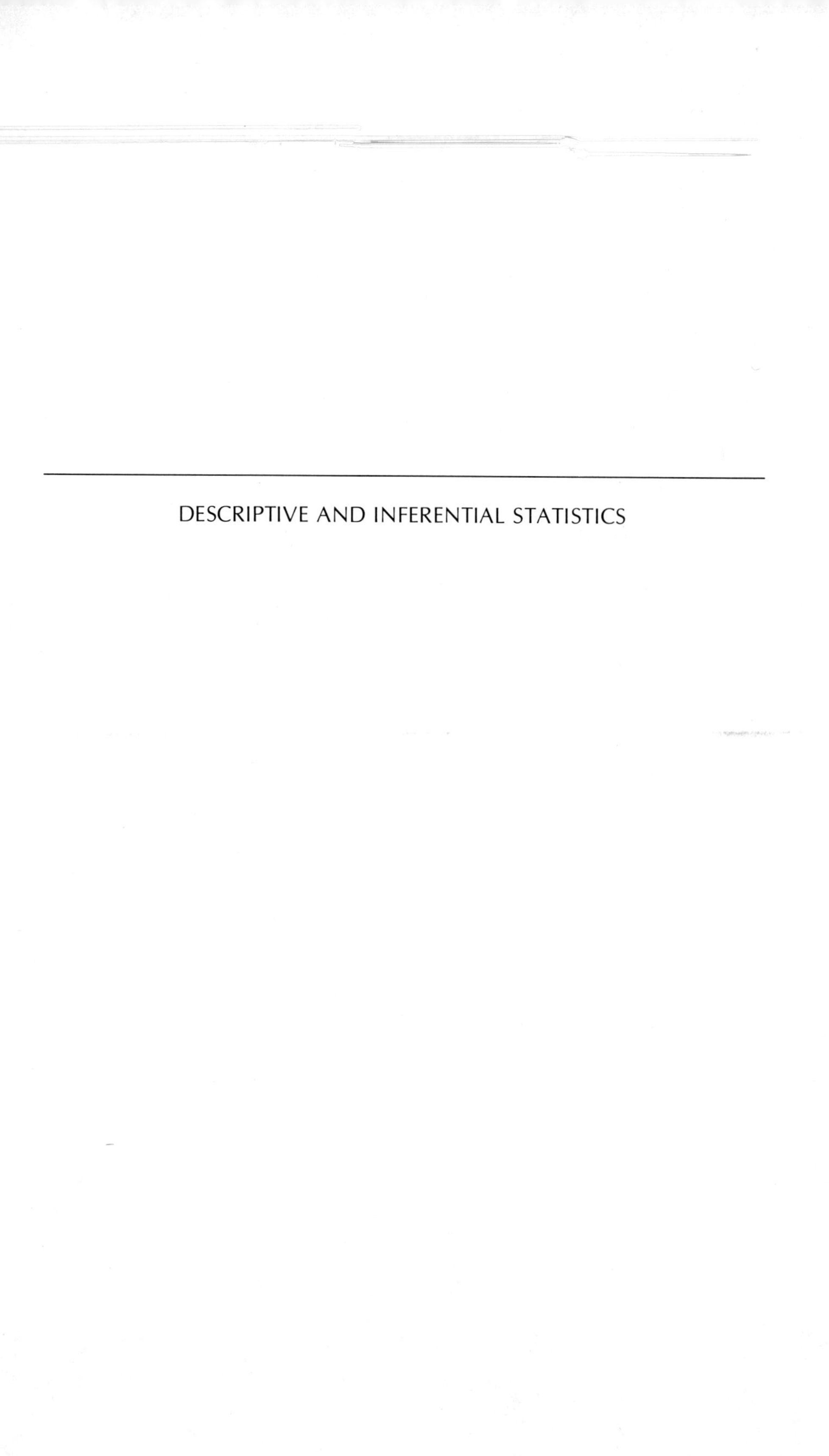

DESCRIPTIVE AND INFERENTIAL STATISTICS

# DESCRIPTIVE AND INFERENTIAL STATISTICS

N. M. DOWNIE
A. R. STARRY
Purdue University

Harper & Row, Publishers
New York, Hagerstown, San Francisco, London

Sponsoring Editor: George A. Middendorf
Project Editor: Cynthia Hausdorff
Designer: T. R. Funderburk
Production Supervisor: Kewal K. Sharma
Compositor: Monotype Composition Company, Inc.
Printer and Binder: Halliday Lithograph Corporation
Art Studio: Vantage Art, Inc.



Library of Congress Cataloging in Publication Data

Downie, Norville Morgan, 1910–
Descriptive and inferential statistics.

Bibliography: p.
Includes index.
1. Statistics. I. Starry, Allan R., 1934–
joint author. II. Title.
HA29.D69 519.5 76-21261
ISBN 0-06-041721-8

# CONTENTS

# IV
## FURTHER TOPICS



## APPENDIXES

# PREFACE

This introductory statistics book is in many ways similar to other statistics books—and in many ways it differs radically from them. The first part of the book discusses descriptive statistics in the classical sense—averages, measures of variability, and the like. The second part, which may be considered the major part, is a detailed introduction to inferential statistics, with major stress being placed on the logic of hypothesis testing and the use of the analysis of variance in such testing. The third part consists of an introduction to the various methods of determining the relationship between variables and an introduction to regression analysis, especially as it concerns the use of statistics in making predictions. The fourth part presents an introduction to distribution-free statistical tests and an llustration of the use of statistics in test theory, construction, and usage.

Emphasis on the use of pocket calculators is one way in which this book departs from tradition. The authors believe that since these calculators have become so inexpensive and are a part of nearly every student's educational paraphernalia, it is logical that the statistics course be geared to them. Every example and exercise in the text can be solved by a simple four-function device, although an automatic square root and certain other features can reduce the student effort required for some problems. The formulas which are presented are the most efficient for use with pocket calculators. By assuming that each student has access to such a device, methodologies using grouped data can be eliminated. Storage limitations of the less expensive machines are circumvented by demonstrating the use of data coding methods. A welcome benefit arising from this approach is a reduction in the step-by-step "cookbook" orientation needed to assure correct solutions and the concurrent opportunity to replace such explanations with expanded discussions of concepts that the student must understand in order to be an intelligent user of statistics.

Students who take introductory courses in applied statistics represent a variety of educational backgrounds and academic interests. Many bring with them only an understanding of elementary algebra, while others will have completed several courses in college level mathematics. While many will be behavioral science or education majors, a sizable number are seeking degrees in such fields as the natural and health sciences, home economics, and industrial management. Some students are 19-year-old undergraduates; others have returned to the university for advanced degrees after several years of experience in their professions. For many, the course will represent their first and last exposure to formal statistical education. These individuals are seeking practical training in a large variety of techniques which may be useful in their future positions. For others, the course is merely the first of several which they will take in the field, and it must provide a solid foundation of fundamental concepts for advanced courses.

Clearly, no single text designed for a one-semester course can satisfy all the needs of such a diverse group. We have attempted to design a text that will meet as many of these needs as possible within the constraints of brevity and readability. Mathematical derivations are limited: A student with minimal mathematical preparation should have no trouble with the examples and exercises. Since a common goal of all students taking the course is to achieve an understanding of how statistical techniques may be applied and interpreted correctly, such matters receive a great deal of emphasis. Examples and exercises draw on easily understood problems in a variety of research fields, both to increase relevancy for nonbehavioral science majors and to broaden students' perceptions with respect to the breadth of statistical applications. Finally, we have attempted to verbalize the rationale underlying several fundamental statistical concepts so that both the mathematically and nonmathematically inclined can achieve a degree of insight and understanding that will be of benefit to them in advanced statistics courses and applied situations.

Some of the ways in which we have attempted to achieve these objectives are quite unique. For example, the subject of mathematical probability, which typically commands one or more chapters in elementary books, receives very limited attention—the minimum required for users of this text. On the other hand, procedures, practices, and concepts surrounding the making of inferences receive much more than usual attention, as does the critical concept of sampling distributions. In the process of trying to pace the introduction and discussion of conceptual material, it was also felt that a departure from conventional topic organization was called for. The reader will notice, for instance, that the chapter on chi-square is not only relatively comprehensive, but precedes the chapter on sampling distributions and estimation. This sequence was adopted because chi-square procedures can be used to illustrate the preceding discussion of inferential techniques without burying the student in nonessential additional material. The authors have used this organization in the introductory course and are satisfied that it is a more efficient arrangement for the student.

While many instructors will find that their students are able to complete the entire book in a one-semester course, others may wish to reduce the number of topics to be studied. We should point out to these individuals that Chapters 1 through 9 represent a sequentially developed instructional core. None of these chapters should be deleted, nor should they be assigned in any other sequence.

The authors are indebted to many people who have had a part in the preparation of this book. We are particularly grateful to Dr. James Price, and to other colleagues who have offered their advice and suggestions. In addition, we greatly appreciate the hundreds of hours of manuscript preparation by a number of very patient and understanding typists, including Diana Haughs, Terri Leonard, Sara McDonald, Suellen Mazzuca, and Caroline Szenina. Particular appreciation goes to Cynthia Hausdorff, project editor, for her most valuable contribution in improving the manuscript during the copy editing and production process.

The authors are deeply grateful to the various authors and publishers who gave us permission to use materials, most of which appear in the various appendixes. We are grateful to the literary executor of the late Sir Ronald A. Fisher, F. R. S., to Dr. Frank Yates, F. R. S., and to the Longman Group, London, for permission to reprint tables in Appendixes D, F, and G from their book *Statistical Tables for Biological, Agricultural, and Medical Research* (6th Edition, 1974).

N. M. Downie
A. R. Starry

# I

# DESCRIPTIVE STATISTICS

# 1

# STATISTICS AND MEASUREMENT

This book is about statistics, important elements in our every day living. When we pick up a newspaper or magazine, we are confronted with all sorts of statistical information. The financial pages of the daily papers present several types of statistical summaries concerning the financial activities of the preceding day. So it is with sports; many kinds of statistics and averages for the different teams and players are presented. The same is also true about the daily weather reports and summaries. In our scientific journals we are confronted with the results of experiments that have been obtained through the application of statistical tests. Also in these journal articles we are given measures of relationships among various variables such as the correlation between intelligence test scores and scores on another test or with academic grades, correlations between physical variables, and relationships among many other known human traits. In these articles we find that statistics are used to summarize and depict data, to average data, to show how data may vary, to show the relationships between sets of data, and to evaluate the results of experiments to see whether the experiment did or did not produce significant results.

The reader is confronted with a mass of statistics, some good, some bad. One of the purposes of this book is to make the user an intelligent consumer of statistics in both popular and scientific contexts. A single course in statistics will not make one a statistician. However, we hope that after using this text, the student will, in addition to knowing how to calculate the simpler statistics, know how to evaluate the use of statistics and to interpret statistical results correctly.

## SAMPLES AND POPULATIONS

In statistical work the student is usually dealing with a *sample* that it is hoped is representative of the larger group from which it was drawn or which it represents. This larger group is called the *population*. We apply various statistical techniques to these data and the resulting values obtained from such studies are called *statistics*. These consist of such values as the number of cases, an average of the scores or measures, a measure of how the scores spread out about the average, and the like. Values such as these, then, that describe samples are statistics and are usually written using Latin letters. The population from which the sample was drawn also has values that describe its properties. Such values are called *parameters* and are usually written using Greek letters. For example,

| | SAMPLE | POPULATION |
|---|---|---|
| **Mean (average)** | $\bar{X}$ | **$\mu$ read mu** |
| **Standard deviation (measure of variability)** | $S$ | **$\sigma$ read sigma** |
| **Number** | $N$ | $n$ |
| **Proportion** | $P$ | $p$ |

As will be seen later we classify certain statistics as descriptive because they describe a sample. Other statistics are called inferential because they are used in making inferences about populations from sample data. This is one of the major uses of modern statistics and the subject of a group of chapters appearing later in the text.

## MEASUREMENT

A *variable,* as used in this book, simply indicates some defined phenomenon or event that is being measured. A variable may take on any value, which is designated as $X$. In statistics we think of numbers obtained by measuring some variable as being points on a line or a *continuum*. For example, the number 13 is considered to cover the distance on a line from 12.5 to 13.4999999, which for practical reasons we round to 13.5. Then 14, the next number, starts at 13.5 and similarly goes to 14.5. Each number, then, has a lower limit and an upper limit. These are known as the *exact limits* of the number. Midway between these two points is the midpoint, which is the number itself. When data are made up of numbers like these, such data are said to be *continuous*. Most of the data we manipulate are of this type. Common examples are measures of height, weight, pressure, and test scores—practically all types of measurements. Continuous data have the advantage of being able to be broken down into smaller and smaller meaningful units. If we say that a boy is 1.5 meters tall, we could also express this as 150 centimeters, 1500 millimeters, and with a finer measuring instrument into even smaller units.

In contrast, as opposed to continuous data we have *discrete* data. With such data, increments are made in whole units, such as 7, 8, 9, where each increase by the whole unit 1. Data like these are exemplified by the number of deaths per week caused by auto accidents, number of letters crossed out per minute in a psychological experiment, number of units assembled per hour, and the like. Discrete data are frequently manipulated statistically similarly to continuous data and so we often end up with statements such as the average number of children per family for a certain group is 2.3.

## TYPES OF MEASUREMENT SCALES

Many statisticians classify measurement scales into four categories: nominal, ordinal, interval, and ratio. As listed, they are in order of increasing complexity. One reason for the concern about types of scales is that when measurements are made using a particular type of scale, certain statistics are best used with that type of data.

In the *nominal* scale, the scale producing the lowest type of measurement, data are merely placed into categories. Individuals can be sorted into two categories: male and female. Or they might be studied on the basis of their political preference: Democrat, Republican, Independent, or Other. From these examples it can be seen that such classifications are common and useful. Unfortunately there is not much that can be done with such data statistically. We can count the frequencies in each category and change these to percents or proportions. Later we shall see that a statistical test may be applied to such data.

The second type of scale is called *ordinal*. With such a scale individuals are placed in order on a continuum such as height. A group of teen-age boys could be ordered or ranked with respect to height, the tallest getting the first position, the next tallest the second position, and so on until all are arranged. Of course, they could also be ordered in the opposite direction. Again there is not much that can be done statistically with such scales chiefly because the distances between individuals on the scales are unequal. Later we shall see that certain correlation coefficients and some special tests of statistical significance are based on data placed in ranks. Ordinal data are ranked data.

The third type of scale is called the *interval* scale. Data obtained with an interval scale are characterized as having equal units of measurement.

The best known example of an interval scale is either the Fahrenheit or centigrade thermometer. On the Fahrenheit scale the distance between 40°F and 60°F is the same as that between 80°F and 100°F, that is, 20 degrees. With such a scale we can say that something is so many units warmer or colder than something else, but nothing more. There has been considerable discussion as to what type of measurement exists in the social sciences. Some consider the results of a test as merely ordinal; that is, on an intelligence test the students are merely ordered from the brightest to the dullest. Others feel that such tests produce interval data because of the manner in which they were constructed. In making these tests the test constructor assumes that each scale is made of equal units. Each item is worth a single point and these points for correct responses are added into a total score. Thus on a test we can say that a student with a score of 90 is 20 points above another student with a score of 70. There is more justification for the manipulation of test scores when we consider them as interval data than when we consider them as ordinal data. The authors subscribe to the viewpoint that much behavioral science data, including scores on standardized tests and measures, provide close approximations to interval scales of measurement.

The fourth and highest type of scale is the *ratio* scale. This is very similar to the interval scale in that it is composed of equal units of measurement, but it differs in that it has an absolute zero as its starting point. A meter stick illustrates such a scale. If we have one measure of 80 centimeters and another of 40, we can say that the first is twice as long as the second or that the second is one-half the length of the first. In other words we have made ratios of our two measurements. Most of the scales used in the physical sciences are ratio scales. Efforts are being made to construct such scales for the social sciences. In the previous paragraph we discussed the ordinary thermometers as being interval scales, that is, 0°F is merely a point on a continuum with no meaning, whereas 0°C is that point at which water freezes under certain conditions. There is a scale used to measure temperatures, the Kelvin scale, that has an absolute zero, the equivalent of −273° centrigrade.

In summary, the nominal scale merely answers the question of how many cases are there in each category. The ordinal scale shows the relative positions of things or individuals on a continuum: who is tallest, who comes next, and so on. Measurements made on an interval scale reveal how many points, degrees, or units one measure is greater than or less than another. Finally, measurements made on a ratio scale tell us how many times larger or smaller one object is than another.

## SOME PROBLEMS IN DEALING WITH NUMBERS

As a general rule we shall, when dealing with whole numbers, carry our results to the nearest tenth. To accomplish this we manipulate our data to the nearest hundredth and then round the final answer to the nearest tenth. For example,

$$77.82 = 77.8$$
$$77.65 = 77.6$$
$$77.75 = 77.8$$

Notice that the last two of the above are exceptions to the general rule. The specific rule here is that when a number ends in 5 and when we are rounding, we drop the 5 when the number preceding it is even (77.65 = 77.6). When the 5 is preceded by an odd number we raise the number preceding the 5 to the next highest digit (77.75 = 77.8). By following this procedure we have a balancing effect. This would not be so if each number ending in 5 were rounded upward; this could cause errors in our work.

When using calculators the student is tempted to copy and record all the digits that the calculator produces for a specific operation, using them merely because they happen to be on the calculator. For example, 1 divided by 7, that is, the reciprocal of 7, is .14285714. If we happen to be dealing with measurements that are made up of two significant digits, then a general rule is to have only two significant digits in the end result rounded as stated above to the nearest tenth. By significant digits, we mean the actual number of digits that we have in our original measurement. If our data consist of two digit numbers such as 76, we have two significant digits. In the long run nothing is gained by using long runs of digits. In the above example .143 would serve adequately as the reciprocal of 7.

In Appendix A the student will find a short review of basic arithmetic and algebraic manipulations. The student should review these and practice them on his calculator, especially if he is long removed from any mathematical work.

## A NOTE ON SYMBOLS

Previously it was pointed out that Greek letters are used for parameters and Latin letters for samples. In addition the following should be noted and learned:

$\Sigma$ sum of
$>$ greater than
$<$ less than
$\neq$ does not equal
$\geqslant$ equal to or greater than
$\leqslant$ equal to or less than

## CHOICE OF A CALCULATOR

We are not recommending any specific brand of calculator. The market for these machines is very unstable, new makes and models appear almost weekly, and competition is apparently driving the prices down. In selecting a calculator, the student should obtain one that, in addition to doing all the basic arithmetical processes, takes square roots, has at least an 8-digit capacity, and accumulates the sum of the squares of numbers without writing down intermediate answers; for example, $6^2 + 4^2 + 9^2 = 133$. Scientific calculators with several internal working registers, addressable storage registers, and automatic scientific notation features are ideally suited to statistical work. However, the examples and problems in this text are workable on the less expensive models. The student should master the technique of making continuous chain calculations when working with formulas so that the maximum number of significant digits is internally stored for each intermediate answer. This has been done with many of the examples used in the book, even though the intermediate answers shown have been rounded to two or three places.

### EXERCISES

For practice with a calculator or to aid in the decision as to which machine to buy, the student should work the problems at the end of Appendix A.[1]

[1] The student may wish to consider the following, or similar, books on pocket calculator usage. Those listed below are much more comprehensive than the typical calculator manual, suggest many useful short-cut techniques, and provide mathematical review material which should prove valuable to some students.

*Basic level*

Roberts, E. M. *Fingertip math.* Dallas: Texas Instruments Learning Center, 1974.

*Intermediate level*

Hunter, W. L. *Getting the most out of your electronic calculator.* Blue Ridge Summit, Pa.: Tab Books, 1974.

*Advanced level*

Gilbert, J. *Advanced applications for pocket calculators.* Blue Ridge Summit, Pa.: Tab Books, 1975.

Smith, J. M. *Scientific analysis on the pocket calculator.* New York: Wiley-Interscience, 1975.

# 2

# FREQUENCY DISTRIBUTIONS, GRAPHS, AND CENTILES

Once data have been collected they have to be put into a form that makes it possible to get information from them. Often we are concerned with the shape that the distribution takes because the shape of the distribution determines what can be done with the data statistically. One of the first things that can be done with data is to set up a *frequency distribution*. This frequency distribution then becomes the basis of a graph that can be made to facilitate the interpretation of the obtained data. The purposes of this chapter are to learn how to construct a frequency distribution and then how to make graphs from this distribution. Finally we shall consider a special type of graph from which centiles are read.

## THE FREQUENCY DISTRIBUTION

As a simple illustration suppose that on a statistics test 10 students obtain the following scores:

25, 24, 23, 23, 22, 22, 22, 21, 20, 19

These scores should be set up and tallied like this:

| SCORE (X) | | f |
|---|---|---|
| 25 | / | 1 |
| 24 | / | 1 |
| 23 | // | 2 |
| 22 | /// | 3 |
| 21 | / | 1 |
| 20 | / | 1 |
| 19 | / | 1 |
| | | $N = 10$ |

Obviously if we had a large number of scores and used a method such as this, it would be more efficient to *group* the data as shown below.

When we have a large sample, a frequency distribution is set using an appropriate interval size. The interval size ($i$) for the previous distribution was 1. In setting up a grouped data frequency distribution, we proceed as follows:

1. The data are ordered as shown in Table 2.1. Doing this is not essential, but it makes the work easier.
2. The difference between the high score and the low score is next determined. This distance is called the *range*. In this case the distance is 147 − 2, which equals 145. Actually, as will be seen later, the range is defined as the high score ($X_h$) minus the low score ($X_l$) plus 1. In this case the range would then be 146.
3. Usage demands that there be between 10 and 20 class intervals (*c.i.*). If there are less than 10 the grouping becomes too coarse and information about the individual scores to a large degree is lost. If there are more than 20 intervals, the amount of work involved raises the question as to whether there was any point in grouping the data. We can decide upon the size of the interval that we are going to use by taking half the distance between 10 and 20, or 15. The range is then divided by 15. In this case we have $146/15 = 9.5$. Rounding this to 10 gives us an acceptable size and one that makes a useful class interval. Note that the most frequently used class interval sizes are 1, 3, 5, and 10.
4. The frequency distribution is then set up with the bottom interval 0–9. The next interval is 10–19 and so on until enough intervals are established to contain all of the scores. It is customary to begin the first interval with a multiple of the interval size, which in this case is 10.
5. The scores from Table 2.1 are then tallied as shown in Table 2.2. These tallies are summarized in the column labeled $f$ for fre-

TABLE 2.1. SCORES OF 100 STUDENTS ON A STATISTICS TEST

| | | | | |
|---|---|---|---|---|
| 147 | 89 | 76 | 63 | 50 |
| 139 | 89 | 75 | 63 | 49 |
| 127 | 87 | 74 | 62 | 49 |
| 120 | 87 | 74 | 62 | 47 |
| 112 | 86 | 73 | 61 | 47 |
| 109 | 85 | 72 | 61 | 46 |
| 109 | 84 | 72 | 60 | 44 |
| 107 | 83 | 70 | 60 | 44 |
| 105 | 82 | 69 | 59 | 44 |
| 105 | 82 | 69 | 59 | 40 |
| 102 | 82 | 69 | 59 | 39 |
| 99 | 80 | 69 | 59 | 37 |
| 99 | 79 | 67 | 58 | 35 |
| 97 | 79 | 67 | 56 | 34 |
| 96 | 79 | 65 | 56 | 33 |
| 96 | 77 | 65 | 56 | 31 |
| 95 | 77 | 65 | 56 | 22 |
| 94 | 77 | 65 | 55 | 18 |
| 93 | 77 | 64 | 55 | 8 |
| 90 | 76 | 63 | 52 | 2 |

quency. The sum of this column ($\Sigma f$) is equal to the number of cases in the distribution ($N$).

6. Another column [column (2)] shows the proportion of cases in each interval. This column is computed by multiplying each entry in the $f$ column by the reciprocal of $N$, or $1/100 = .01$ in the example. The constant or storage feature of many calculators may be used to simplify this task. Since $N = 100$ here, this column does not reveal much. But with other size $N$s it could be very useful.

Observation of one of these intervals may lead the student to conclude that the interval size is 9 and not 10. Take the interval 70–79. Actually this interval has as its lower limit (*ll*) 69.5 and as its upper limit (*ul*) 79.5. The distance between these two points is 10, the interval size. Remember in Chapter 1 we said that any number as we consider it is a point on a continuum with a lower and upper limit. In dealing with intervals we frequently use the mid-point of the interval, which in this case is 74.5—the lower limit (69.5) plus half of the interval size (5).

When we group data that are relatively normal in distribution, we assume that for any interval the frequencies are equally distributed throughout the interval and that the average of the scores in any interval

**TABLE 2.2 FREQUENCY DISTRIBUTION OF THE SCORES OF 100 STUDENTS ON A STATISTICS TEST**

| | | (1) | (2) |
|---|---|---|---|
| $X$ | | $f$ | PROPORTION |
| 140–149 | / | 1 | .01 |
| 130–139 | / | 1 | .01 |
| 120–129 | // | 2 | .02 |
| 110–119 | / | 1 | .01 |
| 100–109 | ~~////~~ / | 6 | .06 |
| 90–99 | ~~////~~ //// | 9 | .09 |
| 80–89 | ~~////~~ ~~////~~ // | 12 | .12 |
| 70–79 | ~~////~~ ~~////~~ ~~////~~ / | 16 | .16 |
| 60–69 | ~~////~~ ~~////~~ ~~////~~ ~~////~~ | 20 | .20 |
| 50–59 | ~~////~~ ~~////~~ /// | 13 | .13 |
| 40–49 | ~~////~~ //// | 9 | .09 |
| 30–39 | ~~////~~ / | 6 | .06 |
| 20–29 | / | 1 | .01 |
| 10–19 | / | 1 | .01 |
| 0–9 | // | 2 | .02 |
| | | $\Sigma f = 100$ | 1.00 |

will equal the midpoint of the interval. As an example consider the interval 30–39, with six cases 39, 37, 35, 34, 33, and 31. The average of these cases is equal to the sum divided by the number of cases, which results in a value of 34.8. If all intervals in several different distributions of test scores were studied, it would be found that for those intervals above the center of the distribution the average of the scores or values in each interval tends to be below the midpoint of the interval and that for the intervals below the center of the distribution, the reverse is true. In this case the obtained value of 34.8 is slightly above the midpoint of the interval, 34.5. These discrepancies are usually slight and have little effect on our data.

## GRAPHS AND GRAPHING

### FREQUENCY POLYGON

In reporting the results of research a frequently encountered graph is the *frequency polygon.* We shall illustrate the construction of such a graph using the frequency distribution set up in Table 2.2. In making frequency polygons certain conventions are usually followed.

1. The frequencies are placed on the perpendicular axis (labeled $y$), the *ordinate*.
2. The scale of numbers is placed on the horizontal axis (labeled $x$), the *abscissa*.
3. The ratio of the $y$ axis to the $x$ axis should be 2 to 3 or 3 to 4. That is, if one is using graph paper with 10 small squares to the inch, the use of 60 of these small spaces on the $y$ axis would require the use of 90 such spaces on the $x$ axis. The use of either of these ratios leads to uniformity with respect to the general shape of the curve, as will be shown later.
4. After the axes have been drawn on the graph paper, appropriate values are placed on the $y$ axis as shown in Figure 2.1. The largest value for $y$ need be no higher than the largest frequency. On the $x$ axis, we could place numbers corresponding to the first number in each of our intervals, such as 0, 10, 20, and so on, or we could place the numbers to correspond to the midpoint of each intervals. Since we assume that for any interval the midpoint is equal to the average of all scores in that interval, it makes sense to place the midpoint values on the $x$ axis and to plot the frequency for each interval above its midpoint.
5. After the points are plotted, they are connected with a straight line. It is usually the custom to anchor the curves at both ends by using a zero frequency for the interval above and below the last intervals in the frequency distribution.

**FIG. 2.1 FREQUENCY POLYGON FOR THE DATA IN TABLE 2.2.**

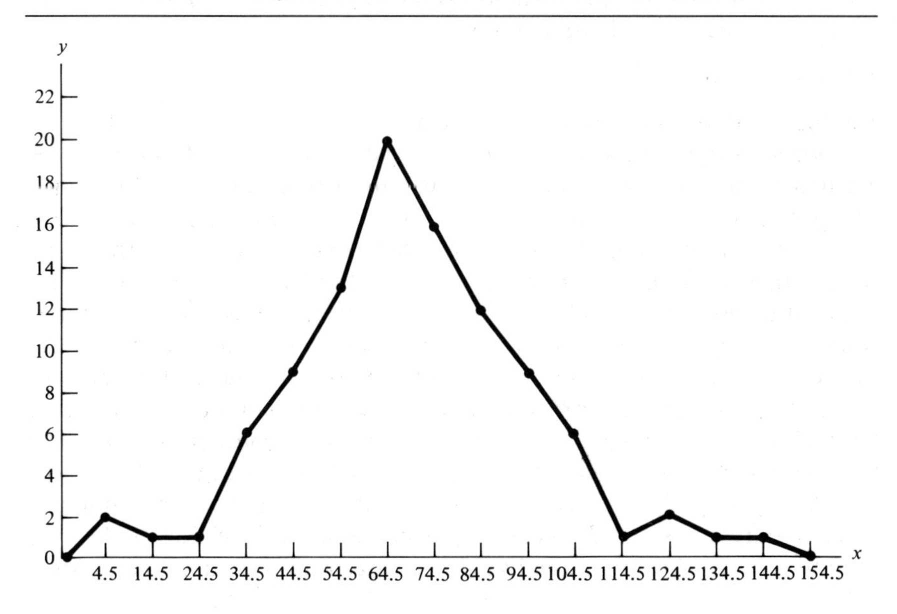

The frequency polygon has many uses in the reporting of research data. One advantage of this type of curve is that several distributions may be plotted on the same axes for comparative purposes by using different colored ink or different types of lines to connect the points.

Occasionally, problems arise when we are plotting distributions in which there are large discrepancies in the size of the number of cases. Sometimes one sample may have so many cases that it will not fit on the same axes as the other samples. The solution to this is to change the frequencies to proportions or percentages and to plot these values instead of the frequencies on the $y$ axis. There are also times when a graph would be more meaningful when plotted in terms of percentages rather than frequencies. As an illustration of this, suppose that we have a five-point scale, called a *Likert scale,* with the following frequencies for a group of seniors and another group of freshmen:

| | POINT | SENIORS $f$ | FRESHMEN $f$ | PERCENT SENIORS | PERCENT FRESHMEN |
|---|---|---|---|---|---|
| **Strongly agree** | **1** | **24** | **60** | **12%** | **15%** |
| **Agree** | **2** | **56** | **120** | **28** | **30** |
| **No opinion** | **3** | **16** | **20** | **8** | **5** |
| **Disagree** | **4** | **40** | **88** | **20** | **22** |
| **Strongly disagree** | **5** | **64** | **112** | **32** | **28** |
| | | $N = 200$ | $N = 400$ | **100%** | **100%** |

When comparing the responses of the two classes, the percentages are much more meaningful than the corresponding frequencies. When plotting these data, the percentages would appear on the $y$ axis and values for the various intervals on the $x$ axis.

## THE HISTOGRAM

Another type of graph frequently encountered in statistical work is the *histogram,* a type of bar graph. In the construction of the histogram, the method followed is very similar to that used in making the frequency polygon. The $x$ and $y$ axes are set up in the same way. The data in Table 2.2 are used in building the histogram shown in Figure 2.2. In this table the frequency of the first interval, 0–9, is 2. A column 2 units high is constructed using the *exact interval limits.* At its base this column extends from −.5 to 9.5, a distance of 10 units. The second interval has a frequency of 1. Therefore a column 1 unit high is constructed between the points 9.5 and 19.5. This procedure is followed until all the intervals have been plotted. Obviously if we were to connect the midpoints of each of the columns we would have the frequency polygon of Figure 2.1.

As a graph, the histogram is not as useful as the frequency polygon because only one set of data can be placed on one pair of axes (unless we wish to plot another set extending downward from the same axes). Also,

FIG. 2.2 HISTOGRAM FOR THE DATA IN TABLE 2.2.
(Exact interval limits have been rounded.)

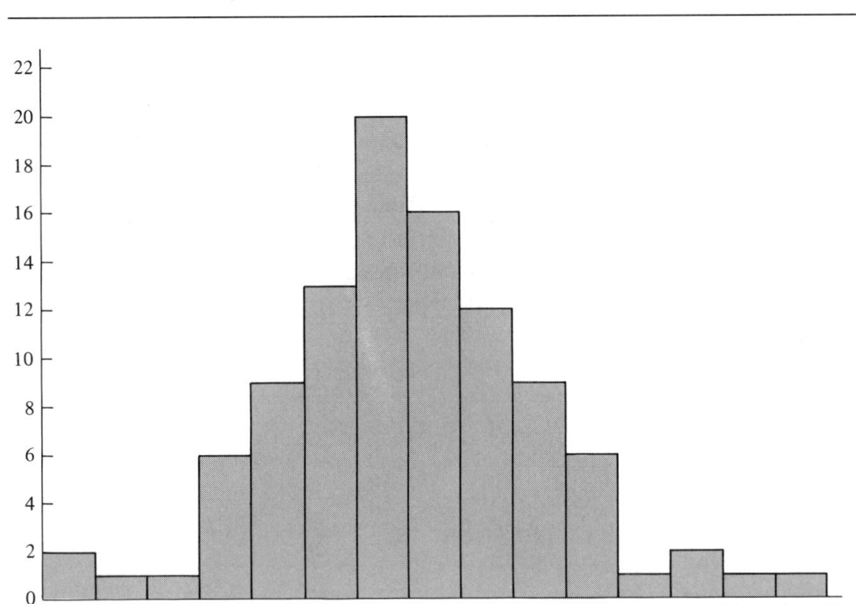


the frequencies could be changed to percentages and the histogram constructed using these. If the columns of such a histogram were separated, the frequently encountered *bar graph* would result.

## CHARACTERISTICS OF DISTRIBUTIONS AND CURVES

Distributions are described in part by the *shape* they take. Three terms are used in describing the shapes of curves and distributions: skewness, kurtosis, and modality.

### Skewness

Skewness is present when a distribution has one tail that extends farther in one direction than does the other tail. When the tail is extended to the right, the distribution is said to be *positively skewed* (Figure 2.3). A distribution of the incomes of a group of adult males for a given state would take this shape. When the tail extends to the left, we have *negative skew*. A curve that represents scores on an easy test would show negative skew (Figure 2.4).

### Kurtosis

Kurtosis is another name for peakedness. When scores on a test tend to pile up near the center of the distribution, we have a peaked or *lepto-*

**FIG. 2.3 A POSITIVELY SKEWED DISTRIBUTION.**

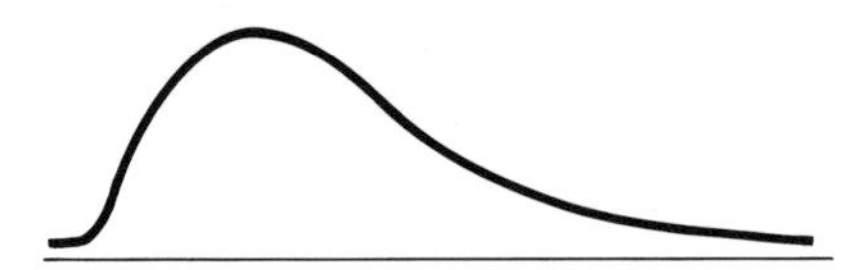

**FIG. 2.4 A NEGATIVELY SKEWED DISTRIBUTION.**

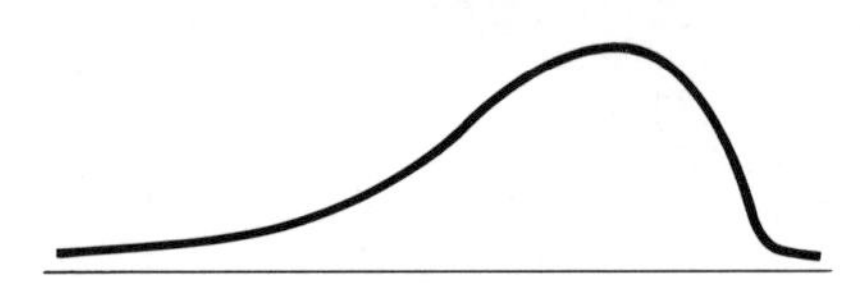

*kurtic* distribution (Figure 2.5). Such a curve occurs when scores obtained by a group on a test are very similar. When we have this condition we say that the group is homogeneous with respect to the trait being measured. As opposed to this condition there is the flat curve, or *platykurtic* distribution. In this distribution there is a wide range of scores with few piling up at the center (Figure 2.6). The group is quite variable on the trait being measured and is said to be heterogeneous with respect to the trait. Between these two conditions we have a moderately peaked curve, described as being *mesokurtic*.

**FIG. 2.5 A LEPTOKURTIC DISTRIBUTION.**

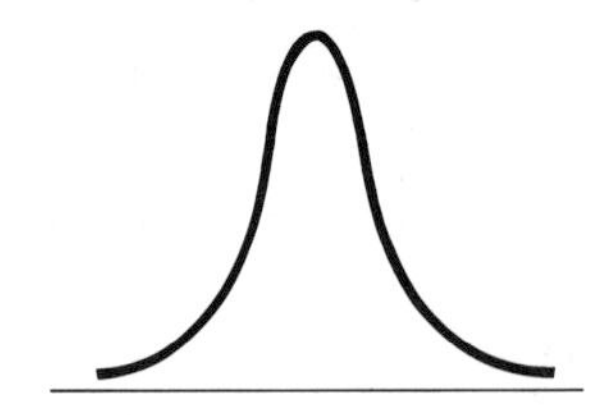

**FIG 2.6 A PLATYKURTIC DISTRIBUTION.**

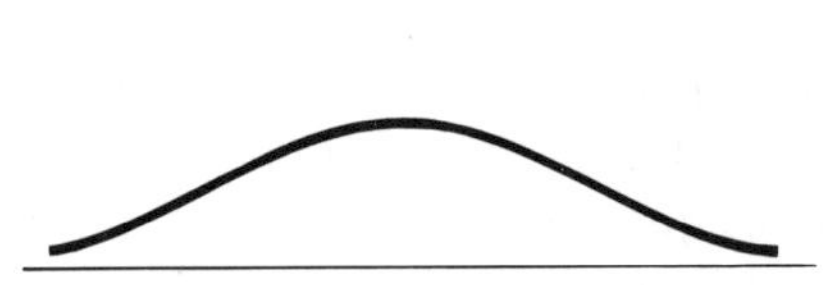

### Modality

A third characteristic of a curve or distribution is modality. Some curves have one peak and are described as being unimodal. Others may have two peaks and are called bimodal (Figure 2.7).

## THE NORMAL CURVE

The *normal* curve, or *bell-shaped* curve as it is frequently called, is the one that is most often encountered in statistical work. Often we assume that a variable is normally distributed in a sample or in the population from which the sample was drawn in order to enable us to compute and use certain statistics or to make certain statistical tests. The normal curve (Figure 2.8) is described as being unimodal, mesokurtic, bilaterally symmetrical, and with zero skew. As can be seen from the accompanying figures, distributions and curves differ considerably in reference to kurtosis and skew.

In the next two chapters we shall consider two other statistics used in describing sets of observations or distributions of samples and populations. These are *averages* and *measures of variability*.

FIG. 2.7 A BIMODAL CURVE.

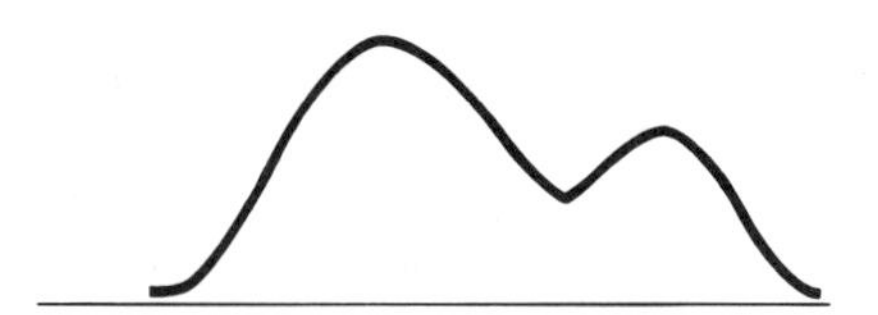

FIG. 2.8 A NORMAL CURVE.

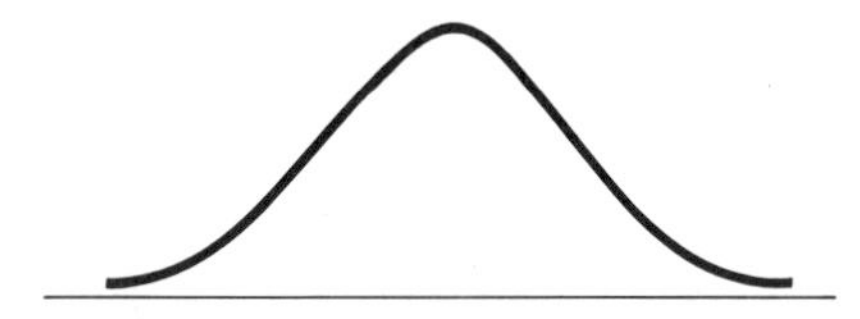

## CENTILES AND THE CUMULATIVE FREQUENCY OR THE CUMULATIVE PERCENTAGE CURVE

Another type of curve that is very useful is the cumulative frequency or cumulative percentage curve. From such a curve we can read centile points or ranks rapidly and efficiently. Before constructing such a curve we shall spend time considering the nature of centiles and how they can be obtained by computation. This is important when one wants one or two centile points, but when a centile distribution is to be set up, such as when one desires a distribution of centile norms for a test, then the only practical way to obtain such statistics is to read them from a cumulative percentage graph.

### CENTILES OR PERCENTILES

A *centile* or *percentile* is defined as that point in a distribution below which a certain percent of the cases fall. Thus the 33rd centile (written $C_{33}$) is that point below which 33 percent of the cases fall. Naturally, 67 percent of the cases are above this point. The *median* (see Chapter 3) is the 50th centile.

In Table 2.3 are the scores of 320 students on a reading test. Suppose we wish to establish a set of centile norms for these scores. First let us find the tenth centile, $C_{10}$, which is defined as that point in the distribution with 10 percent of the cases below it. The first step is to take 10 percent of $N$, which is 320 (.10) = 32. Next we find the point with 32 cases below it. We start by counting up from the bottom and find that at the top of the interval 40–44 we have cut off 18 cases. Hence we have to interpolate in the next interval to obtain our 32 cases. Then

$$
\begin{aligned}
C_{10} &= 44.5 + \frac{32 - 18}{20}(5) \\
&= 44.5 + \frac{14}{20}(5) \\
&= 44.5 + 3.5 \\
&= 48.0
\end{aligned}
$$

In the above, 44.5 is the bottom of the interval in which we interpolate. The numerator of the fraction consists of the difference between the number of cases that we need and the number that we have counted

TABLE 2.3. COMPUTATION OF CENTILES

| | X | f | |
|---|---|---|---|
| | 100–104 | 2 | |
| | 95–99 | 0 | |
| | 90–94 | 4 | |
| | 85–89 | 8 | |
| | 80–84 | 16 | |
| 79.5 ← | | | 30 |
| | 75–79 | 38 | |
| | 70–74 | 48 | |
| | 65–69 | 54 | |
| | 60–64 | 42 | |
| | 55–59 | 40 | |
| | 50–54 | 30 | |
| | 45–49 | 20 | |
| 44.5 ← | | | 18 |
| | 40–44 | 12 | |
| | 35–39 | 4 | |
| | 30–34 | 2 | |
| | | $N = 320$ | |

Any centile may be obtained by the use of this formula when counting up

$$\text{Any centile} = ll + \frac{Np - cf}{f_i}(i) \qquad (2.1)$$

where

$ll$ = lower limit of interpolated interval
$N$ = number of cases
$p$ = proportion equivalent of desired centile
$cf$ = cumulative frequency of cases below the interval in which we are interpolating
$f_i$ = frequency of the interval in which we are interpolating
$i$ = size of the interval

Using this formula, $C_{10} = 48.0$ and $C_{90} = 79.2$, as seen in the text.

up to this point and 5 is the size of the class interval. The denominator is the frequency of the interval in which we are interpolating. This procedure, with a formula for any centile point, is shown in Table 2.3.

Let us take another case, $C_{90}$, that is, the point with 90 percent of the cases below it. We could take 90 percent of $N$ and begin counting up from the bottom. However, it is easier to take 10 percent of $N$ and come down from the top. Thus

$$\begin{aligned} C_{90} &= 79.5 - \frac{32 - 30}{38}(5) \\ &= 79.5 - \frac{2}{38}(5) \\ &= 79.5 - .3 \\ &= 79.2 \end{aligned}$$

Note again that when one works down from the top, a minus sign appears in this equation. The computation done above may be consolidated in formula 2.1.

## OTHER CENTILE POINTS

Certain centile points have special names. $C_{50}$ is the median; $C_{25}$ and $C_{75}$ are called quartile points, $C_{25}$ being $Q_1$, the first quartile point, and $C_{75}$ being $Q_3$, the third quartile point. Also $C_{10}$, $C_{20}$, $C_{30}$, and so on, are referred to as decile points, $D_1$, $D_2$, $D_3$, and so on, respectively.

## THE OGIVE CURVE

As pointed out, the method described above is adequate if we want to find only one or two centile points. However, if we want all of them as when setting up a set of centile norms for a test, it is much easier to construct an *ogive* or cumulative percentage curve and read the centile points from this curve. In building an ogive curve, we first change our frequencies to cumulative frequencies (*cf*), then convert these to cumulative proportions (*cp*), and finally convert these to cumulative percentages, (*c*%). These last values are plotted and the resulting curve from which the centiles are read is the ogive or *S*-shaped curve.
Using the data in Table 2.4, we go through the following steps:

1. We set up column (2), labeled *cf*. This is done by writing down the total number of cases below the top of the intervals. Below the top of the interval 30–34 there are two cases. Below the top of the interval 35–39 there are 6 cases, 2 + 4. By adding the num-

**TABLE 2.4. ESTABLISHING CUMULATIVE VALUES FOR A SET OF DATA**

| | (1) | (2) | (3) | (4) |
|---|---|---|---|---|
| | *f* | *cf* | *cp* | *c*% |
| 100–104 | 2 | 320 | 1.000 | 100.0 |
| 95–99 | 0 | 318 | .994 | 99.4 |
| 90–94 | 4 | 318 | .994 | 99.4 |
| 85–89 | 8 | 314 | .981 | 98.1 |
| 80–84 | 16 | 306 | .956 | 95.6 |
| 75–79 | 38 | 290 | .906 | 90.6 |
| 70–74 | 48 | 252 | .788 | 78.8 |
| 65–69 | 54 | 204 | .638 | 63.8 |
| 60–64 | 42 | 150 | .469 | 46.9 |
| 55–59 | 40 | 108 | .338 | 33.8 |
| 50–54 | 30 | 68 | .212 | 21.2 |
| 45–49 | 20 | 38 | .119 | 11.9 |
| 40–44 | 12 | 18 | .056 | 5.6 |
| 35–39 | 4 | 6 | .019 | 1.9 |
| 30–34 | 2 | 2 | .006 | .6 |
| | $N = 320$ | | | |

ber of cases in each of the succeeding intervals, all the values in column (2) are obtained.

2. Change each of the *cf*'s to a cumulative proportion, *cp*. This is done by taking the reciprocal of 320, .003125, and multiplying each value in column (2) by this reciprocal and rounding the product to the nearest thousandth.
3. Obtain the cumulative percentages (*c*%) by multiplying each value in column (3) by 100.
4. Plot the values in column (4) as shown in Figure 2.9. In plotting these points in an ogive curve, the tally mark is made *above* the point on the *x* axis that corresponds to the *upper limit* of the interval. Then these points are connected by a smooth curve. Notice that in drawing the smooth curve some of the tally marks are on one side of it and some on the other.
5. Read the centiles from the graph. This is illustrated first by finding the median. Since the median is that point with 50 percent of the cases falling below it, we go up the *y* axis to 50, draw a line from that point to the curve, then draw a perpendicular from the latter point on the curve to the base line. The point at which this line meets the base line is the median. As shown of Figure 2.9 this

**FIG. 2.9 OGIVE CURVE FOR THE DATA IN TABLE 2.4.**

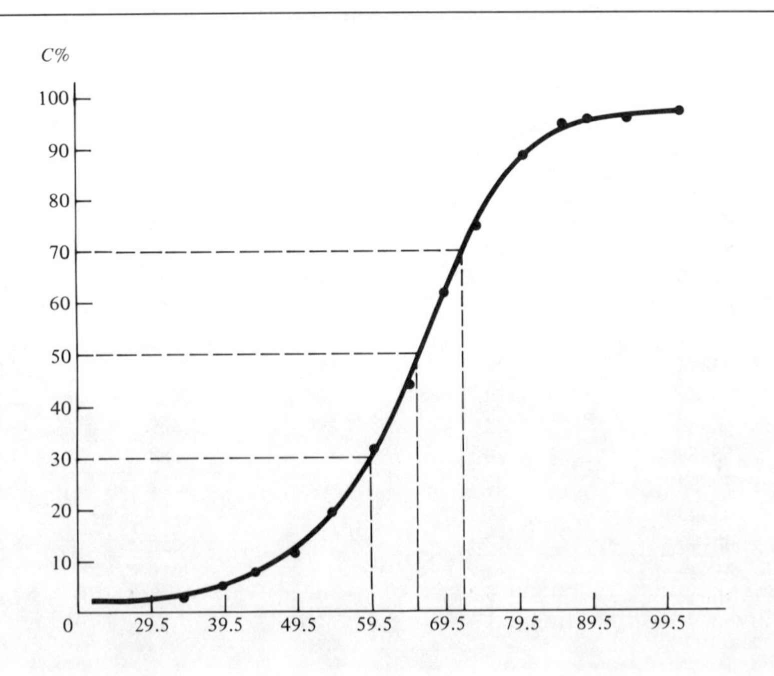

point is approximately 65. Let us see how this compares with the computed median.

$$50 \text{ percent of } 320 \text{ is } 160$$

$$C_{50} = 64.5 + \frac{160 - 150}{54}(5)$$

$$= 64.5 + \frac{50}{54}$$

$$= 64.5 + .9$$

$$= 65.6$$

There is no practical difference between these two values. Similarly $C_{30}$ and $C_{70}$ are found to be approximately 59.5 and 71.0, respectively. If we continued this for all points, we would have a centile point for each score in the distribution and this would be a set of centile norms.

## CENTILE RANKS

*Centile ranks* are very similar to centile points and are often confused with them. A centile rank denotes the percentage of *scores* in a distribution that fall below a specified score. Using the data in Figure 2.9, centile ranks may be obtained by reading the graph in the reverse manner from which it was read for centiles. For example, to find the centile rank of a score of 50, find 50 on the $x$ axis, draw a line from it up to the curve, then another line from this point to the $y$ axis. By doing this we find that a score of 50 has a centile rank of 11. Similarly a score of 85 has a centile rank of 96. Rewriting (2.1),

$$\text{rank} = \frac{100[f_i(X - ll) + (i)(cf)]}{Ni}$$

## USE OF CENTILES

Centiles are frequently encountered in personnel work where tests are used. They are very useful in that they show where an individual falls in the group that took the same test that he did. For example, a student falls at $C_{20}$ on a reading test, $C_{15}$ on a spelling test, $C_{30}$ on a geography test, and $C_{10}$ on an arithmetic test. A statement such as this gives an accurate picture of this student's overall achievement relative to a peer group.

Centiles are not equal units of measurement and should not be averaged or manipulated in any arithmetic fashion. Much of the data used in psychology, education, and the natural sciences, when plotted, approximate the shape of the normal curve. A distribution of centiles, on the other hand, is rectangular. For example, 10 percent of the cases fall below $C_{10}$, an additional 10 percent below $C_{20}$, and so on. When such

FIG. 2.10 RAW SCORE AND CENTILE EQUIVALENTS FOR A SET OF FICTITIOUS DATA.

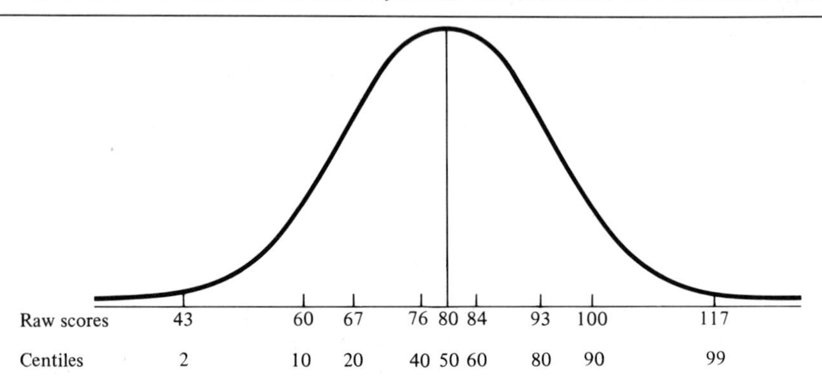


data are plotted, the result is a rectangle made up of ten equal blocks. Figure 2.10 is a curve showing a distribution of raw scores for a set of fictitious data. Notice that, in terms of raw scores, the distance between $C_{50}$ and $C_{60}$ is four raw score points, the dfistance between $C_{60}$ and $C_{80}$ is nine raw score points, and so on. As one proceeds from the center of the distribution toward the tails, the number of raw score points between equal centile points increases. Hence differences between centile points taken near the center of a distribution are small and equal distances in centiles taken in the extremes are large. Thus it may be said that centiles exaggerate differences at the center of the distribution. There is actually no real difference between a person who scores at $C_{44}$ and another who scores at $C_{54}$ on the same test. They are both part of that large pileup of individuals at the center of the distribution. As we shall see, there are better statistics to use with test scores.

## EXERCISES

1. a. What might be an appropriate interval size for the following range of scores?
   (1) 1–12 (2) 22–44 (3) 10–120
   (4) 2.5–3.0 (5) 45–89 (6) 200–402
   b. For each of the above give the exact upper limit, the midpoint, and the lower limit.
2. Below are the scores of 40 students on a statistics test. Construct a frequency distribution for this data. How many different interval sizes might be used?

| | | | | | | | | | |
|---|---|---|---|---|---|---|---|---|---|
| **48** | **36** | **31** | **29** | **27** | **26** | **24** | **23** | **21** | **18** |
| **45** | **35** | **30** | **28** | **27** | **26** | **24** | **22** | **20** | **17** |
| **40** | **32** | **30** | **28** | **27** | **25** | **24** | **22** | **19** | **14** |
| **38** | **31** | **30** | **28** | **26** | **25** | **23** | **21** | **19** | **8** |

3. Using the data in exercise 2
   a. Construct a frequency polygon
   b. On the same axes set up a histogram
4. Convert the frequencies to percentages for the data in exercise 2 and plot the distribution.
5. The following scores were obtained on a final examination in statistics. Construct a frequency distribution, a frequency polygon, and a curve based on percentages.

| | | | | | | | |
|---|---|---|---|---|---|---|---|
| 88 | 74 | 69 | 64 | 57 | 51 | 45 | 33 |
| 85 | 74 | 68 | 63 | 57 | 51 | 45 | 29 |
| 81 | 74 | 68 | 63 | 56 | 50 | 44 | 27 |
| 80 | 72 | 67 | 62 | 55 | 50 | 40 | 26 |
| 79 | 71 | 67 | 61 | 54 | 49 | 39 | 25 |
| 79 | 71 | 66 | 60 | 53 | 49 | 37 | 24 |
| 77 | 70 | 66 | 58 | 52 | 48 | 36 | 23 |
| 77 | 70 | 66 | 57 | 52 | 47 | 35 | 19 |
| 75 | | | | | | | |

6. Given the following distribution of test scores:

| | $f$ |
|---|---|
| 120–129 | 2 |
| 110–119 | 8 |
| 100–109 | 12 |
| 90–99 | 24 |
| 80–89 | 36 |
| 70–79 | 42 |
| 60–69 | 37 |
| 50–59 | 21 |
| 40–49 | 10 |
| 30–39 | 7 |
| 20–29 | 1 |
| | N = 200 |

   a. Find each of the following:
      (1) $C_{90}$ (2) $C_{60}$ (3) $Q_1$ (4) $C_{10}$
      (5) $Q_3$ (6) $C_{50}$ (7) $C_{70}$ (8) $D_3$
   b. Construct an ogive curve for these data and from the curve read the same centiles that you found in problem a above.

# 3

# AVERAGES

After sets of data have been collected, they have to be treated in various ways to simplify their interpretation or to make them meaningful. One of the first things usually done with data is to find an average. An average is simply a single measure that stands for, or represents, a whole series of measures or numbers. The three commonly used averages are the mean ($\bar{X}$), the median (*Mdn*), and the mode (*Mo*).

## THE MEAN

There are several types of means, but the one that is most frequently encountered is the arithmetic *mean*. The equation for this is

$$\bar{X} = \frac{\Sigma X}{N} \tag{3.1}$$

where

$\bar{X}$ is the mean and is read as $X$-bar

$\Sigma X$ is the sum of the scores or of the measures. $\Sigma$ is a capital Greek sigma and is read "sum of" or "summation"

$N$ is the number of cases

Suppose that we have 6 scores: 10, 8, 6, 4, 2, 0. The mean for these scores is then

$$\bar{X} = \frac{10 + 8 + 6 + 4 + 2 + 0}{6} = \frac{30}{6} = 5$$

When scores are numerous, as below, time may be saved by grouping them as shown in Table 3.1.

| | | | | | | | |
|---|---|---|---|---|---|---|---|
| 10 | 8 | 7 | 6 | 5 | 5 | 4 | 3 |
| 9 | 8 | 7 | 6 | 5 | 5 | 4 | 2 |
| 9 | 7 | 6 | 6 | 5 | 4 | 3 | 2 |
| 8 | 7 | 6 | 6 | 5 | 4 | 3 | 1 |

In Table 3.1 the first column, labeled $X$, represents each of the scores and the second column, labeled $f$, is the frequency or number of each score. The third column, $fX$, is the product of the values in the first two columns taken row by row. Then the mean is obtained by this formula:

$$\bar{X} = \frac{\Sigma fX}{N} \tag{3.2}$$

$$= \frac{176}{32}$$

$$= 5.5$$

In the next chapter we shall consider the coding of data, and it will be shown how the work of finding the mean may be shortened.

### THE NATURE OF THE MEAN

To illustrate the nature of the mean we shall use the measures previously listed:

$$\bar{X} = \frac{10 + 8 + 6 + 4 + 2 + 0}{6}$$

$$= \frac{30}{6}$$

$$= 5$$

**TABLE 3.1. OBTAINING THE MEAN FOR GROUPED DATA WHEN THE INTERVAL IS 1**

| $X$ | $f$ | $fX$ |
|---|---|---|
| 10 | 1 | 10 |
| 9 | 2 | 18 |
| 8 | 3 | 24 |
| 7 | 4 | 28 |
| 6 | 6 | 36 |
| 5 | 6 | 30 |
| 4 | 4 | 16 |
| 3 | 3 | 9 |
| 2 | 2 | 4 |
| 1 | 1 | 1 |
| | $\Sigma f = 32$ | $\Sigma fX = 176$ |

Each score is now placed as a point on a continuum as below:

| 0 | 1 | 2 | 3 | 4 | 5 | 6 | 7 | 8 | 9 | 10 |
|---|---|---|---|---|---|---|---|---|---|---|
| ↓ | | ↓ | | ↓ | $\bar{X}$ | ↓ | | ↓ | | ↓ |

with the mean of 5 at the center or at the *fulcrum*. A system such as the one above is said to be in equilibrium because the sum of the distances of the scores from the mean on one side equals the sum of the distances on the other side. In this case the distances are $5 + 3 + 1 = 1 + 3 + 5$. The mean is the only point on this continuum for which this is true. The mean may also be considered as the center of gravity of the distribution.

Another way of looking at this is as follows:

| $X$ | $x$ | $x^2$ |
|---|---|---|
| 10 | 5 | 25 |
| 8 | 3 | 9 |
| 6 | 1 | 1 |
| 4 | −1 | 1 |
| 2 | −3 | 9 |
| 0 | −5 | 25 |
| $\Sigma = 30$ | $\Sigma = 0$ | $\Sigma = 70$ |

In the column labeled $x$, we have the distance of each score from the mean $(X - \bar{X})$. Each small $x$ is the deviation of a score from the mean. When these small $x$'s are summed, the sum is always zero. This then is the mathematical definition of the arithmetic mean—that point in the distribution about which the sum of the deviations is zero. It is noted here that the word *moment* is sometimes used as a synonym for the word *deviation*.

When each of these deviations is squared, $x^2$, the sum of these squares is at a minimum. In other words, the sum of the deviations taken about the mean squared is less than would be the sum of such squares taken about any other point in this distribution. This value of 70, then, is the minimum that the sum of the squares can take for this distribution. Note that the sum of the squares ($SS$) will have extensive use in the rest of this book.

## AVERAGING MEANS

Suppose that there are three means based on the performance of three groups of individuals on the same test. The mean of each group has already been found and later we want to find the mean of the three groups combined, the total or grand mean ($\bar{X}_T$) as it is sometimes called. It is done as follows:

| | $\bar{X}$ | $N$ | $N\bar{X}$ |
|---|---|---|---|
| **Group 1** | **60** | **20** | **1200** |
| **Group 2** | **50** | **40** | **2000** |
| **Group 3** | **40** | **50** | **2000** |
| | | $\Sigma N$ **or** $N_T = 110$ | $\Sigma N\bar{X} = 5200$ |

In the above, each mean is multiplied by its corresponding $N$ and these products are summed. This sum is then divided by the total number of cases.

$$\bar{X}_T = \frac{\Sigma N\bar{X}}{N_T} = \frac{5200}{110} = 47.3 \tag{3.3}$$

In the above, the total sum of the scores, $\Sigma X$, is obtained by writing the formula for the mean differently. We know that

$$\bar{X} = \frac{\Sigma X}{N}$$

Then, clearing the fraction, we have $\Sigma X = N\bar{X}$. So to get the sum of the scores, $\Sigma X$, one multiplies each mean by its corresponding $N$.

## THE MEDIAN (*Mdn*)

The *median* is defined as that point in the distribution with 50 percent of the scores or cases on each side of it. In other words, the median is the midpoint of a distribution. In the previous chapter we learned how to find any centile point. As noted there, the median is the 50th centile. In this chapter we shall rework the procedure for finding the median and discuss its uses as an average.

First we shall obtain the median for a small series of numbers. In the first example we have the following series:

20, 21, 24, 25, 26, 27, 29, 30, 31, 35

The median in this series is 26.5 because this point has an equal number of cases on both sides. 26.5 is the midpoint of this distribution.

In the second example we have this series:

20, 21, 24, 25, 26, 27, 29, 30, 31

Here 26 is the median because it has 4 cases on each side of it. Thus when there is no duplication of scores near the center of the distribution, obtaining the median is a simple task.

For the third example, consider the following series of scores:

20, 21, 22, 22, 22, 23, 28, 33

To get the exact median for this series we proceed as follows. First we divide the number of cases by 2, which in this example results in dividing 8 by 2 and obtaining 4. So the median is the point with 4 cases on each side of it. Counting in from the left we have two cases over to the lower limit of 22, which is 21.5. Beyond the lower limit of 22, we need two more cases, the difference between the total number needed (4) and the number that we have already counted (2). There are three 22's and we assume that they are equally distributed through the interval 22, beginning at 21.5 and going to 22.5. Since we need 2 out of these 3 cases, we must go two-thirds of the way through the interval 21.5–22.5. Since 2/3 is .67, this is added to 21.5 to result in a median of 22.2. This can all be summarized as follows:

$$
\begin{aligned}
Mdn &= 21.5 + \frac{2}{3}(1) \\
&= 21.5 + .67 \\
&= 21.5 + .7 \\
&= 22.2
\end{aligned}
$$

Here is another example:

7, 9, 11, 11, 11, 11, 12, 13, 18, 19, 22

First we divide $N$ by 2, which gives $11/2 = 5.5$. We then start counting from the left. After using the first two numbers we are at the lower limit of 11 which is 10.5. We now need 3.5 more cases $(5.5 - 2)$ of the 4 in the interval of 11. This time we have to go 3.5/4 of the distance through the interval which has limits of 10.5–11.5. Again this is summarized:

$$
\begin{aligned}
Mdn &= 10.5 + \frac{5.5 - 2}{4}(1) \\
&= 10.5 + \frac{3.5}{4}(1) \\
&= 10.5 + .875 \\
&= 11.4
\end{aligned}
$$

As we might expect, the median can also be obtained by working from the other end of the distribution. In this case,

$$
\begin{aligned}
Mdn &= 11.5 - \frac{5.5 - 5}{4}(1) \\
&= 11.5 - \frac{.5}{4}(1) \\
&= 11.5 - .125 \\
&= 11.4
\end{aligned}
$$

It is important to note that, when working from the other end, a minus sign replaces the plus sign. For other samples such as 9, 27, 42, 45, the process is similar. Here the median is 34.5, halfway between the top of 27 and the bottom of 42.

When data are grouped, the process of finding the median is shown in Table 3.2. Here the formula for finding any centile, (2.1), is used.

**TABLE 3.2. FINDING THE MEDIAN FOR GROUPED DATA**

| | X | f | |
|---|---|---|---|
| | 100–109 | 8 | |
| | 90–99 | 12 | |
| | 80–89 | 26 | |
| | 70–79 | 32 | |
| | 60–69 | 38 | |
| 59.5 ← | | | 84 |
| | 50–59 | 33 | |
| | 40–49 | 23 | |
| | 30–39 | 12 | |
| | 20–29 | 8 | |
| | 10–19 | 6 | |
| | 0–9 | 2 | |
| | | $N = 200$ | |

$$Mdn = ll + \frac{Np - cf}{f_i}(i)$$
$$= 59.5 + \frac{(200)(.50) - 84}{38}$$
$$= 59.5 + \frac{(100 - 84)}{38}(10)$$
$$= 59.5 + \frac{16}{38}(10)$$
$$= 59.5 + 4.21$$
$$= 63.7$$

Or coming down from the top

$$Mdn = 69.5 - \frac{(100 - 78)}{38}(10)$$
$$= 69.5 - \frac{16}{38}(10)$$
$$= 69.5 - 5.8$$
$$= 63.7$$

## THE MODE (*Mo*)

The *mode* is defined as the value that appears most frequently in a distribution of scores. In the following:

10, 9, 8, 7, 7, 7, 5, 4, 3, 3, 2, 1

7 is the mode. When data are grouped, the mode is the midpoint of the interval containing the largest number of cases. The mode for the data in Table 3.2 is 64.5.

## USE OF THE THREE AVERAGES

The mode is the least reliable of the three averages. It is properly used with nominal data, and may also be used with ordinal, interval, and ratio data. A statistic is said to be *reliable* when it is stable, varying little from sample to sample.

The median is the appropriate average to use with ordinal data or when a distribution departs from normal. In reality many distributions, when examined, are shown not to be normal. Distributions of salaries are usually positively skewed. Scores on personality tests are often distributed in many ways other than normal. Very often distributions are truncated, that is, they have their terminal intervals labeled, such as $50,000-and-above or 25-and-below. Quite often census data are averaged using the median. Averages found in daily newspapers are often medians.

The mean is used as the average with interval or ratio data when distributions are symmetrical or approximate the shape of the normal curve. The student will note in going through this book that many statistics are based on the mean. Hence if further statistical computations are to be made, it is often essential that the mean be used as the average. This is why means are frequently computed on data that are skewed or are not equal intervals in nature; the researcher has little choice when using certain statistical procedures. To illustrate the use of the different averages with a skewed distribution, consider the following contributions made by 9 office workers to a United Fund campaign:

| $X$ | $f$ | |
|---|---|---|
| 200 | 1 | |
| 50 | 1 | |
| 20 | 1 | |
| 10 | 1 | |
| 7 | 1 | |
| 5 | 3 | |
| 3 | 1 | |
| | $N = 9$ | $\Sigma fX = 305$ |

Finding the averages of the above results in $\bar{X} = \$33.90$, $Mdn = \$7.00$, and $Mo = \$5.00$. Using the mean in this case gives a gross distortion of the typical contribution. Even the median is a bit high. In this case the average that best shows the typical contribution is the mode. There have been many cases where, in order to prove a point, a statistician has used the mean when he should have used one of the other averages, or vice versa.

When a diagram is made showing the three averages in a skewed distribution such as the one above, the relationships shown in Figure 3.1 result. In this figure we see that the mean is pulled into the tail of this positively skewed distribution. The values of the three averages would appear in reverse order in a negatively skewed distribution. Note that in the above distribution if the top contribution had been $200,000, the median would still be $7, the mode $5, and that the mean would have

FIG. 3.1 THE THREE AVERAGES IN A POSITIVELY SKEWED DISTRIBUTION.

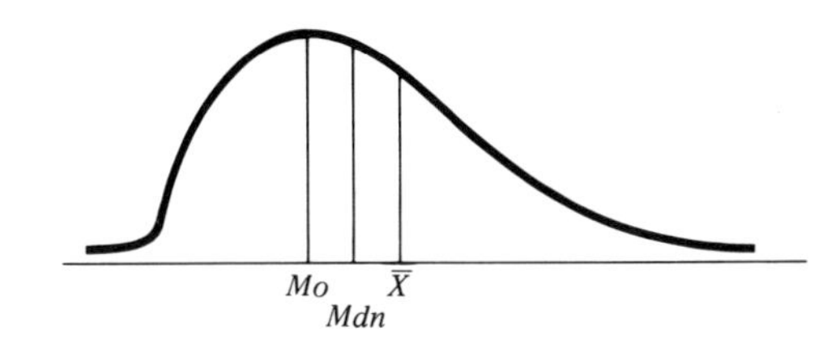


increased drastically. The point is that every single score in a distribution affects the size of the mean. This is not true with the other two averages.

## THE SAMPLE MEAN AND THE POPULATION MEAN

In Chapter 1 we differentiated between sample and population and statistics and parameters. In this chapter we have been considering the sample mean, a descriptive statistic. Frequently a sample value, a statistic, is used as an estimate of a parameter. In later chapters we shall consider the making of a statistical inference. Suffice here to say, the sample mean is an *unbiased* estimate of the population parameter, $\mu$ (mu). That is, if we drew a large number of samples from the same population, the means of these samples would be distributed about $\mu$ and the average of these sample means would be equal to the parameter mean. Therefore, in the long run, a sample mean, while it may be larger or smaller than the population mean, is an unbiased estimate of the parameter. We shall see that this is not true of all statistics.

In summary we can say the following:

The mode is used with nominal data or any distribution when haste is necessary.

The median is used with ordinal data or higher, especially when data depart from normal.

The mean is used with interval or ratio data. It is associated with a symmetrical or normal distribution.

## EXERCISES

1. The following represent the scores of 10 subjects on a series of short tests. Find the mean, median, and mode for each.
   a. 24, 21, 20, 19, 18, 17, 16, 15, 14, 12
   b. 22, 20, 19, 18, 18, 18, 18, 16, 16, 10
   c. 28, 27, 27, 26, 26, 26, 26, 24, 24, 22
   d. 30, 30, 30, 29, 29, 29, 29, 28, 28, 28
   e. 32, 30, 30, 29, 29, 24, 24, 24, 23, 22

2. The following are the scores of 15 students on a test. Find the three averages for these scores.

| | | |
|---|---|---|
| 760 | 620 | 685 |
| 740 | 580 | 670 |
| 740 | 570 | 665 |
| 710 | 570 | 520 |
| 690 | 540 | 515 |

3. On a test of psychomotor skills the following results were obtained. Find the three averages for these scores.

| $X$ | $f$ |
|---|---|
| 18 | 1 |
| 17 | 2 |
| 16 | 0 |
| 15 | 3 |
| 14 | 7 |
| 13 | 12 |
| 12 | 18 |
| 11 | 24 |
| 10 | 16 |
| 9 | 8 |
| 8 | 4 |
| 7 | 3 |
| 6 | 2 |
| 5 | 0 |
| 4 | 1 |
| 3 | 0 |
| 2 | 1 |
| | $N = 102$ |

4. Given the following data, what is the mean when all three samples are combined?

| | $\bar{X}$ | $N$ |
|---|---|---|
| (1) | 42 | 40 |
| (2) | 38 | 20 |
| (3) | 40 | 45 |

# 4

# VARIABILITY

Knowing the average scores of three groups on a test gives us some information about the groups, but not enough to be of much use. As an illustration, suppose that the mean of each of three groups on an achievement test is 90. From this we might conclude that the three groups are very similar in achievement. This may or may not be the case, for before anything definite may be said about the three groups, we must know how the scores of each group spread about the mean. In other words, we must know something about the *variability* of the groups. Suppose that in the first group (A) with a mean of 90, the lowest score is 20 and the highest is 160. In the second group (B) whose mean is 90, the low score is 80 and the high score 100. In the third group (C), also with a mean of 90, the scores range from 50 to 130. Examination of the range of the three groups shows that they are indeed quite different in reference to their performance on this test (see Figure 4.1). It follows then that along with an average score for each group we must also have a measure of the variability or dispersion of the scores about the mean. In this chapter we shall discuss three such measures of variability.

## THE RANGE

The *range* is defined as the high score minus the low score plus one, or plus one of the smallest units of measurement.

$$\text{Range} = (X_h - X_l) + 1$$

For the data in the above paragraph, the ranges are 141, 21, and 81, respectively.

FIG. 4.1 THREE DISTRIBUTIONS WITH THE SAME MEAN.

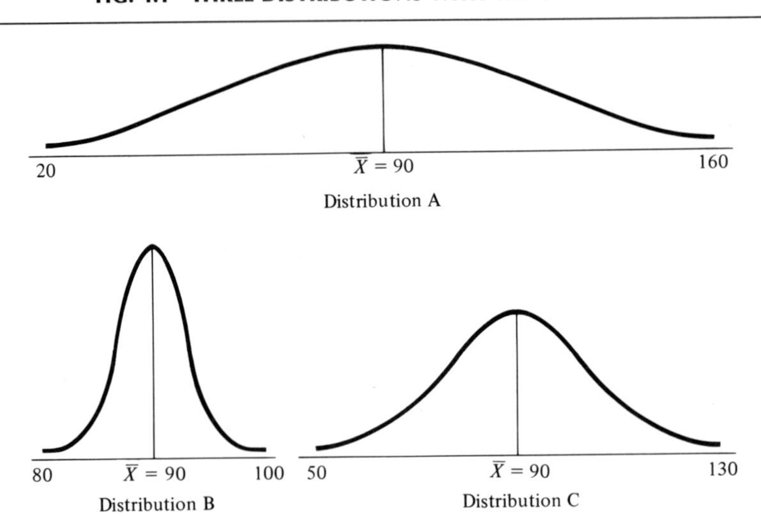


As a statistic the range leaves much to be desired. For example, in the first case where the scores run from 20 to 160, suppose that the only low score is 20 and that the next score in order of size is 50. Then, if there were no low score of 20, the range would now be 111. Thus one low score is responsible for an increase of 30 points in the range. When a single score produces a difference like this, we say that the result is unreliable. The range then tends to be an *unreliable* statistic.

Sometimes the reliability of the range is improved by lopping off the top and bottom 5 percent of the scores and using the middle 90 percent in determining the range. It is obvious that such a statistic, being rid of the high and low cases, is a more reliable statistic than the actual range.

## THE STANDARD DEVIATION AND VARIANCE

By far the most important measure of variability is the *standard deviation,* or its square, the *variance*. The standard deviation of a sample ($S$) is defined as

$$S = \sqrt{\frac{\Sigma x^2}{N}} \tag{4.1a}$$

where

$$x = (X - \overline{X}) \quad \text{and} \quad x^2 = (X - \overline{X})^2$$

The standard deviation of a population ($\sigma$) is defined in the same way:

$$\sigma = \sqrt{\frac{\Sigma x^2}{n}} \tag{4.1b}$$

As a simple illustration we have

| | x | $x^2$ |
|---|---|---|
| $X$ | $(X - \bar{X})$ | $(X - \bar{X})^2$ |
| 10 | 5 | 25 |
| 8 | 3 | 9 |
| 6 | 1 | 1 |
| 4 | −1 | 1 |
| 2 | −3 | 9 |
| 0 | −5 | 25 |
| $\Sigma = 30$ | $\Sigma = 0$ | $\Sigma = 70$ |

$$\bar{X} = \frac{\Sigma X}{N} = \frac{30}{6} = 5$$

Then

$$S = \sqrt{\frac{\Sigma x^2}{N}}$$

$$= \sqrt{\frac{70}{6}}$$

$$= \sqrt{11.6667}$$

$$= 3.4$$

Since it can be shown that

$$\Sigma x^2 = \Sigma X^2 - \frac{(\Sigma X)^2}{N} \tag{4.2}$$

is the sum of squares, a more direct solution follows:

| $X$ | $X^2$ |
|---|---|
| 10 | 100 |
| 8 | 64 |
| 6 | 36 |
| 4 | 16 |
| 2 | 4 |
| 0 | 0 |
| $\Sigma = 30$ | $\Sigma = 220$ |

$$\Sigma x^2 = \Sigma X^2 - \frac{(\Sigma X)^2}{N}$$

$$= 220 - \frac{(30)^2}{6}$$

$$= 220 - \frac{900}{6}$$

$$= 220 - 150$$

$$= 70$$

This value of 70, called the *sum of the squares* and frequently denoted by the letters *SS*, is identical with that obtained by the other method.

Combining equations (4.1a) and (4.2) produces a more direct formula for obtaining the standard deviation.

Equation (4.1a) $$S = \sqrt{\frac{\Sigma x^2}{N}}$$

Equation (4.2) $$\Sigma x^2 = \Sigma X^2 - \frac{(\Sigma X)^2}{N}$$

$$S = \sqrt{\frac{\Sigma X^2 - (\Sigma X)^2/N}{N}}$$

$$= \sqrt{\frac{1}{N}\left[\Sigma X^2 - \frac{(\Sigma X)^2}{N}\right]}$$

$$= \sqrt{\frac{1}{N}\left[\frac{N\Sigma X^2 - (\Sigma X)^2}{N}\right]}$$

$$= \frac{1}{N}\sqrt{N\Sigma X^2 - (\Sigma X)^2} \tag{4.3}$$

## THE STANDARD DEVIATION AS A BIASED ESTIMATE OF THE PARAMETER STANDARD DEVIATION

A statistic is an unbiased estimate of its parameter if the mean of a large number of sample values equals the parameter value. For reasons that will not be discussed here, the sample standard deviation is a biased estimate of the parameter standard deviation. This bias is such that the sample standard deviation tends to be an underestimation of the parameter standard deviation. To estimate the population standard deviation, $s$ as defined below should be used

$$s = \sqrt{\frac{\Sigma x^2}{N - 1}} = \sqrt{\frac{N\Sigma X^2 - (\Sigma X)^2}{N(N - 1)}} \tag{4.4}$$

The use of $N - 1$ in the above formula tends to eliminate this bias and it is apparent that when $N$ is small such a correction is necessary when the sample standard deviation is used as an estimate of the parameter standard deviation. For the present, we shall use $S$ as obtained in formula (4.3) for the standard deviation in describing the variability of a known group. Later, when we are concerned with inferential statistics, we shall use formula (4.4) to produce unbiased estimates of the parameter.

## THE VARIANCE

The standard deviation squared is called the *variance*:

$$S^2 = \frac{\Sigma x^2}{N} \quad \text{or} \quad s^2 = \frac{\Sigma x^2}{N - 1} \tag{4.5}$$

Since we are dividing the sum of the squares by the number of cases, we are obtaining the *mean square.* The term mean square (*MS*) is often used as a name for the variance.

## STANDARD DEVIATION AND THE NORMAL CURVE

Standard deviation units are useful in describing relationships associated with the normal curve. Suppose that with data that approximate a normal distribution we compute the mean and standard deviation and find them to be 73.3 and 20.2, respectively. If one standard deviation, 20.2, is taken on each side of this mean, notice that approximately 68 percent of the area of the curve has been cut off (see Figure 4.2). Instead of 68 percent of the area, we might say that 68 percent of the cases are included in the designated area because we assume that the cases are equally spread over the area of the curve. From Figure 4.2 we see then that the points that include 68 percent of the cases are 53.1 and 93.5. Sometimes it is stated that approximately two-thirds of the cases are included in the area defined by the mean plus and minus one standard deviation. When two standard deviations (often called *sigma* units) are measured off on each side of the mean, approximately 95 percent of the area is included. Finally, three standard deviations taken on each side of the mean will include practically all of the area or cases, there being only 13 in 10,000 cases remaining in the area on each side of the mean beyond three standard deviation units.

## THE RANGE AND STANDARD DEVIATION

The percentages and values described in the above paragraph are true only when there is a large number of cases, say 500 or more. As the number of cases decreases, a smaller number of standard deviations is needed

FIG. 4.2 RELATIONSHIP BETWEEN STANDARD DEVIATION UNITS AND THE NORMAL CURVE.

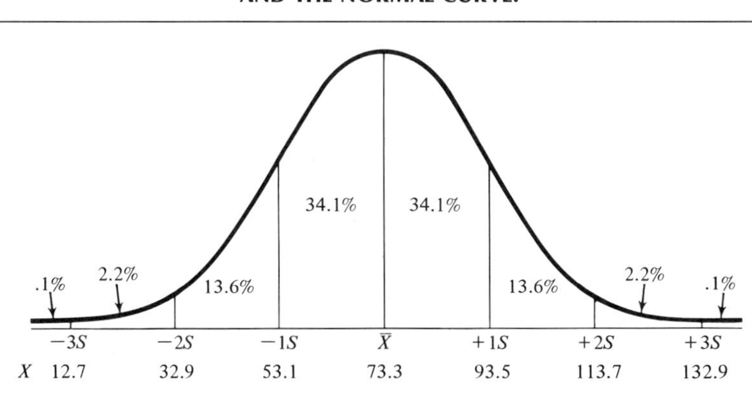

to cover the range, as is shown in the following table.[1] Therefore, when the number of cases is small, it is important that we consider this.

| $N$ | NUMBER OF STANDARD DEVIATIONS INCLUDED IN THE RANGE |
|---|---|
| 5 | 2.3 |
| 10 | 3.1 |
| 25 | 3.9 |
| 30 | 4.1 |
| 50 | 4.5 |
| 100 | 5.0 |
| 500 | 6.1 |
| 1000 | 6.5 |

## AVERAGING STANDARD DEVIATIONS

When two or more standard deviations are to be averaged, it is done by the following formula:

$$S_T = \sqrt{\frac{N_1(\bar{X}_1^2 + S_1^2) + N_2(\bar{X}_2^2 + S_2^2)}{N_1 + N_2} - \bar{X}_T^2} \tag{4.6}$$

where

$S_T$ = standard deviation of the combined groups
$N_1, N_2$ = the number in each group
$\bar{X}_1, \bar{X}_2$ = the mean of each group
$\bar{X}_T$ = the average or weighted mean of the two groups [see (3.3)]
$S_1, S_2$ = the standard deviations of the two groups

The above formula may be modified when there are more than two groups by putting additional terms in both the numerator and denominator.

## CODING DATA

When numbers are large and a small pocket calculator is being used, it is almost imperative that data be coded to meet the capacities of the calculator. Below we shall discuss the effects of coding by addition and subtraction and then the effects of multiplication and division on both the means and standard deviations.

[1] Adapted from L. W. C. Tippett. "On the Extreme Individuals and the Range of Samples from a Normal Population." *Biometrika,* 1925, *17,* 386.

CODING BY ADDITION OR SUBTRACTION OF A CONSTANT

First we consider the effect on the mean and the standard deviation resulting from adding or subtracting a constant from each score in a distribution. This is illustrated below:

| $X$ | $x$ | $x^2$ | $X + 10$ | $x$ | $x^2$ |
|---|---|---|---|---|---|
| 10 | 4 | 16 | 20 | 4 | 16 |
| 8 | 2 | 4 | 18 | 2 | 4 |
| 6 | 0 | 0 | 16 | 0 | 0 |
| 4 | −2 | 4 | 14 | −2 | 4 |
| 2 | −4 | 16 | 12 | −4 | 16 |
| $\Sigma X = 30$ | $\Sigma x = 0$ | $\Sigma x^2 = 40$ | $\Sigma = 80$ | $\Sigma = 0$ | $\Sigma = 40$ |

**For the original data:**

$$\bar{X} = \frac{\Sigma X}{N} = \frac{30}{5} = 6$$

$$S = \sqrt{\frac{\Sigma x^2}{N}} = \sqrt{\frac{40}{5}} = \sqrt{8} = 2.83$$

**For the coded data:**

$$\bar{X} = \frac{\Sigma X}{N} = \frac{80}{5} = 16$$

$$S = \sqrt{\frac{40}{5}} = \sqrt{8} = 2.83$$

From the above we see that when a constant is added to each score in a distribution, the same constant is added to the mean, but the standard deviation is not affected. Hence, in cases such as this the coded mean has to be uncoded to get the real mean. Similar results are obtained when a constant is subtracted from each score in a distribution, but the coded mean is less than the actual mean of a size equal to the constant.

CODING BY MULTIPLICATION AND DIVISION

The effects of multiplying each score in a distribution by a constant are shown below:

| $X$ | $x$ | $x^2$ | $2X$ | $x$ | $x^2$ |
|---|---|---|---|---|---|
| 20 | 4 | 16 | 40 | 8 | 64 |
| 18 | 2 | 4 | 36 | 4 | 16 |
| 16 | 0 | 0 | 32 | 0 | 0 |
| 14 | −2 | 4 | 28 | −4 | 16 |
| 12 | −4 | 16 | 24 | −8 | 64 |
| $\Sigma X = 80$ | $\Sigma x = 0$ | $\Sigma x^2 = 40$ | $2X = 160$ | $\Sigma x = 0$ | $\Sigma x^2 = 160$ |

$$\bar{X} = \frac{\Sigma X}{N} = \frac{80}{5} = 16$$

$$S = \sqrt{\frac{\Sigma x^2}{N}} = \sqrt{\frac{40}{5}} = \sqrt{8} = 2.83$$

$$\bar{X} = \frac{\Sigma X}{N} = \frac{160}{5} = 32$$

$$S = \sqrt{\frac{160}{5}} = \sqrt{32} = 5.66$$

(*Note:* $2 \times 2.83 = 5.66$)

From the above we see that when each score in a distribution is multiplied by a constant, *both* the mean and standard deviation are multiplied by the same constant. Hence, in this case both of the obtained statistics have to be divided by the same constant to obtain the real mean and standard deviation. Similarly it can be shown that when each score in a distribution is divided by a constant, both the mean and standard deviation are divided by the same constant.

The student will save much time and effort if the above rule is applied whenever possible. For example, suppose that one has a group of scores ranging from 760 to 520. These would best be coded by subtracting 500 from each score as it is entered in the calculator. Or suppose we have a group of measures such as .13, .18, .24, and so on. Coding here would consist of removing the decimal points, that is, multiplying each score by 100. The student must remember that, when data are coded, the answers often have to be uncoded depending on the type of coding carried out.

## A SUGGESTED MODEL FOR OBTAINING THE STANDARD DEVIATION

In Table 4.1 the suggested model for computing a mean and standard deviation is shown. First, all the values in column (1) have had the con-

**TABLE 4.1. MODEL FOR COMPUTING MEAN AND STANDARD DEVIATION**

| (1) | (2) | (3) |
|---|---|---|
| $X$ | $X - 35$ | $(X - 35)^2$ |
| 58 | 23 | 529 |
| 56 | 21 | 441 |
| 52 | 17 | 289 |
| 50 | 15 | 225 |
| 48 | 13 | 169 |
| 48 | 13 | 169 |
| 46 | 11 | 121 |
| 45 | 10 | 100 |
| 44 | 9 | 81 |
| 44 | 9 | 81 |
| 42 | 7 | 49 |
| 40 | 5 | 25 |
| 38 | 3 | 9 |
| 37 | 2 | 4 |
| 35 | 0 | 0 |
| | $\Sigma = 158$ | $\Sigma = 2292$ |

$$S = \frac{1}{N}\sqrt{N\Sigma X^2 - (\Sigma X)^2}$$

$$= \frac{1}{15}\sqrt{15(2292) - (158)^2}$$

$$= .07\sqrt{34380 - 24964}$$

$$= .07\sqrt{9416}$$

$$= .07(97.04)$$

$$= 6.79 = 6.8$$

Coded mean:

$$\bar{X}_c = \frac{158}{15}$$

$$= 10.5$$

Uncoded mean:

$$\bar{X} = 10.5 + 35$$

$$= 45.5$$

stant of 35 subtracted from them. The remainders appear in column (2) and as far as the formula is concerned these coded values are now the $X$ values. Then each number in column (2) is squared and the squares placed in column (3). Columns (2) and (3) are then summed and these sums are entered in formula (4.3) as shown. Note that, since we coded by subtraction, only the obtained mean has to be uncoded to get the real mean.

## THE QUARTILE DEVIATION OR SEMI-INTERQUARTILE RANGE

A measure of dispersion that is used when data are skewed, or in some other way depart from normal and the median is used as the average, is the *quartile deviation* ($Q$). This is defined as

$$Q = \frac{Q_3 - Q_1}{2} \tag{4.7}$$

where

$Q_3$ = the third quartile
$Q_1$ = the first quartile

The method of obtaining this statistic for grouped data is illustrated in Table 4.2.

**TABLE 4.2 COMPUTATION OF THE QUARTILE DEVIATION FOR GROUPED DATA**

| | X | f | |
|---|---|---|---|
| | 120–129 | 1 | |
| | 110–119 | 4 | |
| | 100–109 | 8 | |
| | 90–99 | 12 | |
| 89.5 ← | | | 25 |
| | 80–89 | 18 | |
| | 70–79 | 22 | |
| | 60–69 | 26 | |
| | 50–59 | 24 | |
| 49.5 ← | | | 35 |
| | 40–49 | 18 | |
| | 30–39 | 8 | |
| | 20–29 | 7 | |
| | 10–19 | 2 | |
| | | $\Sigma = 150$ | |

First obtain $Q_1$:

$$25 \text{ percent of } 150 = \frac{150}{4} = 37.5$$

$$Q_1 = 49.5 + \frac{2.5}{24}(10)$$
$$= 49.5 + 1.0$$
$$= 50.5$$

$$Q_3 = 89.5 - \frac{12.5}{18}(10)$$
$$= 89.5 - 6.9$$
$$= 82.6$$

$$Q = \frac{Q_3 - Q_1}{2}$$
$$= \frac{82.6 - 50.5}{2}$$
$$= \frac{32.1}{2}$$
$$= 16.05 = 16.0$$

When the data are ungrouped, the process is as shown below.

| | | | $X$ | | | | |
|---|---|---|---|---|---|---|---|
| 40 | 37 | 28 | 25 | 19 | 15 | 9 | 6 |
| 39 | 36 | 28 | 22 | 17 | 12 | 8 | 6 |
| 39 | 30 | 27 | 20 | 17 | 11 | 8 | 4 |
| 39 | 29 | 27 | 18 | 16 | 11 | 7 | 2 |
| 38 | 28 | 26 | 18 | 16 | 10 | 7 | 1 |

In this distribution there are 40 scores. First the values are ordered by magnitude, as shown. Then we take 25 percent of $N$ or divide $N$ by 4 to obtain the number of scores below $Q_1$. In this case this number is 10. Counting up from the bottom we find $Q_1$ to be 9.5. Counting down from the top 10 scores takes us to the bottom of 29 where we have to interpolate to get the tenth score.

$$Q_3 = 28.5 - \frac{1}{3}$$
$$= 28.5 - .3$$
$$= 28.2$$

$$Q = \frac{Q_3 - Q_1}{2}$$
$$= \frac{28.2 - 9.5}{2}$$
$$= \frac{18.70}{2}$$
$$= 9.35 = 9.4$$

In any distribution, one quartile deviation taken on each side of the median will include 50 percent of the cases. When $N$ is large it is also true that eight quartile deviations taken on each side of the median include practically all of the cases.

## WHEN THE VARIOUS MEASURES OF VARIABILITY ARE USED

1. Range. This is the least reliable of the measures and is used only when one is in a hurry to get a measure of variability. It may be used with ordinal, interval, or ratio data.
2. Standard Deviation and the Variance. The standard deviation is used whenever a distribution approximates a normal distribution. As will be shown it is the basis for much of the statistics that come later in this book. As the most reliable measure of variability it is used with interval and ratio data.

3. Quartile Deviation. This statistic is used when the median is used as an average; that is, when data depart noticeably from normal. It is also the measure to be used with ordinal data.

## EXERCISES

1. For the following find $\bar{X}$, $S$, $s$, and $Q$.
   a. 20, 19, 18, 16, 15, 14, 13, 12, 8
   b. 34, 34, 30, 29, 28, 25, 20, 18, 12, 8
   c. 40, 38, 36, 32, 24, 20, 18, 10
   d. 10, 10, 10, 10, 10, 10, 10, 10
   e. 12, 11, 11, 11, 9, 9, 9, 9, 8, 7, 2

2. On a statistics test a group of students obtained the following scores. Find both $S$ and $\bar{X}$. Code your data and use formula (4.3).

| | | | |
|---|---|---|---|
| 41 | 34 | 31 | 24 |
| 40 | 34 | 31 | 23 |
| 38 | 33 | 29 | 20 |
| 35 | 32 | 29 | 16 |
| 35 | 31 | 28 | 14 |

3. Below are the scores of 10 students on a quiz. Find $\bar{X}$, $S$, and $S^2$.

   **86, 54, 58, 78, 58, 68, 77, 78, 84, 88**

4. Two statistics classes were given the same statistics test with the following results. What is the standard deviation of the two groups combined?

| | $\bar{X}$ | $S$ | $N$ |
|---|---|---|---|
| Class 1 | 44 | 15 | 60 |
| Class 2 | 46 | 10 | 40 |

5. The following scores were obtained on a short test of psychomotor skill. Find $Q$ and evaluate the use of $Q$ for these data.

| | | | | |
|---|---|---|---|---|
| 18 | 17 | 15 | 13 | 12 |
| 18 | 16 | 14 | 13 | 12 |
| 17 | 16 | 14 | 12 | 11 |
| 17 | 15 | 14 | 12 | 11 |

6. The following are the scores of a group of high school seniors on the verbal test of the SAT. Compute $\overline{X}$ and $S$.

| | | | | | |
|---|---|---|---|---|---|
| 780 | 740 | 710 | 630 | 580 | 520 |
| 760 | 740 | 690 | 620 | 550 | 520 |
| 760 | 730 | 690 | 610 | 550 | 510 |
| 750 | 730 | 680 | 610 | 540 | 500 |
| 750 | 730 | 640 | 580 | 520 | 490 |

# 5

# STANDARD SCORES AND THE NORMAL CURVE

Among the most useful of statistics are standard scores. The basic standard score, called $z$, is defined as

$$z = \frac{X - \bar{X}}{S} = \frac{x}{S} \tag{5.1}$$

where all symbols are as previously defined. To illustrate the computation of standard scores, let us assume a test with $\bar{X} = 38$ and $S = 7.4$. Some of the scores obtained with this test are the following:

| (1) | (2) | (3) |
|---|---|---|
| $X$ | $z$ | TRANSFORMED SCORE |
| **51** | **1.76** | **68** |
| **40** | **.27** | **53** |
| **38** | **.00** | **50** |
| **30** | **−1.08** | **39** |
| **20** | **−2.43** | **26** |

By use of formula (5.1) we obtain the $z$ scores shown in column (2). Note that the mean of the test (38) has an equivalent $z$ score of zero.

Values above the mean have positive $z$ scores and those below the mean are negative. Actually $z$ scores are raw scores transferred into standard deviation units. The score of 51 has a $z$ score of 1.76. This indicates that it is 1.76 standard deviation units above the mean. If a large distribution is made it will be noted that $z$ scores have a mean of zero and a standard deviation of one. With a large sample, practically all the $z$ scores are contained within a band made up of the mean $\pm 3$ $z$-score values (Figure 5.1). It should be pointed out here that changing a distribution of scores that is negatively skewed does not make the distribution normal. The $z$ scores will have the same shape in their distribution as did the original data. Since $z$ scores are usually decimals and since in most distributions about half of them are negative, they are frequently transformed into other types of standard scores that contain no decimals and are all positive. This linear transformation is brought about in the following way: The obtained $z$ score is multiplied by the new standard deviation and this product added to the new mean:

$$z\ (\text{new } S) + \text{new } \bar{X} = \text{transformed score} \tag{5.2}$$

Suppose we take the first score in our previous distribution and transform it to a standard score with a mean of 50 and a standard deviation of 10. The $z$ score for this raw score is 1.76.

$$\begin{aligned} 1.76(10) + 50 &= 17.6 + 50 \\ &= 67.6 = 68 \end{aligned}$$

In this manner the other $z$ scores have been transformed.

In Figure 5.1 some of the most frequently encountered standard scores appear. In this figure CEEB stands for scores of tests used by the college boards, the verbal part of the Scholastic Aptitude Test (SAT-V) or the mathematics part of the Scholastic Aptitude Test (SAT-M). The

**FIG. 5.1 DISTRIBUTION OF THE VARIOUS TYPES OF STANDARD SCORES.**

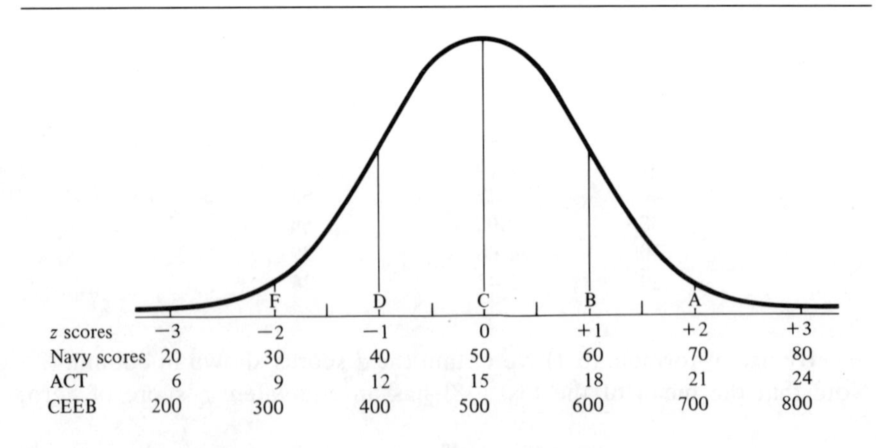

Graduate Record Examination (GRE) also uses a mean of 500 and a standard deviation of 100. ACT in the same figure represents the American College Testing program. ACT scores have a mean of 15 and a standard deviation of 3. Some intelligence tests such as the Wechsler Adult Intelligence Scale (WAIS) have a mean of 100 and a standard deviation of 15. There really is no limit to the types of standard scores that can be transformed from the basic $z$ score.

One of the major uses of standard scores is in the manipulation of the results of educational and psychological tests. As shown, they are easily made. They are also easily comprehended by the user after a little experience with them. They have another decided advantage in that they can be averaged and treated mathematically in other ways. In this respect they differ drastically from centiles. Whenever scores on different tests, quizzes, assignments, and the like are to be averaged, the scores on each measure should first be converted to $z$ scores and an average $z$ ($\bar{z}$) computed for each student. Were this not to be done, each score would be automatically (and perhaps erroneously) weighted in the average according to the relative size of the standard deviation of its respective distribution.

If an instructor wishes to give different weights to two or more tests or to weight such things as laboratory grades, each element that goes into the average should be changed into standard scores and these weighted. Suppose that an instructor gives three hourly examinations and a final and that he wishes the final examination to count one-half of the total grade. The weighting is carried out as follows:

$$\frac{z_1 + z_2 + z_3 + 3z_f}{6}$$

where

$z_1, z_2, z_3$ = the standard scores a student has obtained on each of the hourly tests

$z_f$ = the student's $z$ score on the final, taken 3 times—(it is weighted by a factor of 3)

6 = The number of elements to be accounted for in the final weighting scheme

Similarly a $z$ score for laboratory work or for an average of short quizzes could be weighted into the final score. $z$ scores that have been transformed may be weighted in the same manner as above (provided that each has been transformed with the same formula).

Standard scores also make it possible to compare an individual's performance on different tests; the scores of which have been transformed into standard scores. Suppose that on an aptitude battery a job applicant has the following standard scores:

| | |
|---|---|
| Mechanical | 62 |
| Clerical | 48 |
| Spatial | 36 |
| Computational | 42 |
| Block design | 30 |

Since all of these have a mean of 50 and a standard deviation of 10, it is very easy to determine the strong and weak points of this applicant. As the student progresses through this book he will note that there are many other uses for standard scores.

## THE NORMAL CURVE

In Chapter 2 we described the normal curve as having no skew, possessing mesokurtosis, being bilaterally symmetrical, and having one mode, with this maximum height being at the center. Also in the normal curve, the mean, median, and mode are identical. Finally the normal curve is said to be *asymptotic;* that is, the tails never touch the base line but extend to infinity in both directions.

Actually there are many normal curves, each based on a specific number of cases, a particular mean, and a particular standard deviation. There is a mathematical formula for the normal curve, but its use is beyond the scope of this book. We shall be concerned here with the so-called unit normal curve that has a mean of zero, a standard deviation of one, and with an area equal to one. The values that appear in Appendix C have been obtained by solving the equation for the normal curve for a large number of $z$ score values. In the rest of this chapter we shall be concerned with the use of Appendix C, showing its application to various statistical problems. As we shall note later, another name for Appendix C is the normal probability table.

### AREAS BETWEEN POINTS UNDER THE NORMAL CURVE

As was pointed out in Chapter 4, standard deviation units measured from the mean cut off certain proportions of the area of the normal curve or a certain percentage of the cases. The student should review Figure 4.2. First we shall consider the area of the curve or the number of cases between different points of the curve. Suppose that we have a distribution based on 500 cases with a mean of 78 and a standard deviation of 12. Various questions may now be posed. For example, how many cases would you expect to fall above a score of 100 in this distribution that we are assuming to be normal? First we find the $z$ score for the raw score of 100.

$$z = \frac{X - \bar{X}}{S} = \frac{100 - 78}{12} = \frac{22}{12} = 1.83$$

FIG. 5.2 PROPORTION OF CASES IN A NORMAL CURVE ABOVE A SCORE OF 100.

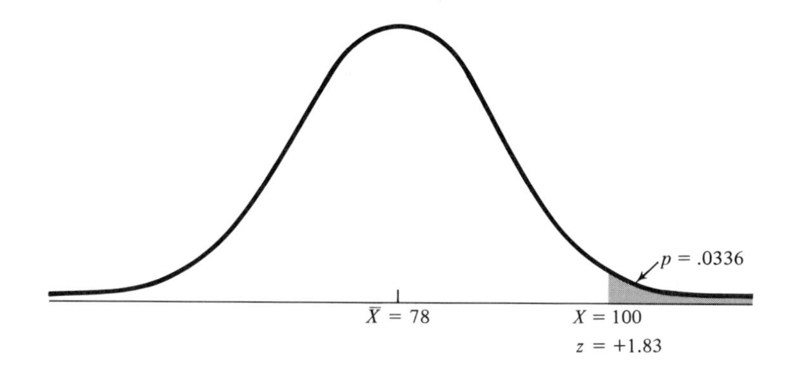


Next we go to Appendix C, enter it for a $z$ score of 1.83, and, for the area in the smaller portion, find a proportion ($p$) of .0336 (Figure 5.2). Multiplying this by 500 gives 16.8 cases.

We could also ask another question using the same data: What is the probability of an individual getting a score of 100 or more on this test? Since .0336 of the area of the normal curve is above this score of 100, we can say that the *probability* of a person who takes this test getting a score of 100 or more is 336 in 10,000, or to make it a bit more meaningful, slightly over 3 in 100.

Next we shall examine the area between two points on the curve. Suppose that for the same data we wish to find the proportion and number of cases that fall between scores of 90 and 100. We have already found the $z$ score for $X = 100$. This $z$ score, which we shall now call $z_1$, is 1.83. Now we compute the second $z$ score:

$$z_2 = \frac{X_2 - \bar{X}}{S} = \frac{90 - 78}{12} = \frac{12}{12} = 1.00$$

These two $z$ scores are shown in Figure 5.3. There are several ways of getting the desired area from Appendix C. We could get the area from the mean to each of the $z$ scores and subtract these areas or we could take the two smaller areas cut off by the $z$ scores and subtract these. Let us first follow the former procedure.

| | AREA FROM $\bar{X}$ TO $z$ | |
|---|---|---|
| $z_1 = 1.83$ | **.4664** | |
| $z_2 = 1.00$ | **.3413** | **(subtract)** |
| | **.1251** | |

Or by the second method:

FIG. 5.3 PROPORTION OF CASES BETWEEN TWO POINTS IN A NORMAL CURVE.

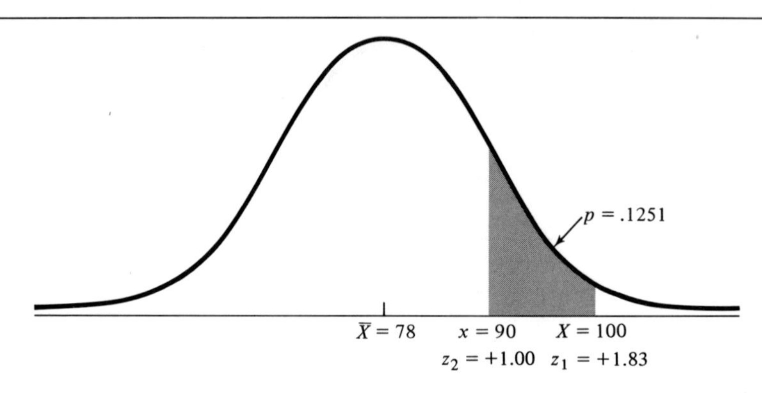


| | PROPORTION IN THE SMALLER AREA | |
|---|---|---|
| $z_2$ = **1.00** | **.1587** | |
| $z_1$ = **1.83** | **.0336** | **(subtract)** |
| | **.1251** | |

Thus .1251 of the area of the curve, or 500 × .1251 cases, or 62.55 = 63 cases fall between these two points. In terms of probability we can say that the probability of a person's score on this test falling between 90 and 100 is 125 in 1000 or 12.5 in 100. If the points with which we were concerned were on different sides of the mean, we would add the obtained areas rather than subtract them.

## FINDING CENTILES FROM THE NORMAL CURVE

Suppose, using the same data, $\bar{X} = 78$, $S = 12$, we wish to find the 40th centile ($C_{40}$). Recall that the 40th centile is that point in the distribution with 40 percent of the cases below it. For this operation we enter Appendix C looking for the standard score that has 40 percent of the cases below it. From Appendix C we see that the $z$ score that divides the area of the curve into a 40–60 split is .25. Since this centile is below the mean, the $z$ score is negative, −.25 (Figure 5.4). The next step is to convert this $z$ score of −.25 to a raw score, $X$. This is done as follows:

$$z = \frac{X - \bar{X}}{S}$$

$$zS = X - \bar{X}$$

transposing,

$$X = zS + \bar{X}$$

For the given data

$$\begin{aligned} X &= (-.25)12 + 78 \\ &= -3.00 + 78 \\ &= 75 \end{aligned}$$

Hence for these data, $C_{40}$ is 75.

With little effort $C_{60}$, the complement of $C_{40}$, may be obtained. As noted above a $z$ score of .25 divides the area of the curve into approximately a 60–40 split. Thus if we use the positive value of .25 in the above equation we will have $C_{60}$:

$$\begin{aligned} X &= (+.25)12 + 78 \\ &= 3.00 + 78 \\ &= 81 \end{aligned}$$

In addition to finding centile scores or points, Appendix C may be used in obtaining centile ranks. A centile rank was defined as that score with a certain percent of the cases below it. Again for the same data suppose we wish to obtain the centile rank for a score of 98 (Figure 5.5):

$$z = \frac{X - \bar{X}}{S} = \frac{98 - 78}{12} = \frac{20}{12} = 1.67$$

From Appendix B we see that a $z$ score of $+1.67$ has .9525 of the area in the larger portion or below it. Hence an $X$ of 98 has a centile rank of 95.

FIG. 5.4 OBTAINING CENTILE POINTS FROM THE NORMAL CURVE.

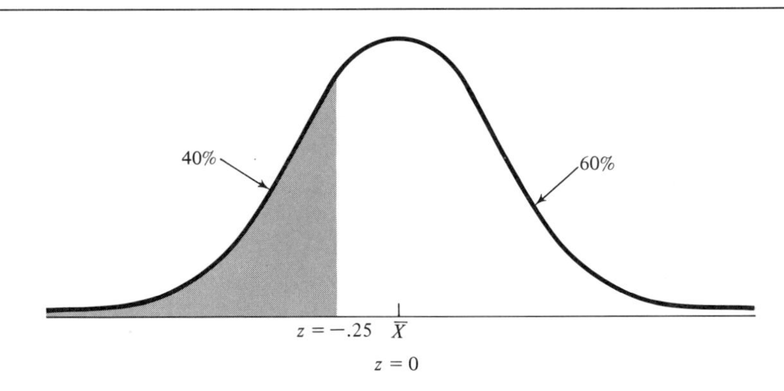

FIG. 5.5 OBTAINING CENTILE RANKS FROM THE NORMAL CURVE.

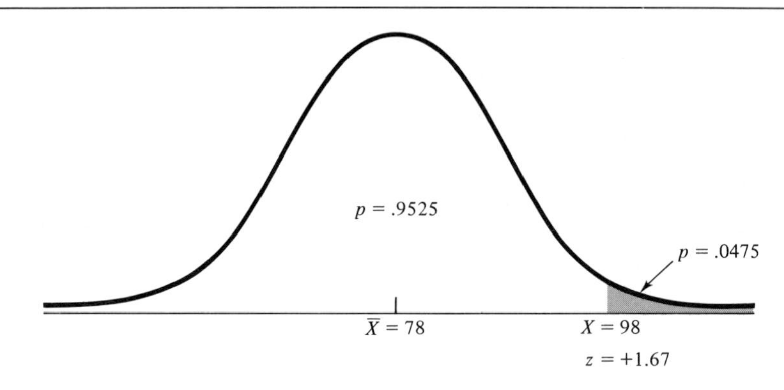


## EXERCISES

1. Given the following data related to a test: $\bar{X} = 94$, $S = 14$, $N = 800$.
   a. Convert each of the following scores to a $z$ score:

      (1) 94 (2) 80 (3) 106
      (4) 64 (5) 130 (6) 71

   b. Now change each of the $z$ scores obtained in part a to a standard score with a mean of 500 and $S = 100$.
   c. Change each of the scores in part a to an ACT score with $\bar{X} = 15$ and $S = 3$.
2. Below are the scores of a student on three tests. Also are presented the $\bar{X}$ and $S$ of each test:

| TEST 1 | TEST 2 | TEST 3 |
|---|---|---|
| **80** | **75** | **12** |
| $\bar{X} = 90$ | $\bar{X} = 60$ | $\bar{X} = 6$ |
| $S = 12$ | $S = 14$ | $S = 2$ |

   a. What is his average standard score on these tests? For standard scores, use $\bar{X} = 50$, $S = 10$.
   b. Suppose that the test scores were to be weighted 2, 4, and 1. What is his average weighted standard score?
3. A naval recruit scores as follows on a series of tests all of which have a mean of 50 and standard deviation of 10.

| | |
|---|---|
| Clerical ability | 62 |
| Mechanical comprehension | 54 |
| Finger dexterity | 38 |
| General information | 50 |
| Spatial visualization | 41 |

a. What is his average score on this test battery?
b. What inferences can be made about this man from his performance on this test battery?

4. Given a normal distribution with $\bar{X} = 70$, $S = 15$, and $N = 700$:
   a. How many individuals would you expect to find above each of these scores in this distribution?

      80, 48, 110

   b. How many scores would you expect to find between the following pairs of scores?

      80–110, 60–90, 85–100

   c. In part a what is the probability of an individual obtaining a score higher than each of the scores listed there?
   d. In part b what is the probability of an individual getting a score between 80 and 110?
5. Using these data: $\bar{X} = 60$, $S = 12$, $N = 500$, find the following centile points:

   $C_{90}$, $C_{10}$, $C_{80}$, $C_{20}$

6. In the distribution used in Exercise 5, what is the centile rank of a score of 50?
7. With large samples which of the following do you think would approximate a normal distribution?
   a. The weights of individual apples in a bushel of apples of the same kind from the same tree.
   b. Scores on a test of artistic ability.
   c. The income of a group of 40-year-old women in any state.
   d. The life of automobile tires guaranteed for 30,000 miles.
   e. The distribution about the population mean of the means of a large number of samples of size 15 drawn from a normally distributed population.
   f. The distribution about the population mean of the means of a large number of samples of the same size drawn from a slightly skewed population.
   g. The measurements made by a large group of individuals in measuring the intensity of the light of a particular star.
   h. The heights of 1000 randomly selected men.

# II

# INFERENTIAL STATISTICS

# 6

# POPULATIONS, SAMPLES, AND INFERENCE

In Part I, statistics and parameters were studied which serve to describe and summarize sample or population data distributions. Part II will be concerned with ways in which statistics computed on sample data may be used to make inferences and test hypotheses about populations. To develop an understanding of this process, we must first take a more detailed look at the nature of populations, samples, and the concepts which underlie statistical inference.

## POPULATIONS

Before any type of meaningful research can be conducted, the investigator must define the population being studied as thoroughly as possible. In general, a *population* is a systematically defined group or set of life forms, objects, or events. The things that make up populations, their *members,* may consist of such diverse phenomena as seventh grade students, American elm trees, protozoa, white mice, hydrogen atoms, automobile accidents, and so forth. A little reflection on the part of the student will demonstrate that populations are generally dynamic—constantly changing in both character and composition across time. Populations of seventh grade students, for example, change nearly 100 percent in composition from one school year to the next and the physical and mental characteristics of their members are undergoing continuous change. Thus, *time* is a critical dimension and must receive attention in population definitions.

A defined population is merely a collection of members. To form a data distribution for a population, each of its members must be observed or measured on some *criterion* variable. The $X$ values obtained are *entries* in the population data distribution. Individual seventh graders, then, are members of a population of seventh grade students. The scores of these students on a standardized reading comprehension test (a criterion variable) would be the $X$ values (entries) used to develop one of the many possible data distributions for this defined population.

Populations are conceived as being either finite or infinite in size. A population of accidental home fires occurring in Chicago during 1975 is of finite size because it contains a limited number of members (separate home fires). At the other extreme, the population of hydrogen atoms in the earth's atmosphere contains a virtually infinite number of members. Even though the populations with which most researchers work would seem to be finite, and appropriate statistical procedures have in fact been developed for working with finite populations, for the sake of simplicity the procedures covered in subsequent chapters of this book will be based on the assumption that all populations being studied are so large that they can be considered infinite in size. In normal practice the errors thus introduced are negligible.

Let us look at a theoretical study involving only population data. Suppose that an Internal Revenue Service researcher hypothesizes that the average reported gross income of taxpayers in the recession year of 1974 was less than the 1973 average. He proceeds to plan a study to test this hypothesis. Two populations are thus involved: (A) U.S. taxpayers, 1973 and (B) U.S. taxpayers, 1974. The criterion variable to be measured on the population members is reported gross income. Each entry or $X$ value in the population data distribution is a gross income figure for an individual taxpayer. Since this researcher has access to computerized tax records, he can instruct his computer programmers to generate the median income for all of population A and the median income for all of population B. The resulting medians are *parameters* of the respective populations. He would then be able to compare directly the relative magnitudes of the two parameters without the need for any type of inferential technique. Either the 1974 average is less than that for 1973 (hypothesis supported) or it is not (hypothesis unsupported). A decision could thus be made regarding the hypothesis using only descriptive parameters. In other words, inferential statistical techniques are of value only when sampling is involved and relevant population parameters are not available.

## SAMPLES

In few instances will a researcher have access to data for entire populations. Consider the tasks, for instance, of (1) determining the standard deviation of the height of every mature pin oak tree growing on the North American continent during 1977, (2) collecting data on the deterioration rate of every square foot of interstate highway, or (3) determining any measureable effects of radiation on the world's human population during a particular period of heightened sunspot activity. In the vast majority of cases data on entire populations are impossible to amass and the researcher must be content with estimates computed on samples. Even in situations where population data can seemingly be used, the researcher will often wish to generalize his findings to future population members—making inferences across time. Imagine a college registrar who wants to develop a formula for predicting students' grade-point averages at his institution from their high school ranks. Even if he uses all of the currently enrolled students in the study, they comprise only a portion of the population of interest (present and future students). His formula, then, will actually be developed with sample data and applied to future members of the population.

The primary requirement of a sample is that it be *representative* of the defined parent population on every relevant dimension. In an attitudinal study, for example, this means that the same range and types of attitudes must exist in the experimental sample as exist in the population of interest. Similarly, in a field study of California soil fertility, the sample plots on which tests are conducted must be completely representative of the soil composition, growing conditions, and fertilization combinations to be found in the statewide soil populations. Obviously sampling is no simple task, yet it is one that is extremely crucial if the results of studies are to have any meaning. Only a brief introduction to the topic of statistical sampling can be provided here.

## RANDOM SAMPLING PLANS

Most commonly used inferential techniques rest on the assumption of a random sampling plan, with population members chosen for inclusion in samples on a purely random basis.

### Simple Random Samples

In *simple random* sampling each population member has an equal chance of being drawn for the sample and every possible sample which could be constructed has an equal chance of being used in a particular study. If a defined finite population is known to have 100,000 members ($n$ = 100,000), the chances of any individual member being selected to make up a single sample of size 1 ($N$ = 1) are the same, and equal to 1/100,000, or .00001. Were a sample of size $N$ = 100 to be randomly drawn from this population, the chances that any particular member will be selected are 100/100,000, or .001. Many everyday drawings for prizes follow a random sampling plan. Entrants' names or ticket numbers are placed in a container and picked out blindly to determine the winners. This procedure should be random enough, but if the contents of the container are not thoroughly mixed the drawing could be biased, as was the case with the 1970 Office of Selective Service draft lottery. Capsules containing birth dates (month and day) were apparently put into a container according to months, with January dates going in first, February dates second, and so on, through December dates. Although the capsules were supposedly well mixed, a disproportionately high number of dates for months later in the year were picked out. If the person(s) doing the drawing had been instructed to select capsules from random positions within the container, this fiasco would have been averted, even given the inadequate mixing of capsules. Needless to say, the cause of this bias in the procedure was eliminated before the 1971 lottery was conducted.

A similar situation might arise if a plant worker were told to determine the contents of a 500-gallon drum by drawing off a sample in a gallon container. Suppose that the drum actually contains 90 percent crude oil and 10 percent water. If the spigot used is located near the bottom of the drum, the sample drained off would contain only water. The worker would make an erroneous conclusion about the population of contents in the drum because of an unrepresentative sample.

When properly conducted, the simple "hat-draw" technique is a perfectly legitimate method of constructing random samples. Random number lists may also be generated by computer or from tables of random numbers (see Appendix N). In using such lists to construct a sample, the researcher must first assign a unique number to every member of the defined population. The random number list is then used to identify through corresponding numbers those population members that will make up the sample.

Stratified Random Samples A *stratified random* sampling plan differs from simple random sampling in that the population is first grouped into relevant strata and random sampling is then performed within each stratum, or subpopulation. It is particularly useful in those situations where the various strata differ substantially in size. Suppose that a researcher wishes to sample the attitudes of American and foreign students at a large university and finds that there are 20,000 of the former and only 200 of the latter in the university population; a ratio of 100 to 1. A simple random sample of 100 students would result in a sample with approximately 99 Americans and 1 foreigner. Because it is foolish to expect that any individual foreign student could represent the views of all 200 in that subpopulation, a stratified random sample is employed. The researcher divides the population into the two relevant strata, American and foreign, and randomly selects, say, 75 Americans and 25 foreigners to constitute a total sample of 100 for study. It should be noted that the resulting sample is unlikely to be representative of the attitudes of the entire population of 20,000 students, as would be expected with a simple random sample. If such representativeness were desired, each subsample statistic would have to be *weighted* in proportion to the relative size of the parent subpopulations.

Cluster Samples When dealing with large populations, it becomes difficult or impossible to identify each member and to insure that the sample obtained is truly random. One sampling plan designed to minimize this problem while still providing an approximation to random sampling is called *cluster sampling*. With this procedure subpopulations are defined that are similar to the total population and to one another with respect to relevant dimensions; in essence, each constitutes a mini-population. Intact clusters are then selected at random to make up the sample; or, members of each of the selected clusters may be randomly chosen for the sample. To illustrate, think of a study in which normative data are needed for a standardized test of automotive mechanics. Union locals to which experienced mechanics belong could be identified and constitute the clusters. A number of clusters (locals) could then be chosen at random. By contacting the selected locals for names of their members, a random sample of mechanics could be drawn from this subpopulation of clusters. Many such techniques or combinations of techniques are available to the researcher who wishes to approximate closely a random sample but who is unable to assemble a membership directory for an entire large population.

Errors in Stratified and Cluster Sampling Obviously, approximations to a simple random sampling plan are only as good as the judgment of the researcher who defines the relevant strata or clusters. It is unfortunate, but in sampling a researcher's mistakes are not easily detected and the nonrepresentative or biased samples which may result are typically assumed to be close approximations to those which would have resulted with simple random sampling. Some actual examples of the errors that can be made will be mentioned to demonstrate that the researcher must approach approximation procedures with great caution.

The Literary Digest poll of 1936 predicted that Landon would defeat Roosevelt as based on a sampling of voters with listed telephone numbers or registered automobiles. Because these two clusters did not adequately represent the entire voting population, the resulting sample was biased. Similarly, political pollsters, with the lower socioeconomic groups underrepresented in their voter samples, predicted that Dewey would be overwhelmingly victorious in 1948. A more recent poll indicated that the Equal Rights Amendment would be passed by the voters of a particular state. The Amendment was later soundly defeated. A postmortem revealed a flaw in the complex and rigorous sampling plan—a combination of stratified, cluster, and area sampling. Although the approximations to a random sampling of voter's places of residence (area sampling) proved to be adequate, the interviewers sent to these specific residences to contact voters were making their rounds between 8:00 A.M. and 5:00 P.M. on weekdays. A disproportionate number of interviews, consequently, was with unemployed women who were apparently more disposed to vote for the Amendment than employed members in the same households. The final stage of sampling, therefore, was nonrandom and biased.

## NONRANDOM SAMPLING PLANS

Purposive Samples The old adage, "As goes Maine, so goes the nation," refers to the historical fact that Maine voters usually preferred the same presidential candidate on their early election dates as would later be elected by the rest of the country's voters. We might say that Maine voters constituted a mini-population for Presidential election purposes. When a researcher has a great deal of confidence in such a group (cluster), he may be tempted to study only it and generalize to the entire population. This is an example of *purposive sampling*; the clusters are purposely, rather than randomly, selected. The chances for error are greatly increased over the cluster sampling plan described above, but the magnitude of the researcher's task is reduced considerably.

Quota Samples In *quota* sampling plans, the relative proportion of various population subgroups (strata) are determined and a sample systematically constructed to have the same subgroup proportions, usually without regard to random selection within the strata. Thus, a researcher may be satisfied if his sample contains the same percentage of individuals as the population in terms of sex, age, and educational levels, even though random selection was not practiced. The student should understand that this logic is sound only if every stratum that might have some bearing on the results of the study has been identified. In addition, only through sheer luck will nonrandom selection of sample members within each substratum result in a representative sample from the subgroups. Perhaps the most justifiable use of the quota technique is for checking the adequacy of a quasi-random sampling procedure, such as drawing every *n*th name from a telephone book. A quota comparison which indicates that the proportion of obtained sample subgroups (for example, proportions by sex) differs considerably from that of the parent population suggests that an uncontrolled source of bias has been present in the sampling procedure used.

Incidental Samples The type of sampling plan employed in a great percentage of studies is *incidental sampling*. Actually, an incidential sample involves no systematic plan at all. The researcher takes any population members that are readily available and includes them in the sample. Thus, a teacher may use students in his classes or school; a drug researcher may use locally available paid subjects; or physicians may use patients in their practice. The flaw in incidental sampling should be apparent to the student. Although the population of interest may be well defined and include the members of the sample, the extent to which the sample is representative of the population is virtually unknown.

Errors in Nonrandom Sampling  Samples in behavioral and medical research that are composed of volunteers are not only incidental samples, but are often obviously biased when compared with nonvolunteer samples that might be drawn from the same population. For example, suppose that an administrator at a private four-year college in the East decides to conduct a study of the attitudes of the student body regarding the use of marijuana and develops a questionnaire survey which is mailed to a random sample of students. Only 60 percent of the students in the sample complete and return the questionnaire. The sample can only be described as incidental. Is it reasonable to generalize these results to the entire student body (the defined population); that is, to assume that nonrespondents hold the same views toward the use of this drug as respondents? Such a projection is particularly hard to defend in this case. The person conducting the survey has two options: (1) he can attempt to increase the response rate to nearly 100 percent without introducing additional sources of bias; or (2) he can redefine the population about which inferences are to be made. These options should be pursued in order. If the first is unsuccessful, the population would then be *redefined* as that proportion of the student body willing to participate in attitudinal studies concerning marijuana. It is only to this population that the results could safely be inferred. Any attempt to make generalizations to similarly defined students in other colleges and universities is, of course, also indefensible unless data from that population were obtained.

At this point the student may begin to have reservations about the adequacy of sampling in most studies to which he has been exposed. Such enlightened skepticism is appropriate, for no matter how exotic the statistical techniques employed, the results of any study are only as good as the representativeness of the groups on which measurements are made.

### SAMPLING WITH REPLACEMENT

Sampling may be carried out *with* or *without replacement.* In sampling with replacement, each member drawn for a sample is returned to the population before the next draw is made. In sampling without replacement, a member that is drawn is not returned, thus reducing the size of the population each time by one member. Most research studies are conducted with samples that have been drawn without replacement. Nevertheless, to simplify the presentation in this book we will adopt the assumption that sampling is being conducted *with* replacement. This assumption is permissible to make for the typical research situation because the ratio of sample size to population size ($N/n$) is usually extremely small.

## SIZE OF SAMPLES

In later discussions sample size will be associated algebraically with the size of the errors in random sampling. At this point only some general observations will be made to help establish a perspective for the importance of sample size as compared to sample representativeness and its relationship to population size.

Some individuals who conduct research studies erroneously believe that large sample sizes can somehow compensate for sample bias or lack of representativeness. This is definitely untrue. The extent of error in nonrandom samples is generally incalculable via statistical methods. A random sample, no matter how small, is of greater value for making population inferences than a large incidental sample or one based on questionable strata or clusters.

It may seem reasonable to suspect that the larger the defined population, the larger the sample required. Again, this is generally untrue. Much more important is the extent to which the sample *represents* the population in all those dimensions relevant to the study. For instance, a new treatment for arthritis may, in fact, be beneficial to 46 percent of those suffering from this ailment, have no effect on 53 percent, and precipitate serious side effects in 1 percent. If the treatment is tested on only 35 patients selected randomly from a population, it is probable that the adverse reaction which affects only 1 out of every 100 patients would not show up at all. Since the importance of being able to identify categories of patients who should not receive a treatment is obvious, medical researchers must strive to study as large and as representative an experimental sample as possible before any new treatment is released for general usage. At the other extreme, consider the entomologist who wishes to describe the physical dimensions of a population of honey bees. Since worker bees and drones from each colony are monomorphic, or physically near-identical, the scientist needs only to select a few of each type from every colony in the defined population to have a representative sample.

The common element in these diverse examples is the extent of population homogeneity/heterogeneity with respect to the variable(s) under investigation. Large samples are not needed to study relatively homogeneous population data distributions, such as the size of drones in bee colonies or the molecular structure of distilled water, but are essential when investigating those that are heterogeneous and complex, such as the effects of a medical treatment on humans or the causes of traffic accidents.

## INFERENCE

When a researcher projects the results of a study conducted on an experimental sample to the defined parent population, he is making an inference. The inference may simply involve estimation of a parameter with a statistic, for example, using $\bar{X}$ as an unbiased estimate of $\mu$; or it may be a statistical decision that certain populations differ with respect to a measured variable, for example, that individuals receiving biofeedback training (the experimental population) can influence their blood pressure to a greater extent that those untrained in these techniques (the control population). In all such cases the researcher is interested in making decisions about defined populations, not about the samples employed in the research.

### STEPS IN MAKING INFERENCES

A sequence of steps is followed in making inferences and testing hypotheses. In general, these steps are:

1. Identification of variable(s) of interest
2. Definition of population(s) to be studied
3. Identification of appropriate statistical techniques
4. Determination of sample size(s)
5. Formulation of the decision model
6. Sampling
7. Data collection
8. Statistical analysis
9. Decision making and inference

The relative sequence of steps 1 through 4 is not particularly critical, but all must precede step 5. Brief attention has already been given to several of these steps and others will be illustrated in the following chapter. Before we proceed further, however, the subjects of hypothesis testing, decision-making errors, levels of significance, and decision rules must be given brief consideration.

### DECISION MODELS

The decision model (step 5) is made up of four components: the null hypothesis ($H_0$), the alternate hypothesis ($H_1$), the level of significance ($\alpha$), and the decision rule ($DR$).

Hypothesis Forms In everyday language, an hypothesis is a guess or hunch that is stated without proof. In statistical inference, hypotheses are complete statements of all the possible outcomes of an investigation with respect to each variable that is to be studied, and are formulated in a manner suitable for testing via statistical procedures. The researcher bases his decisions regarding the hypotheses on the results of these procedures. In general, the *null hypothesis* ($H_0$) may be thought of as a statement of no difference between populations that are being compared, although we will see shortly that it can have additional implications. *Alternate hypotheses* ($H_1$) are generally statements of differences between populations. Notice that both are statements concerning populations, not samples.

Consider an investigation of two methods, symbolized as methods C and D, of teaching first-year algebra in a secondary school system. The criterion variable consists of a standardized algebra achievement test administered at the completion of the course. The population of interest is defined as present and future students taking first year algebra in this school system. There are three possible outcomes to this study:

1. Method C is equal to method D (C = D)
2. Method C is superior to method D (C > D)
3. Method D is superior to method C (D > C)

The first outcome forms the basis of the null hypothesis of no difference ($H_0$). The alternate hypothesis ($H_1$) will be formed from outcomes 2 and 3 involving differences.

It is important to realize that measurements of achievement will be taken on samples of students and not samples of methods. The methods of instruction that are being studied differentiate students in the population into two groups. As a consequence, there are actually two populations, one consisting of students in this system taught by method C, and the other, those taught by method D. The hypothesis statements could be written something like this:

$H_0$: Students in the population taught by method C will demonstrate equal achievement to those in the population taught by method D.

$H_1$: Students in the population taught by method C will demonstrate either less or more achievement than those in the population taught by method D.

Notice that outcomes 2 and 3 have been combined into a single alternate hypothesis. This is known as a *nondirectional* hypothesis form. Although its implications for statistical tests cannot yet be explored, $H_1$ contains no indication by the researcher as to which method will result in greater student achievement. A nondirectional alternate hypothesis includes all the possible outcomes of difference and is used whenever a researcher has no substantial reason to believe that a particular outcome will result from a planned study. A *nondirectional statistical test* (sometimes called a *two-tailed test*) will result from this hypothesis form.

Suppose, on the other hand, that the literature has shown method C to be superior to method D in other school systems where both have been tried. It would then be to the researcher's advantage to use a *directional* hypothesis form, with an alternate hypothesis which clearly states only one predicted outcome:

$H_0$: Students in the population taught by method C will demonstrate achievement equal to or less than those in the population taught by method D.

$H_1$: Students in the population taught by method C will demonstrate greater achievement than those in the population taught by method D.

The directional alternate hypothesis statement ($H_1$) presents only that particular research outcome expected by the researcher. $H_0$ now includes both a statement of no difference and the other remaining outcome, which in this case is the possibility of C being inferior to D. All possible outcomes are always divided among the two hypothesis statements. A *directional statistical test* (sometimes called a *one-tailed test*) will result from this hypothesis form.

When an investigator makes a statistical test on any set of hypothesis statements, he is actually only deciding whether to reject or not reject the null hypothesis, $H_0$. Let us assume that this school researcher elected to use the nondirectional form and a subsequent statistical test on sample data causes him to reject the null hypothesis. Having made this decision, he has no choice other than to accept the alternative hypothesis that populations taught by the two methods differ ($C > D$ or $C < D$). No other inference is fully justified, however. He cannot now look at the sample data, see which sample had the highest average achievement score, and proclaim with any stated degree of statistical certainty that method C or D will work best in the defined population. Being able to say only that the methods differ may be sufficient in some studies, but in an applied situation such as this a decision may have to be made as to which method the school system should adopt. What are the options of our researcher in this instance? First, he could delay making a choice between the methods until a second study with new samples can be conducted, this time using a directional hypothesis form in which $H_1$ hypothesizes greater achievement with the method which was superior in the first study. If, as expected, $H_0$ can again be rejected with a stated degree of certainty, he can now accept $H_1$ as being true in the population and feel confident in adopting this teaching method for the first-year algebra classes in the school system. This is the best option for him to pursue. If he had reviewed the literature and had been aware of this method's superiority in other systems, of course, he could have used the directional form in the first place and saved a great deal of work. The second, but less desirable, option would be to go ahead and adopt the method associated with the greatest sample achievement in the original study. While his decision between the methods is quite likely to be correct, he forfeits his right to make any statistical statement concerning the probable accuracy of the decision by choosing this option. As will be seen later, this means that he cannot report that he has rejected the null hypothesis at some given level of significance. Thus, this solution is best reserved for applied settings in which an immediate decision must be made following the rejection of $H_0$ with a nondirectional test. The final option, which may already have come to the student's mind, would be to simply rewrite the nondirectional hypothesis statements of the original study into the "appropriate" directional form so that the alternate hypothesis predicts the known outcome of the study. Unfortunately, this option is methodologically unsound and should never be pursued. Formulation or reformulation of any part of the decision model after observing the data violates the assumptions on which test probabilities are based.

Let us now consider the possibility that the researcher's results indicate no difference between achievement for the two teaching methods. Consideration of this outcome will help to demonstrate why the nondirectional form is best to use when a researcher has no particular reason to expect a specific outcome, for if $H_0$ cannot be rejected, the researcher can then feel free to adopt whichever teaching methods he might prefer for reasons such as relative cost, teacher time involved, and so on. However, if he had initially used the directional form and could not reject $H_0$, he is left with the possibility that the methods are equal or that one method is superior to the other. He might now decide to run another study with a nondirectional hypothesis form to determine if the methods were unequal and perhaps be forced to conduct a third study with the remaining directional form if the results now suggested that one method might actually be superior.

In lieu of these potential problems, some guidelines might be set forth. If a researcher has reason to believe that a particular outcome will result, a directional test should be used. When no such evidence is available, a pilot study should first be conducted with a small, representative sample. If the pilot study suggests that population differences exist, appropriate directional hypotheses can be formulated for the main study. If population differences are not suggested, the main study should be conducted using a nondirectional test.

"Accepting" vs. "Not Rejecting" $H_0$ When a null hypothesis cannot be rejected, it has not necessarily been demonstrated to be true. A null hypothesis is never "proven." An example should help to clarify this matter. Investigations are currently underway to determine whether life exists on other planets. Scientists place sensors in satellites or soft-landing probes to ascertain the presence of life forms or sufficient environmental conditions for life as we know them. If life forms are perceived by these sensors, the null hypothesis of no life on that planet can be rejected. However, if life forms are not perceived, it does not prove that no life exists. The forms may be located in other areas of the planet that the sensors have not scanned or they may be of some unusual nature which the sensors were not designed to perceive. Instead of accepting a null hypothesis, therefore, we simply *do not reject* the assumption that it is correct. Through our own experience and that of other researchers we may, in time, draw the conclusion that a particular null hypothesis we are testing does in fact represent the true state of affairs in the population, even though statistics cannot directly provide us with a probability statement indicating the extent to which we can be confident of our decision.

Errors in Decision Making The decision made by a researcher about populations will be either correct or incorrect, depending on the accuracy with which his study has predicted the true state of affairs in the population.

Continuing with the same example as before, let us theoretically assume that in the population of interest there are no differences in algebra achievement between students taught by method C and those taught by method D ($H_0$ is true). If, on the basis of results with sample data, our researcher decides that he cannot reject the nondirectional hypothesis of no difference ($H_0$), he has made a correct decision. On the other hand, if the sample results should cause him to reject $H_0$, he has committed a decision error. This type of mistake, rejecting $H_0$ on the basis of sample results when it is actually true in the population, is called a *Type I* or $\alpha$ *error.* In practice, of course, the researcher will seldom or never find out if such a mistake has been made.

Now let us theoretically assume that there *are* achievement differences in the population between students taught with method C and those taught with method D ($H_0$ false). If, on the basis of sample results, the researcher decides to reject the nondirectional null hypothesis, he has made a correct decision. However, if the results of his study should cause him to not reject $H_0$, he has made a second type of decision error. The mistake of not rejecting $H_0$ on the basis of sample results when it is actually false in the population (when real differences do exist) is called a *Type II,* or $\beta$ *error.* Type I and II errors are summarized in Table 6.1. Although the state of affairs in the population will be unknown to the researcher, $H_0$ is actually true or false. The researcher must make a decision based on sample data to reject or not reject $H_0$. This decision will either be correct (no error) or incorrect (Type I or II error). Both directional and nondirectional hypothesis forms are subject to these errors.

Level of Significance The probability of making a Type I error is defined as $\alpha$, the *level of significance.* This level is an essential component of the decision model and, as with the other components, is specified by the

**TABLE 6.1. TYPE I AND TYPE II ERRORS**

| DECISION BASED ON SAMPLE DATA | State of Affairs in the Population | |
|---|---|---|
| | $H_0$ TRUE | $H_0$ FALSE |
| **Reject $H_0$** | **Type I error ($\alpha$)** | **Correct decision No error** |
| **Do not reject $H_0$** | **Correct decision No error** | **Type II error ($\beta$)** |

researcher in advance of data collection. The most commonly used levels of $\alpha$ are .10, .05, .01, and .001, although any values less than .50 are permissible. An $\alpha$ level of .05 simply indicates that the probability of making a Type I error by chance alone is .05, or that a Type I error will be made 5 times in 100 when $H_0$ is actually true. If the $\alpha$ level is set at .001, Type I error will occur less frequently, only 1 time in 1000. Thus, the researcher can minimize the probability of making a Type I error (rejecting $H_0$ when it is true in the population) by setting $\alpha$ at a very low level. As he does so, however, the probability of making a Type II error, $\beta$, is increased. As *$\alpha$ decreases, $\beta$ increases,* and vice versa. Although $\alpha$ is under the researcher's direct control, $\beta$ is not. Given a particular sample size and research design the probability of making a Type II error can be reduced only by increasing the probability of making a Type I error; that is, by setting $\alpha$ at a relatively high level. The relationship between $\alpha$ and $\beta$ errors will be discussed in more detail in Chapter 9.

The concept of *probability* is being used here to describe the likelihood of observing some defined event, just as it is in everyday communications. We are all familiar with probability in connection with weather forecasting (for example, the probability of rain is 20 percent), for predicting the tendency of a fair coin to land heads up (50 percent), or in describing the odds that a single card drawn from an ordinary playing deck will be the queen of hearts (1 in 52, or .019). Probability has to do with the inexact. Our best guess is that 10 heads will be observed when a fair coin is tossed 20 times, for this is the most likely event given an underlying probability of .50. We would not be surprised, however, if as few as 6 or as many as 14 heads appear. The event would be passed off as something that might happen rather frequently by chance alone. On the other hand, we might be willing to reject the assumption that the coin is fair if only 1 head is observed in 20 tosses, citing as our justification the low probability of such an extreme event.

In making statistical tests we set an $\alpha$ level and calculate an observed value of a test statistic on sample data. The $\alpha$ level enables us to establish a range in which the observed value—the event—is expected to fall. If the observed value falls within this range, the null hypothesis is not rejected (correctly or erroneously). If the observed value is extreme and falls outside of this range, the probability of its having occurred by chance is less than the stated $\alpha$ and the null hypothesis is rejected (again, correctly or erroneously).

Minimizing Crucial Errors *Magnitudes* of all possible errors may be reduced by using the largest sample sizes possible, but $\alpha$ and $\beta$ errors cannot be eliminated in hypothesis testing. Therefore, preliminary planning on the part of the researcher is necessary in order to define and, when possible, minimize those errors that are relatively *crucial* in a particular investigation.

If the directional hypothesis form can be used, the crucial error can often be controlled by writing the hypotheses in such a way that the crucial error becomes the Type I error. Let us see how this might be accomplished in the educational research problem we have been studying in this section. The researcher would first define the possible errors regarding the choice between teaching methods C and D. These are (1) deciding that C is better when it is not and (2) deciding that D is better when it is not. Next, the relative impact of the two errors are compared, taking into consideration the costs, teacher time, probable acceptance of the two methods, and instructional benefits. Let us assume that the adoption of method D on a system-wide basis would require a tremendous investment in expensive new instructional equipment, while method C could be instituted with existing facilities. The most expensive, or crucial, error would thus be to decide that method D is better when it actually is poorer than or equal to C in the population. This error could now be defined as the Type I error via appropriate structuring of the hypothesis statements:

$H_0$: Students in the population taught by method D will demonstrate achievement equal to or less than those taught by method C.

$H_1$: Students in the population taught by method D will demonstrate greater achievement than those taught by method C.

Rejecting $H_0$ would result in acceptance of $H_1$ and adoption of the expensive teaching method D. Writing the hypothesis in the above manner allows the researcher to control the probability of erroneously rejecting $H_0$ and committing a Type I error, as he can set $\alpha$ at a sufficiently low level, say .01 (only 1 chance in 100 that $H_0$ will be rejected when it is actually true). Setting $\alpha$ low has the negative effect of increasing the probability of making a $\beta$ error, now associated with not rejecting the equality or superiority of method C when, in the population, D is actually superior. This is acceptable to the researcher, for he has already decided that it is the less crucial of the two errors.

Unfortunately, this procedure cannot be applied to nondirectional tests, for the hypothesis statements can only be written in one way. If the crucial error happens to be the Type I error, it can be controlled by setting $\alpha$ at a relatively low level. If the crucial error happens to be the Type II error, however, the researcher can only try to minimize it by using large samples and/or setting $\alpha$ at a relatively high level. Until the student has acquired sufficient background to decide for himself what is considered a "low" or "high" $\alpha$ level, perhaps some tentative guidelines are in order. The authors suggest that the .05, .10, and larger $\alpha$ levels might be considered high, and .01, .001, and smaller levels low; at least for the time being.

Examples used in subsequent chapters will include consideration of crucial errors in setting up hypotheses and choosing appropriate $\alpha$ levels. What may seem complex and unclear to the student at this stage should prove to be simple as the procedure becomes more familiar and routine.

Decision Rules The last component of the decision model is the *decision rule* (*DR*). As with the other components, it is developed before sample data are collected. The decision rule sets forth a course of action that the researcher will follow subsequent to statistical analysis of sample data. Its generalized form is as follows:

> DR: Reject $H_0$ if the observed value of the test statistic is greater than (or sometimes less than) some stated critical value(s) of that statistic. Otherwise, do not reject $H_0$.

In other words, the researcher will compute an observed value of some test statistic on sample data. If that value is, say, greater than a stated *critical value,* which he will look up in a table, his decision will be to reject $H_0$. If the observed value is less than the critical value, he will not reject $H_0$.

Several steps must be completed before this generalized decision rule form can be made specific to a particular study. These include identification of an appropriate statistical test (step 3) and determination of the critical value. The latter requires utilization of the stated $\alpha$ level and, with some tests, the degrees of freedom for the problem. These matters will be taken up in more detail in conjunction with the introduction of the simple, but very useful, chi-square tests in the next chapter.

## EXERCISES

1. Carefully define three populations of your choice. Indicate whether they are finite or infinite.
2. Indicate which sampling plan you would use and how you would proceed to draw samples for research studies from each of the three populations you have defined in exercise 1.
3. Suppose that each of the 100 members of a small population are listed by number (1–100). Using the random number table in Appendix N, select and list 12 random numbers to be used in drawing a simple random sample. What is the probability of a particular population member being included in the sample?
4. What is the difference between cluster and stratified random sampling plans?
5. A pollster working for a political candidate interviews 50 people at shopping centers two months before the election and predicts a landslide victory for his client. The candidate is later defeated. List several possible explanations for this error.
6. In a research journal, magazine, or newspaper, locate a study conducted on a sample.
   a. Is the population defined? If so, give the definition. If not, give the definition which seems appropriate for the study.
   b. What type of sampling plan was apparently used?
   c. Was the sampling procedure adequate? Why or why not?
7. Mentally formulate a simple research investigation in your field.
   a. Write a set of nondirectional hypothesis statements ($H_0$ and $H_1$) for this study.
   b. Rewrite these hypotheses into a directional form.
8. Define the Type I and Type II errors for the nondirectional and directional hypothesis statements you provided in exercises 7a and 7b.
9. Decide which error in question 8 is most crucial.
   a. Write a set of directional hypothesis statements for your investigation which puts this crucial error directly under the researcher's control (as defined by $\alpha$).
   b. State two ways in which you could attempt to minimize $\beta$ at the same time.
10. Consider why the probability of obtaining a one-spot with a six-sided die is 0.167.

# 7

# CHI-SQUARE TESTS

The simplicity of chi-square tests and the lack of restrictive assumptions regarding their use make them ideal vehicles for illustrating the elements of inference discussed in the previous chapter.

The chi-square ($\chi^2$) statistic, which serves as the foundation of all chi-square tests, is computed with frequencies—or proportions that can be converted to frequencies—within defined categories. The basic formula for $\chi^2$ is

$$\chi^2 = \sum \frac{(O - E)^2}{E} \tag{7.1}$$

where $O$ is an observed cell frequency and $E$ is the theoretical or expected frequency for that cell.

For making decisions about populations, observed values of $\chi^2$ computed on sample data are compared with appropriate critical values found in Appendix D. Without going into further detail, let us use this test statistic in a demonstration of the decision-making process.

Suppose we have speculated that Eisenhower dollar coins, with their heavy weight and wafer construction, might be biased in favor of landing primarily in either a heads-up or tails-up position when tossed. We would expect "fair" coins to land heads 50 percent of the time and tails 50 percent of the time. Thus, if we could obtain and toss a random sample of Eisenhower coins our "theory" could be tested. Some limited experience with tossing coins and common sense tells us that the chances (probability) of obtaining exactly 50 percent heads and 50 percent tails with any sample of coins, no matter how fair they might be, are not very great. Therefore, a test is needed to help us decide whether an observed deviation from 50/50 merely represents chance sampling fluctuation or is indicative of true bias in the population of Eisenhower coins. Since the chi-square test can be used to provide decision-making information of this type, we can proceed to plan and conduct this experiment.

The format used for presenting the study in Table 7.1 roughly follows the steps for making inferences listed in Chapter 6 and may be adapted to most statistical testing procedures. Notice that the first step, identification of the variable of interest, was encompassed in the original speculation which led to the study and has now been defined as the criterion variable. The population to which we wish to make inferences is then carefully described. The sample description includes the number of members to be studied and the way in which they are to be selected (randomly).

Formulation of the decision model begins by stating testable hypotheses about the population. These statements are nondirectional, for

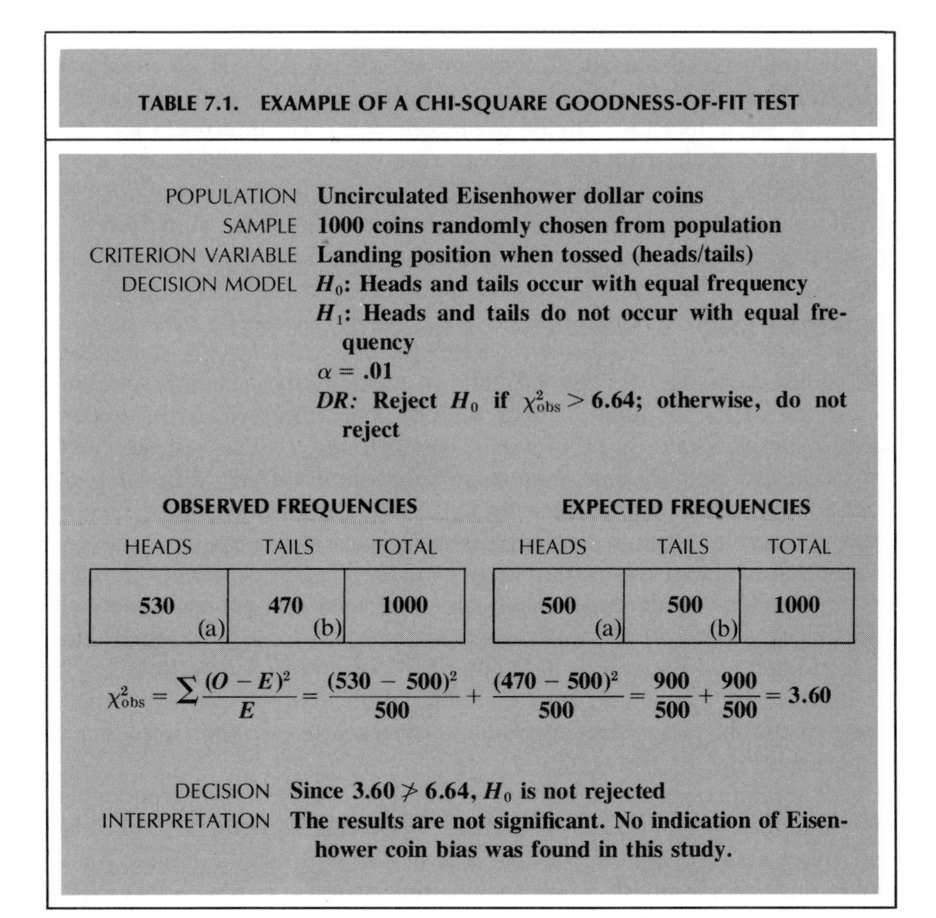

TABLE 7.1. EXAMPLE OF A CHI-SQUARE GOODNESS-OF-FIT TEST

POPULATION **Uncirculated Eisenhower dollar coins**
SAMPLE **1000 coins randomly chosen from population**
CRITERION VARIABLE **Landing position when tossed (heads/tails)**
DECISION MODEL **$H_0$: Heads and tails occur with equal frequency**
**$H_1$: Heads and tails do not occur with equal frequency**
**$\alpha = .01$**
***DR:* Reject $H_0$ if $\chi^2_{obs} > 6.64$; otherwise, do not reject**

OBSERVED FREQUENCIES

| HEADS | TAILS | TOTAL |
|---|---|---|
| 530 (a) | 470 (b) | 1000 |

EXPECTED FREQUENCIES

| HEADS | TAILS | TOTAL |
|---|---|---|
| 500 (a) | 500 (b) | 1000 |

$$\chi^2_{obs} = \sum \frac{(O-E)^2}{E} = \frac{(530-500)^2}{500} + \frac{(470-500)^2}{500} = \frac{900}{500} + \frac{900}{500} = 3.60$$

DECISION **Since $3.60 \ngtr 6.64$, $H_0$ is not rejected**
INTERPRETATION **The results are not significant. No indication of Eisenhower coin bias was found in this study.**

we are not predicting a specific outcome; for example, heads will occur more frequently. The $\alpha$ level is set at a relatively low level of .01 for two reasons. First, the very large sample size will effectively minimize $\beta$; and second, deciding that Eisenhower coins are biased when they actually are not seems to be the more crucial of the two errors. The decision rule, *DR*, states that we will reject $H_0$ if our observed value of $\chi^2$ is greater than the critical value of 6.64 (procedures for determining critical values will be taken up later).

After sampling is completed, the 1000 coins are each tossed and the frequencies of heads and tails are recorded in an *observed* data table. Next, an *expected* data table is constructed which is identical in form to the observed table but which lists theoretical frequencies based entirely on the expected outcome of the study when $H_0$ is true. With 1000 coins, the null hypothesis implies that approximately 500 heads and 500 tails should have been obtained. The chi-square test will help us decide if the discrepancy between the observed and expected outcome is merely sampling fluctuation or an indication of population bias.

Formula (7.1) is used to generate an observed value of chi-square ($\chi^2_{obs}$) for these data. Since this value is not greater than the critical value ($3.60 \ngtr 6.64$), a decision is made to *not reject* $H_0$. We interpret this result carefully, for $H_0$ has not been proven true (although unlikely, a Type II error might have been committed). However, our interest in the possibility of Eisenhower coins being biased has certainly been diminished, if not extinguished.

## CONSIDERATIONS IN THE USE OF CHI-SQUARE TESTS

Chi-square tests may be applied without making any assumptions about the shape of the population data distributions from which the experimental samples have been drawn. They belong to a class of statistical tests that are called *nonparametric,* or distribution-free. More of these nonparametric tests will be covered in Chapter 15. Because of its freedom from restrictive assumptions and its computational simplicity, the chi-square test is one of the more widely used inferential techniques in many research fields. Students in the behavorial sciences, for example, will find that it is particularly well-suited to the analysis of attitudinal data. Unfortunately, chi-square is also one of the most misused techniques, as pointed out by Lewis and Burke (1949), Marascuilo (1972), and others. Some of the mistakes frequently made by researchers when applying $\chi^2$ will be mentioned in this chapter.

To demonstrate further how the $\chi^2$ statistic is based on frequencies within defined categories, consider the observed frequency data in Table 7.2. Here, a single random sample was first drawn from a defined population and its members were then categorized according to sex and

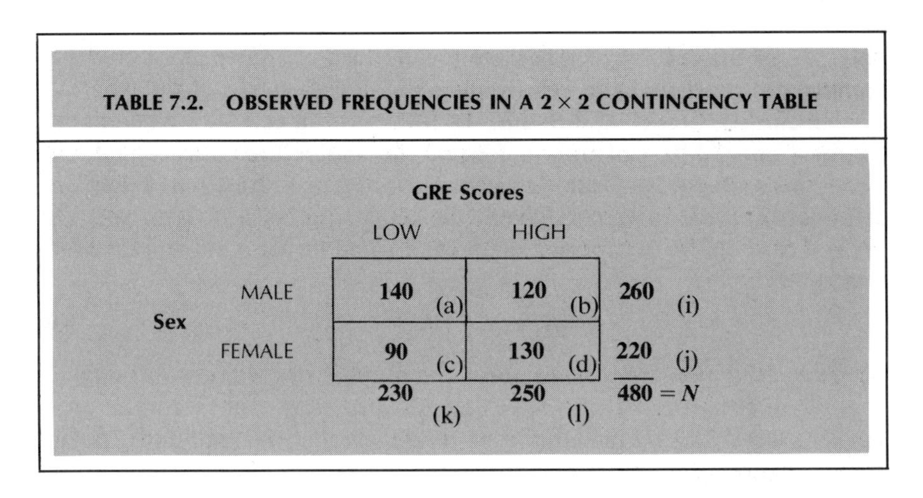

**TABLE 7.2. OBSERVED FREQUENCIES IN A 2 × 2 CONTINGENCY TABLE**

| | | GRE Scores: LOW | GRE Scores: HIGH | |
|---|---|---|---|---|
| Sex | MALE | **140** (a) | **120** (b) | **260** (i) |
| | FEMALE | **90** (c) | **130** (d) | **220** (j) |
| | | **230** (k) | **250** (l) | **480** = $N$ |

whether they scored "high" or "low" on the Graduate Record Examination. This resulted in a 2 × 2 table (2 rows and 2 columns) with 4 cells. The entries in the cells are frequencies. Thus, we see that 220 of the candidates were women, and of these 90 scored "low" on the GRE and 130 scored "high." These types of tables are often called *contingency tables*. In order to classify individuals according to the 4 cells or categories, two measures were made on sample members after they were drawn; one on the variable of sex and another on the variable of GRE scores. Since the researcher has exercised no control over either, both are *criterion* variables. Notice that the sex variable corresponds to discrete data on a nominal scale, while the GRE score variable corresponds to continuous data on an ordinal or interval scale of measurement and has been arbitrarily dichotomized into two categories. The chi-square test may, in fact, be used with any scale of measurement, including the nominal scale. When dealing with a continuous variable on an ordinal, interval, or ratio scale, the data are categorized into two or more groups that have been defined in advance by the researcher. In the above example, the mean of the normative population provided by the test publisher was used to divide the obtained GRE scores into the categories of "high" and "low." Obviously, these categories must make sense to the researcher with regard to the type of population inference he is attempting to draw.

### DEGREES OF FREEDOM

For most statistics, the concept of *degrees of freedom* (*df*) is directly related to sample size. For chi-square, degrees of freedom is based on the number of *cells* in the contingency table. Look at Table 7.2 again. The cell frequencies sum to $N = 480$, as do each set of *marginal* frequencies ($260 + 220 = 480$ and $230 + 250 = 480$). If the marginal frequencies and only one of the cell frequencies are given in a 2 × 2 table, we could still compute all the remaining cell frequencies. For example, if the marginal frequencies (i), (j), (k), and (l) and the frequency in cell (a) are given; cell (b) must equal 120, cell (c) must equal 90, and cell (d) must equal 130. Thus, this and any 2 × 2 table has only *one* degree of freedom. When one cell is set and the marginals given, the remaining cells are also set. The general formula for computing degrees of freedom for a chi-square problem is

$$df = (r - 1)(c - 1) \tag{7.2}$$

where $r$ is the number of rows and $c$ the number of columns in the table.

When either $(r - 1)$ or $(c - 1)$ equals *zero,* however, we must cross out that term. The *df* for a table with one row and five columns, for instance, would be computed as follows:

$$df = \cancel{(1 - 1)}(5 - 1) = 4$$

The *df* for Table 7.2 would be $(2 - 1)(2 - 1) = 1$.

## COMPUTING EXPECTED FREQUENCIES

With certain types of chi-square tests, such as the goodness-of-fit problem in the first example, the expected frequencies are predetermined according to some scientific law or theory. In all other problems involving chi-square, however, the expected cell frequencies must be computed from the observed data table. Let us use the data in Table 7.2 as an example. Here, an expected table is needed which gives the cell frequencies expected when there is no relationship between sex and GRE scores ($H_0$ true). These entries must be proportional to the marginal frequencies. This is not a trial-and-error task when the marginals are unequal. The solution is to multiply the two marginals associated with a particular cell and divide this product by the total frequency, $N$. First, the observed table is reproduced as shown below, leaving out the cell frequencies.

| | | |
|---|---|---|
| $\frac{(230)(260)}{480} = 124.58$ (a) | $\frac{(250)(260)}{480} = 135.42$ (b) | 260 (i) |
| $\frac{(230)(220)}{480} = 105.42$ (c) | $\frac{(250)(220)}{480} = 114.58$ (d) | 220 (j) |
| 230.0 (k) | 250.00 (l) | 480 = $N$ |

Next, multiply the marginals for cell (a) and divide by $N$; $a = ki/N$. The formula for cell (b) is $li/N$, and so on for the remaining cells. The expected entries must sum to equal the marginals. Thus, for a table such as this with $df = 1$, only one cell needs to be computed with this procedure. The rest may be obtained through subtraction.

## ASSUMPTIONS OF CHI-SQUARE TESTS

Use of the $\chi^2$ statistic assumes that the table categories are *mutually exclusive* and *exhaustive*; that is, measures on each member of a sample may be assigned to one and only one cell. We should not, therefore, use the chi-square tests described here in situations where measures on a sample member will fall into more than one cell, as with a test-retest or repeated-measures design. Such misuse is all too common. In addition, the above assumption implies that an appropriate cell must exist for each measure. Thus, the researcher must not throw out obtained measures merely because they do not match the categories in the table or because the number of frequencies in a particular cell is too large or too small for his liking (another common mistake).

A second assumption of $\chi^2$ is that the members of the sample or samples are *randomly* and *independently* drawn from the population(s) of interest. By independence, we mean that the selection of any particular sample member has no influence on the probability of selecting any other member. Finally, the $\chi^2$ statistic approximates the theoretical distributions in Appendix D only when the frequencies in the cells are fairly large; at least 10 per cell in a table with 1 degree of freedom and at least 5 per cell in larger tables. More will be said about this last requirement in later sections.

## DETERMINATION OF CRITICAL VALUES

If $\chi^2$ statistics were computed on thousands of large independent samples randomly chosen from a defined population, the frequency distribution of these observed values would approximate one of the theoretical distributions of $\chi^2$ provided in Appendix D. Each degree of freedom is associated with a unique theoretical distribution. To find a critical value of $\chi^2$ in Appendix D, we must first know the degrees of freedom for our problem and specify an $\alpha$ level at which the test is being conducted. For the example in Table 7.1, $df = 1$ and $\alpha = .01$. The intersection of 1 and .01 shows a critical value of 6.635.

The chi-square statistic has a theoretical range of zero to infinity.

## CHI-SQUARE FORMULAS

Formula (7.1) presented in the introductory paragraphs is a general formula for computing the $\chi^2$ statistic which may be used with any size data table. A mathematically equivalent formula, which may be more efficient for some pocket calculators, is

$$\chi^2 = \sum \frac{O^2}{E} - N \qquad (7.3)$$

where $N$ is the total number of frequencies in the data table. The use of this formula will be illustrated in a later example.

A third formula is computationally simple, but is appropriate *only* for $2 \times 2$ tables:

$$\chi^2 = \frac{N(ad - bc)^2}{i\,j\,k\,l} \tag{7.4}$$

where $a$, $b$, $c$, and $d$ are cell frequencies as identified in Table 7.2, and $i$, $j$, $k$, and $l$ are marginal frequencies. It is important that the cell frequencies be identified in the same way as they have been in Table 7.2. Expected frequencies are not needed and rounding errors are reduced when this formula can be applied.

Positive values of $\chi^2$ will always be produced with these three formulas. Thus, with all the chi-square tests which will be presented in this chapter, the researcher will base his decision about rejecting $H_0$ on whether or not the observed $\chi^2$ value is *greater* than some critical value. This makes it impossible to test directional hypotheses. Therefore, we will only consider chi-square as a *nondirectional* test.

## TYPES OF CHI-SQUARE TESTS

One of the reasons for the frequent misuse of $\chi^2$ is the failure of researchers to define their problems adequately, resulting in faulty experimental designs and incorrect interpretations of test results. We will discuss three types of chi-square tests for various problems: *goodness-of-fit*, *homogeneity*, and *independence*. All may be used to provide the researcher with decision-making information about populations and all are based on the extent of the discrepancy between the observed sample frequencies and some expected or theoretical set of population frequencies.

### GOODNESS-OF-FIT TESTS (ONE SAMPLE, ONE OR MORE CRITERION VARIABLES)

A $\chi^2$ goodness-of-fit test is appropriate to determine if an obtained set or distribution of sample frequencies corresponds to some theoretical distribution. It might be used, for example, to decide whether an obtained sample data distribution could have been drawn from a population data distribution that is normal in shape, or to decide if the results of an experiment are consistent with those expected according to some scientific model, law, or theory. We conduct this test by first obtaining a table of observed frequencies on criterion variable categories with a single random sample and then constructing a table of expected frequencies that are consistent with the theoretical distribution or the scientific model. The data for a goodness-of-fit problem can always be presented in a one-way

[(1 × $c$) or ($r$ × 1)] table, regardless of the number of variables that are involved. The procedures discussed below can be generalized for use with any size table.

One Criterion Variable  The coin-tossing experiment in Table 7.1 is an example of a goodness-of-fit test with one criterion variable (landing position). The expected frequencies were predetermined by our definition of fair coins in the null hypothesis and the test was conducted to evaluate the goodness-of-fit between the distribution of frequencies in the observed table and the distribution expected under the mathematical laws of chance. Since $H_0$ could not be rejected, we could describe these results as being *nonsignificant*. It would be advisable for the student to review the example at this point.

Two Criterion Variables  In some goodness-of-fit problems, measures on two or more uncontrolled criterion variables are obtained. The expected frequencies in each combination of categories (cells) are generated from some theoretical distribution. Confusion may be avoided if such studies are always set up in a one-way [(1 × $c$) or ($r$ × 1)] table. Let us briefly consider an illustration of a problem with two criterion variables.

A Mendelian cross of a sample of peas is carried out by a geneticist with the criterion variables of color (yellow or green) and skin texture (smooth or wrinkled). Four combinations are possible as a result of the crossing: yellow-smooth (YS), yellow-wrinkled (YW), green-smooth (GS), and green-wrinkled (GW). According to Mendel's law, YS combinations should occur 9 times as frequently as GW; while YW and GS peas should occur 3 times as frequently as GW combinations. This theoretical distribution is used to construct the frequencies in a 1 × 4 expected data table. Since the sample on which the measures are to be taken has 960 members (peas), the scientist would expect to find that the possible combinations occur with approximately these frequencies: GW = 60, GS = 180, YW = 180, and YS = 540. These expected frequencies and the actual frequencies observed with the sample are shown below:

**Observed Frequencies**

| GW | GS | YW | YS | TOTAL |
|---|---|---|---|---|
| 50 | 200 | 160 | 550 | 960 |

**Expected Frequencies**

| GW | GS | YW | YS | TOTAL |
|---|---|---|---|---|
| 60 | 180 | 180 | 540 | 960 |

The degrees of freedom for this table is

$$(r - 1)(c - 1) = (1 - 1)(4 - 1) = 3$$

Formula (7.1) is computed as follows:

$$
\begin{aligned}
\chi^2_{\text{obs}} &= \sum \frac{(O - E)^2}{E} \\
&= \frac{(50 - 60)^2}{60} + \frac{(200 - 180)^2}{180} + \frac{(160 - 180)^2}{180} + \frac{(550 - 540)^2}{540} \\
&= \frac{100}{60} + \frac{400}{180} + \frac{400}{180} + \frac{100}{540} \\
&= 6.30
\end{aligned}
$$

Formula (7.3) could also be used:

$$
\begin{aligned}
\chi^2_{\text{obs}} &= \sum \frac{O^2}{E} - N \\
&= \frac{50^2}{60} + \frac{200^2}{180} + \frac{160^2}{180} + \frac{550^2}{540} - 960 \\
&= 41.667 + 222.222 + 142.222 + 560.185 - 960 = 966.296 - 960 \\
&= 6.30
\end{aligned}
$$

Notice that both of these formulas involve the computation of a term for each cell in the data table. Since this is a $1 \times 4$ table, formula (7.4) cannot be applied.

If we assume that the geneticist decided in advance on an $\alpha$ level of .10, the critical value of $\chi^2$ determined with Appendix D would be 6.25. The null hypothesis that the observed distribution fits the distribution expected with Mendel's law would be rejected, for $\chi^2_{\text{obs}}$ is greater than the critical value. When $H_0$ is rejected with a test, researchers often label the results as being *significant.* Keep in mind, of course, that the chances of making a Type I error were 1 in 10 with this decision ($\alpha = .10$).

## TESTS OF HOMOGENEITY (TWO OR MORE SAMPLES, ONE CRITERION VARIABLE)

The chi-square statistic is frequently used to determine if two or more populations are homogeneous; that is, if their data distributions are similar with respect to a particular criterion variable. As an illustration, we may wish to find out if the attitudes of Democrats and Republicans in a particular state are similar with respect to the initiation of local flood control projects. Two completely separate random samples would be drawn; one from the population of Democrats and another from the population of Republicans. The criterion variable is attitude, as expressed in favorable-unfavorable categories. If the chi-square test of homogeneity indicates that the attitudes of the two groups do not differ significantly, the two populations may be viewed as being homogeneous, or essentially the same with respect to the particular attitude measured. This test is sensitive to any differences in the populations (shape, central tendency, or variability) that affect the proportions in the categories. While the samples drawn from each population in a test of homogeneity need not be identical in size, it is highly recommended that they be similar in magnitude whenever possible.

Chi-square tests of homogeneity are identical to the parametric two-sample $t$ tests under certain conditions, as will be discussed in Chapter 9. They are useful alternatives to these latter tests when the more stringent parametric assumptions cannot be met.

Two Samples To illustrate the chi-square test of homogeneity for two populations, let us suppose that a forester wishes to make a comparison between the annual growth rates of douglas fir seedlings in the states of Oregon and Washington (see Table 7.3 for this example). Two populations are involved; trees growing in Oregon and trees growing in Washington. Sampling is accomplished by independently and randomly selecting 100 trees from each population for study. The growth rate criterion variable has been arbitarily dichotomized into the categories of "less than 5 inches" and "5 inches or more."

The hypothesis statements again must be nondirectional. $H_0$ states that the two populations are essentially homogeneous, or do not differ with respect to the criterion measure. Because the Type II error seems rather crucial to this investigator, he sets $\alpha$ at .05 and finds the critical value to be 3.84 for $df = 1$.

Not all the trees in the samples survived the year; reducing sample A to 96 and sample B to 94. Those remaining were classified according to measured growth and the observed frequency table constructed. Formula (7.4) has been used to compute $\chi^2$ for this $2 \times 2$ table, eliminating the task of computing expected frequencies. A glance at the magnitudes of the intermediate terms generated, however, will quickly demonstrate the difficulty of applying this formula when $N$ is large. Unless the hand calculator available has an automatic scientific notation feature, formula (7.4) should be reserved for fourfold tables with $N$'s of less than 100.

The observed $\chi^2$ is greater than the critical value and $H_0$ is rejected at the .05 level. Although the data seem to support the conclusion that the growth rate in Oregon was greater than in Washington, the forester cannot make this specific decision at a known probability level with a nondirectional chi-square test. If such an alternate hypothesis could have been stated in advance and if other assumptions could have been met, a directional $t$ test may have supported this specific conclusion.

Three or More Samples Let us now briefly consider a larger $4 \times 3$ chi-square test of homogeneity. A nationwide marketing survey is conducted to determine whether or not regional differences exist in preferences among three styles of men's jackets. Random samples of 60 potential buyers are separately drawn from each of the four regions (populations) of interest. The criterion variable is measured by asking each member

**TABLE 7.3. EXAMPLE OF A CHI-SQUARE TEST OF HOMOGENEITY**

POPULATIONS **Three-year-old douglas fir seedlings during the period 7/1/76 to 7/1/77**
**A: Growing in Oregon**
**B: Growing in Washington**

SAMPLE A **100 seedlings randomly chosen from population A**
SAMPLE B **100 seedlings randomly chosen from population B**
CRITERION VARIABLE **Growth rate during period: $<$ 5 inches/$\geq$ 5 inches**
DECISION MODEL **$H_0$: Growth rates of population A and B are equal**
**$H_1$: Growth rates of population A and B are unequal**
**$\alpha = .05$**
***DR:* If $\chi^2_{obs} > 3.84$, reject $H_0$; otherwise, do not reject**

**OBSERVED FREQUENCIES**

| | GROWTH RATE $<$ 5 in. | $\geq$ 5 in. | |
|---|---|---|---|
| SAMPLE A | **9** (a) | **87** (b) | **96** (i) |
| SAMPLE B | **23** (c) | **71** (d) | **94** (j) |
| | **32** (k) | **158** (l) | **190** $= N$ |

$$\chi^2_{obs} = \frac{N(ad - bc)^2}{i\,j\,k\,l} = \frac{190[(9)(71) - (87)(23)]^2}{(96)(94)(32)(158)}$$

$$= \frac{190[639 - 2001]^2}{45625344} = \frac{352458360}{45625344} = 7.73$$

DECISION **Since 7.73 > 3.84, reject $H_0$**
INTERPRETATION **Distributions of growth rates as categorized in this study are not equal. Growth rate is not the same in these two populations of seedlings.**

to indicate the one style personally preferred (ratings or ranked data would have violated the assumptions of chi-square tests). Observed and expected frequency data are given in Table 7.4. The null hypothesis states that regional differences do not exist. The expected frequencies were again obtained by using the marginals from the observed table. Expected frequencies for every cell in a particular column are the same in this table because all the row marginals are equal. Cells in the style A column, for example, all have expected frequencies of (41)(60)/240 = 10.25. Notice that the expected table is consistent with the null hypoth-

**TABLE 7.4. OBSERVED AND EXPECTED FREQUENCIES FOR A 4 × 3 CHI-SQUARE TEST OF HOMOGENEITY**

**Observed Frequencies**

| | STYLE A | STYLE B | STYLE C | |
|---|---|---|---|---|
| EAST | **6** (a) | **32** (b) | **22** (c) | **60** |
| MIDWEST | **10** (d) | **21** (e) | **29** (f) | **60** |
| SOUTH | **22** (g) | **15** (h) | **23** (i) | **60** |
| WEST | **3** (j) | **38** (k) | **19** (l) | **60** |
| | **41** | **106** | **93** | **240** |

**Expected Frequencies**

| | STYLE A | STYLE B | STYLE C | |
|---|---|---|---|---|
| EAST | **10.25** (a) | **26.50** (b) | **23.25** (c) | **60** |
| MIDWEST | **10.25** (d) | **26.50** (e) | **23.25** (f) | **60** |
| SOUTH | **10.25** (g) | **26.50** (h) | **23.25** (i) | **60** |
| WEST | **10.25** (j) | **26.50** (k) | **23.25** (l) | **60** |
| | **41** | **106** | **93** | **240** |

eses of no differences in style preference by region, even though styles B and C are more popular overall than style A.

The formula for $\chi^2_{obs}$ will contain 12 terms; one for each cell:

$$\chi^2_{obs} = \sum \frac{(O - E)^2}{E}$$

$$= \frac{(6.0 - 10.25)^2}{10.25} + \frac{(32.0 - 26.5)^2}{26.5} + \cdots + \frac{(19.0 - 23.25)^2}{23.25}$$

$$= 34.93$$

The degrees of freedom for the problem is $(4 - 1)(3 - 1) = 6$. Suppose the researcher has chosen to use an $\alpha$ level of .01. The critical value, as found in Appendix D, would then be 16.81. Since $\chi^2_{obs}$ is greater than this critical value, $H_0$ is rejected. A significant $\chi^2$ has been obtained for the table taken as a whole. The researcher feels confident in making the inference that preference for men's jacket styles differs by region.

*Post Hoc* Procedures for Homogeneity Tests Suppose, now, that the above researcher desires to obtain additional information from the data in Table 7.4. For example, is style A less acceptable than styles B and C for the population of all regions taken as a whole (involving comparisons of the marginal frequencies)? Is style B preferred over style C in the Eastern population (involving a comparison between two cell frequencies)? Are style preferences of the East and West populations the same (involving comparisons of certain row frequencies)? No matter how tempting it might be to make such additional *post hoc* comparisons with isolated sets of data, the researcher must remember that this would entail selected reanalyses of data already used in the overall chi-square test of regional differences. He would be in violation of assumptions of the test and erroneous decisions could result. We will come up against this same issue in Chapter 10 with regard to analysis of variance tables. The researcher has only three acceptable courses of action in this situation. The first is to make the desired analyses by using an extension of Scheffé's method of multiple contrasts, which will not be explained here but may be found in Marascuilo (1971). The second is to draw new samples, collecting fresh data for $\chi^2$ analyses of each comparison or contrast of interest as suggested by inspection of the overall table. The third is to forego any additional tests of significance and simply make observations as to which cells or sets of cells have contributed the most to the significant overall $\chi^2_{obs}$ for the table. For instance, cells (g) and (h) are two that particularly stand out. The $\chi^2$ term for cell (g) is $(22.0 - 10.25)^2/10.25 = 13.5$, while the term for (h) is $(15.0 - 26.50)^2/26.50 = 5.0$. Together they account for over half of the magnitude of $\chi^2$ for the entire table. Upon observing this, the researcher might infer that style A could be marketed more successfully in the Southern region than style B. He is not in a position to determine the probability of this decision being right or wrong, for the specific null hypothesis of no difference between the relative preference of the Southern population for styles A and B has not been tested at a given $\alpha$ level. If a wrong decision could result in serious financial repercussions, additional statistical evidence should be gathered by drawing a new sample from the Southern population and collecting data in the form of a $1 \times 2$ observed table to test the hypothesis of no difference in preference for styles A and B.

### TESTS OF INDEPENDENCE (ONE SAMPLE, TWO CRITERION VARIABLES)

Unlike the one-sample goodness-of-fit test, the one-sample *test of independence* always involves measures on *two* variables for each sample member and an independent theoretical distribution is not used, per se, to construct expected frequencies. It is sometimes referred to as a test of *association* and is related to the correlational analysis procedures which will be covered in later chapters. The independence test sample is composed of members that have been randomly drawn from a *single* population of interest. Therefore, populations are not being compared for homogeneity. Instead, the test is used to determine if measures on two criterion variables are either independent or associated with one another within a particular population. Examples of investigations in which we might wish to test the independence of variables might include studies of possible association between the retail price of TV sets and their picture quality, frequency of lightning flashes and amount of rainfall, number of books people claim to read and their level of education, weight of mammals and brain size, the duration of tiltmeter anomolies and earth tremor magnitudes, and so on.

Computationally, the test of independence is identical to the test of homogeneity. One of the most obvious distinctions between them is that the researcher determines the size of each sample in the test of homogeneity, thereby setting in advance the total frequency for the problem and the marginal totals of either the rows or the columns. In the test of independence only the total frequency (the single sample size, $N$) is known; the marginal row and column totals are random variables for which frequencies are not determined until after the sample is drawn and data have been collected. The student should now be able to recognize that the problem previously described in Table 7.2 would be analyzed with a test of independence.

Consider the question of whether attitudes of a city's residents toward the construction of a new civic center (the first variable) are independent of attitudes toward construction of a new city hall (the second variable). If the two variables are positively associated, people who are in favor of the center would also tend to favor the city hall. If the variables are negatively associated, people in favor of the center would tend to oppose the city hall. If the variables are in fact independent, people in favor of the center would be split in their attitudes toward the city hall; half approving and half disapproving. In other words, when the attitudes are independent, knowledge of a person's attitude toward one construction project tells us nothing about his attitude toward the other project.

Let us carry out this example further, as shown in Table 7.5. A single population would be defined and one sample drawn. Sample members would respond to two attitudinal questions; one concerning the center and one concerning the city hall. The observed table would be constructed by assigning each member to a particular cell of a fourfold table which matches the two responses given. The marginal frequencies in the observed table are again used to calculate expected frequencies for these tests. Notice how the expected table is proportional with respect to the marginals. Although the respondents tend to be opposed to both construction projects, the expected table still depicts a situation consistent with $H_0$. Of those 132 in favor of project A, for example, about 43 percent would favor project B. Of those 218 opposed to project A, again about 43 percent would be in favor of project B.

Formula (7.3) has been used to compute the observed value of $\chi^2$. Computations with formula (7.4) have also been provided to demonstrate

**TABLE 7.5. EXAMPLE OF A CHI-SQUARE TEST OF INDEPENDENCE**

| | |
|---|---|
| POPULATION | **Adult residents of City Z, June 1975** |
| SAMPLE | **350 individuals randomly chosen from population** |
| CRITERION VARIABLES | **A: Attitude toward construction of new city hall (favor/oppose)**<br>**B: Attitude toward construction of new civic center (favor/oppose)** |
| DECISION MODEL | **$H_0$: Attitudes A and B are independent**<br>**$H_1$: Attitudes A and B are associated**<br>$\alpha = .05$<br>*DR:* **Reject $H_0$ if $\chi^2_{obs} > 3.84$; otherwise, do not reject** |

**OBSERVED FREQUENCIES**

| | | Attitude B FAVOR | Attitude B OPPOSE | | |
|---|---|---|---|---|---|
| **Attitude A** | FAVOR | 47 (a) | 85 (b) | 132 | (i) |
| | OPPOSE | 102 (c) | 116 (d) | 218 | (j) |
| | | 149 (k) | 201 (l) | 350 | |

(table continues)

Table 7.5 (*Continued*)

EXPECTED FREQUENCIES

| | | Attitude B | | |
|---|---|---|---|---|
| | | FAVOR | OPPOSE | |
| Attitude A | FAVOR | 56.19 (a) | 75.81 (b) | 132 (i) |
| | OPPOSE | 92.81 (c) | 25.19 (d) | 218 (j) |
| | | 149 (k) | 201 (l) | 350 = N |

$$\chi^2_{\text{obs}} = \sum \frac{(O)^2}{E} - N$$

$$= \frac{(47)^2}{56.19} + \frac{(85)^2}{75.81} + \frac{(102)^2}{92.81} + \frac{(116)^2}{125.19} - 350$$

$$= 39.31 + 95.30 + 112.10 + 107.48 - 350 = 354.20 - 350$$

$$= 4.20$$

$$\chi^2_{\text{obs}} = \frac{N(ad - bc)^2}{ijkl} = \frac{350[(47)(116) - (85)(102)]^2}{(132)(218)(149)(201)} = 4.21$$

DECISION **Since 4.20 > 3.84, reject $H_0$**

INTERPRETATION **Attitudes A and B are associated (are not independent) in the defined population. The strength of this association should be computed with an index of association**

that it was not absolutely necessary to construct the expected data table for this 2 × 2 problem, although the availability of expected values facilitates the interpretation of results. The decision to reject $H_0$ at a previously determined $\alpha$ level of .05 leads the researcher to conclude that the attitudes are related in the population. It appears that this association is negative, for there is a weak tendency for those who favor one project to oppose the other. The magnitude of this relationship should be determined with a *post hoc* index of association.

Tests of independence are not limited to fourfold tables. If the response categories in the above study had been defined as "Favor," "Undecided," and "Oppose," for example, a 3 × 3 table would have resulted.

Formula (7.1) or (7.3) would have been used to compute $\chi^2$ and the results would still be interpreted in terms of the independence or association of the two criterion variables in the population.

*Post Hoc* Procedures Tests of independence merely indicate the presence of association; not its magnitude. The sensitivity of $\chi^2$ to any degree of association increases as a function of sample size. Thus, a significant $\chi^2$ value might be observed with a large sample even though the relationship between variables is so weak that it has limited or no practical meaning. When the null hypothesis of independence has been rejected, an *index of association* should be computed in order to obtain a numerical estimate of the strength of the relationship. These indices include the phi coefficient, Cramér's statistic, and the contingency coefficient; all of which are discussed in Chapter 14.

## SMALL EXPECTED FREQUENCIES IN CHI-SQUARE TESTS

One of the assumptions specified for $\chi^2$ was that each expected cell frequency be at least 5 or 10. All the examples we have used have met this requirement, which is essential to insure that the distribution of the observed frequencies around each expected frequency can approach normality in shape. Unfortunately, satisfying this assumption is sometimes difficult in practice. The expected frequencies in tests of homogeneity and independence depend on the marginals of the observed tables; data over which the researcher has no direct control. Consider the following observed frequency table for a test of independence between the number of traffic accidents during a one-year period and use of tranquilizers for a sample of licensed drivers within a particular state.

| | | Traffic Accidents | | | |
|---|---|---|---|---|---|
| | | 0 | 1 | ≥2 | |
| Tranquilizer Usage | REGULARLY | **5** (a) | **8** (b) | **7** (c) | **20** |
| | SOMETIMES | **13** (d) | **16** (e) | **11** (f) | **40** |
| | NEVER | **102** (g) | **26** (h) | **12** (i) | **140** |
| | | **120** | **50** | **30** | **200** |

The expected value of cell (c) would be (20)(30)/200 = 3. Since this does not satisfy the requirement of a minimum expected frequency of 5 in a table with more than 1 degree of freedom, the researcher may consider collapsing the traffic accident categories of "1" and "2 or more" into a single category labeled "1 or more." Some statisticians would approve of this solution. Others would take the position that deliberate manipulation of established categories of criterion variables should be avoided whenever possible. The researcher may also think of the possibility of throwing out all the data associated with the category of "2 or more." This alternative is clearly unacceptable, as it violates the assumption that all the observed data be used. Violations of this type are frequently found in the literature, particularly with respect to elimination of the middle or "average" categories of criterion variables.

The authors suggest the following rules with regard to combining or eliminating categories after data collection:

1. In $\chi^2$ tests of homogeneity, *population* categories may be combined or dropped if necessary to achieve adequate expected cell frequencies, providing that the new combined categories or set of remaining categories make sense with respect to the experimental problem.
2. In any type of $\chi^2$ test, categories of *criterion* variables should never be thrown out of the analysis, and should be combined with other categories only as a last resort to achieve adequate expected cell frequencies.

Sound solutions to this problem are available. The preferred solution for any size or type of $\chi^2$ table is to increase the $N$'s in the overall sample(s). In the above example only 3 of every 200 frequencies would be expected to fall in cell (c). Consequently, the researcher would have to sample another 150 or so members of the population in order to achieve an expected frequency of 5. This solution may, unfortunately, be too costly or too time consuming to be practical in many circumstances. The effects of violating the minimum cell size requirement are related to the table size ($df$) and the relative number of small frequencies involved. Since the particular table under consideration is large ($df = 4$) and only one cell is in violation, the effects of using the original data as obtained would probably be minimal; the theoretical $\chi^2$ distributions should still be adequately approximated. If the researcher could not collect more data his best solution would thus be to go ahead and conduct the test with the original set of 200 cases rather than to combine or eliminate categories of the criterion variables.

Small expected frequencies in 2 × 2 tables present less of a problem, as simple methods are available to handle this situation. The first set of methods are *correction for continuity* adjustments to the formula for $\chi^2_{\text{obs}}$ as developed by Yates, Pirie and Hamden, and others. The Pirie-Hamden formula for 2 × 2 tables is:

$$\chi^2_{\text{obs}} = \frac{N[|ad - bc| - .5]^2}{i\,j\,k\,l} \qquad (7.5)$$

where the terms are as defined for formula (7.4).

This formula differs from (7.4) only in that the value of .5 is subtracted from the absolute difference, $ad - bc$, before squaring. Use of this formula is illustrated in Table 7.6. Determination of critical values and all other procedures remain unchanged.

The authors suggest that the Pirie-Hamden formula be used for computing the observed value of $\chi^2$ whenever the total frequency, $N$, is greater than 20 and one or more of the expected frequencies is between 3 and 10.

When any expected frequency is as small as 2 in a table with $df = 1$, solutions which do not involve $\chi^2$ are available to the researcher. Although the procedures will not be covered here, Fisher's exact method is recommended in this situation or whenever the total frequency is 20 or less. A discussion of this method may be found in several texts, including McNemar (1969).

**TABLE 7.6 COMPUTATION OF CHI-SQUARE WHEN THE CORRECTION FOR CONTINUITY FORMULA IS APPROPRIATE**

**Observed Frequencies**

| | | |
|---|---|---|
| 3 (a) | 9 (b) | 12 (i) |
| 7 (c) | 13 (d) | 20 (j) |
| 10 (k) | 22 (l) | 32 = $N$ |

$$\chi^2_{obs} = \frac{N[|ad - bc| - .5]^2}{ijkl}$$

$$= \frac{32[|39 - 63| - .5]^2}{(12)(20)(10)(22)}$$

$$= \frac{32(24 - .5)^2}{52800}$$

$$= \frac{17672}{52800}$$

$$= .33$$

## EXERCISES

1. On a true-false test, a student answers 40 out of 70 items correctly. Does this score differ from chance (35 correct) at the .05 level?
2. A six-sided die is rolled 120 times and a one-spot appears 40 times. Is the die fair? Use an $\alpha$ level of .001.
3. During a 30-day period, a person driving through a particular intersection in a northerly direction encounters a red light 90 times, a yellow light 10 times, and a green light 30 times. The street department claims that the light is red 50 percent, yellow 10 percent, and green 40 percent of the time for traffic proceeding in that direction through the intersection. Formally outline a study, using examples in this chapter as illustrations, to determine if these observations are compatible with the street department's claim. State your decision at the .01 level and interpret the results.
4. During the decade of the 1960s, suppose that the Department of Agriculture categorized total domestic beef cattle production by breed (A, B, and Miscellaneous) and by region (East and West of the Mississippi). The following percentages were obtained within categories:

| East | | | West | | | |
|---|---|---|---|---|---|---|
| A | B | MISC | A | B | MISC | |
| 10%. | 14% | 6% | 21% | 34% | 15% | 100% |

A present-day researcher has categorized nationwide production by breed and region with a current sample of 1000 animals. These observations, in frequencies, are:

| East | | | West | | | |
|---|---|---|---|---|---|---|
| A | B | MISC. | A | B | MISC. | |
| 117 | 118 | 42 | 263 | 339 | 121 | 1000 |

Test for goodness-of-fit at the .05 level between this sample distribution and the USDA's. Interpret the results. (Be sure to convert the expected table to frequencies for an $N$ of 1000.)

5. A random sample of white mice, the progeny of several generations trained in running mazes, is compared with a random sample whose ancestors were untrained to determine if success in learning to run a maze differs for the underlying populations. The observed data are given below:

| | SUCCESSFUL | UNSUCCESSFUL | |
|---|---|---|---|
| UNTRAINED | 18 | 22 | 40 |
| TRAINED | 23 | 17 | 40 |
| | 41 | 39 | 80 |

Formally outline this study and test the null hypothesis at the .10 level. Use both formulas (7.3) and (7.4) to compute $\chi^2_{\text{obs}}$.

6. Samples of three species of insects are placed in a container and a prescribed amount of insecticide is added. After 60 minutes the number of each species surviving is counted. Does the insecticide differ in its effectiveness according to species at the .001 level?

| | Insect Species | | |
|---|---|---|---|
| | A | B | C |
| SURVIVING | 21 | 65 | 43 |
| NOT SURVIVING | 179 | 135 | 157 |
| | 200 | 200 | 200 |

7. One random sample of patients is drawn from hospital Y and another sample from hospital Z, both over the same one-year period from 1/1/76 to 1/1/77. The individuals are categorized as to the length of their stay.

| | | Length of Stay in Days | | | | |
|---|---|---|---|---|---|---|
| | | ≤1 | 2–3 | 4–7 | ≥8 | |
| **Hospital** | Y | 9 | 55 | 31 | 15 | 110 |
| | Z | 2 | 48 | 53 | 17 | 120 |

a. Formally outline this study, testing the hypothesis that length of stay, as categorized, does not differ between hospitals at the .05 level.
b. If *post hoc* tests are justified, indicate what you would test and how you would proceed.

8. Random samples of American cougar, African lion, and cheetah cubs are acquired for a behaviorial study. All are handled daily while young. At 3 years of age they are classified as to aggressive behavior toward people.

| | AGGRESSIVE | UNPREDICTABLE | NONAGGRESSIVE | |
|---|---|---|---|---|
| COUGAR | 8 | 8 | 4 | 20 |
| LION | 10 | 7 | 3 | 20 |
| CHEETAH | 2 | 10 | 8 | 20 |

a. Formally outline this study, testing the hypothesis of no behavioral differences between breed of cat at the .01 level.
b. Explain why the assumptions for the $\chi^2$ test have probably been satisfied.
c. Would *post hoc* tests be justified? Why or why not?

9. A study is conducted with a sample of red delicious apples to determine if the variables of coloration and uniformity of shape are associated with one another.

| | | **Uniformity** | |
|---|---|---|---|
| | | GOOD | POOR |
| **Coloration** | GOOD | 11 | 7 |
| | POOR | 12 | 14 |

Test for independence at the .10 level. Use the correction for continuity formula if appropriate.

10. The relationship between fraternity/sorority membership and future career plans is being studied at university Y using a random sample of students.

| | **Career Plans** | | |
|---|---|---|---|
| | GRADUATE/ PROFESSIONAL SCHOOL | WORK | OTHER |
| MEMBER | 28 | 21 | 5 |
| NONMEMBER | 44 | 74 | 28 |

a. Formally outline this study, testing for independence at the .05 level.
b. What *post hoc* procedures would be appropriate?

# 8

# SAMPLING DISTRIBUTIONS AND ESTIMATION

In the previous chapter we indicated that if the chi-square statistic for a given problem were to be computed on a large number of independent random samples drawn from a defined population, the resulting frequency distribution would approximate one of the theoretical distributions of $\chi^2$ in Appendix D. Such distributions of sample statistics are called *sampling distributions*. While it was possible to demonstrate the use of $\chi^2$ tests without exploring the concept of sampling distributions, at least a basic understanding is needed before parametric procedures for making tests and constructing confidence interval estimates can be given. The following discussion will center around the concept of sampling distributions composed of sample means.

## SAMPLING DISTRIBUTIONS

Use of the mean and standard deviation as descriptive statistics requires that the data being described meet the standards of equal-interval scales and that the data distribution is at least symmetrical and unimodal; or, where the normal probability table is to be employed, approximately normal in shape. For purposes of making inferences from sample statistics to population parameters, on the other hand, these standards are questioned not with respect to the empirically obtained sample data but rather to the underlying and typically unknown population of data from which the sample was drawn.

Let us consider a purely hypothetical investigation being independently conducted by three naive and somewhat neurotic researchers on the amount of time required for male college students in a certain country to run a mile. There are approximately 1.2 million able-bodied male students in this population. Running time, the criterion variable, is measured in seconds. Because of the nature of the variable under study, there is some reason to believe that the population data would tend to be unimodal. The distribution would not be normal, however, for practical ceilings would exist on minimum and maximum times obtained and common sense indicates that a positive skew is quite likely to be observed.

Although the sample size selected for any real study is restricted by economic, time, and other considerations, let us assume for illustrative purposes that all three researchers have the facilities and support needed to collect data on a sample of any size from this population. The first researcher decides to capitalize on this unique opportunity and makes the necessary arrangements to sample the running time of every healthy male college student in the country. Figure 8.1 shows the distribution of raw $X$ values (measures on the criterion variable) that he obtains. Since every possible population member has been included in this sample, the sample is the population. Thus, the statistics obtained in this very unusual case are not estimates of parameters, but actually are the parameters; $\bar{X} = \mu$, $S = \sigma$, $N = n$. No inferences need be made here. Note that the population of $X$ values is positively skewed.

For an illustration of the opposite extreme of sampling, let us assume that the second investigator decides initially to take a single sample of

FIG. 8.1 POSITIVELY SKEWED DISTRIBUTION OF MILE RUNNING TIME IN SECONDS.

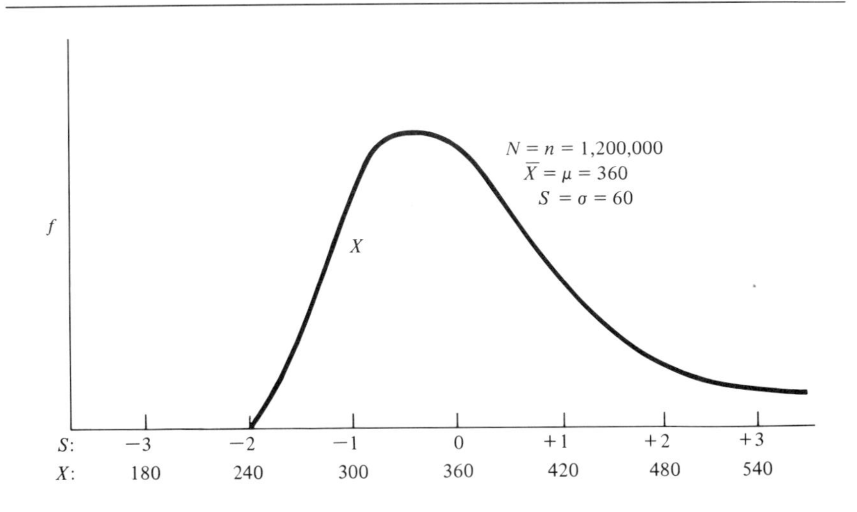

size 1. The decision is to draw one student on a purely random basis from this population, measure his running time, and generate estimates of the population parameters. This single $X$ value is found to be 341 seconds. Thus, $\overline{X} = 341/1 = 341$, the best available estimate of $\mu$. He discovers that no estimate of $\sigma$ can be made, since $s$ cannot logically be computed on a sample of one case. Suppose now that this investigator decided to continue taking new random samples of size 1, computing the mean of each sample of one case, until he has sampled and obtained a measure on every member in the population. The distribution of mean values resulting from this rather bizarre decision would be called a *sampling distribution of means,* where $N_{\overline{X}}$, the number of cases in each sample, is 1 (see Figure 8.2). It should be apparent to the reader that if every subject is measured only once (sampling without replacement from a finite population), this particular sampling distribution of mean values would in fact be identical to the population of $X$ values obtained by the first researcher. Comparison of Figures 8.1 and 8.2 will show this to be true; only the notation differs. Notice that the scale of measurement in Figure 8.2 is now labeled $\overline{X}$ and that the standard deviation remains at 60 seconds.

Sampling distributions are a type of population data distribution. The parameters of a sampling distribution composed of means are:

$\mu_{\overline{X}}$ = mean of the sampling distribution
$\sigma_{\overline{X}}$ = standard deviation of the sampling distribution, more commonly called the *standard error* of the mean
$N_{\overline{X}}$ = the sample size used in computing each mean in the distribution

**FIG. 8.2 POSITIVELY SKEWED SAMPLING DISTRIBUTION OF MEAN RUNNING TIME; $N_{\overline{X}} = 1$.**

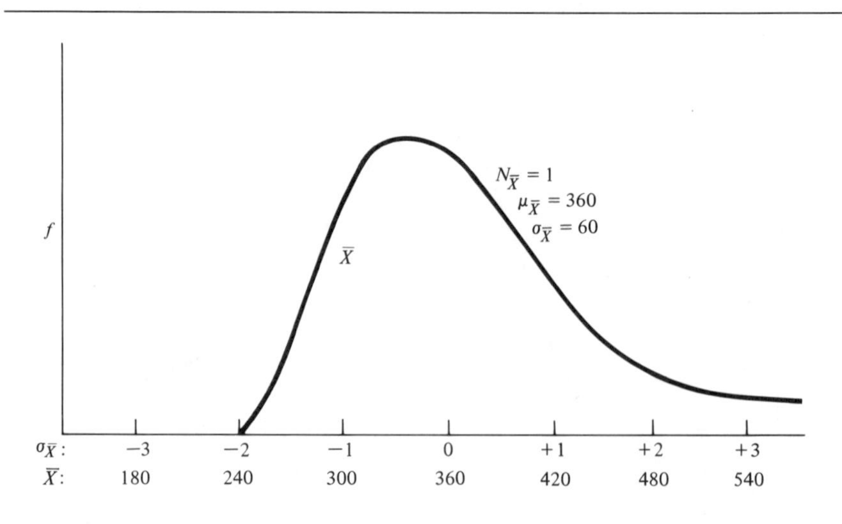

No symbol is needed for the total number of means (entries) in the sampling distribution, for this quantity is usually considered to be infinite. This number happens to equal $n$ in the second example only because every possible sample of size $N_{\bar{X}} = 1$ was drawn without replacement (without putting each sample member drawn back into the population before drawing the next sample) from a finite population. The reader should be advised at this point that sampling distributions are ordinarily constructed on a theoretical basis, with the assumption that a very large number (approaching infinity) of samples of size $N_{\bar{X}}$ are drawn with replacement—putting each sample back into the population before drawing the next.

Now let us assume that the third researcher chooses to measure the running time of a single sample of size 4, randomly drawn from the population. The following results are obtained:

| PERSON | $X$ | |
|---|---|---|
| 1 | 365 | $N = 4$ |
| 2 | 439 | $\bar{X} = 355.8$ |
| 3 | 282 | $S = 56.6$ |
| 4 | 337 | $s = 65.3$ |

Since this researcher is working independently, he does not know the actual parameters of this population as computed by the other investigators. His best available estimate of $\mu$ is the mean of this sample, 355.8, and his best estimate of $\sigma$ is 65.3. He is quite understandably concerned about the accuracy of these estimates and the amount of confidence that he can place in them. To get some indication of the stability and accuracy of his statistics, the researcher decides to draw two more random samples of size 4 from the population for comparison.

| **Sample B** | | | **Sample C** | | |
|---|---|---|---|---|---|
| PERSON | $X$ | | PERSON | $X$ | |
| 1 | 423 | $N = 4$ | 1 | 348 | $N = 4$ |
| 2 | 284 | $\bar{X}_B = 362.0$ | 2 | 411 | $\bar{X}_C = 357.5$ |
| 3 | 349 | $S_B = 52.1$ | 3 | 385 | $S_C = 47.0$ |
| 4 | 392 | $s_B = 60.2$ | 4 | 286 | $s_C = 54.2$ |

As might be expected, statistics computed from the three sets of data were not identical, fluctuating from sample to sample. Our knowledge of $\mu$ as computed by the other researchers enables us to see that errors are present in all of the sample estimates he has obtained. The researcher is dissatisfied with the magnitude of these fluctuations, and now proceeds to demonstrate his neurotic tendencies by drawing tens of thousands of samples of size 4 from this population, computing $\bar{X}$ for each. He plots a frequency polygon of these sample means, as shown in Figure 8.3, and computes the mean and standard deviation for the resulting distribution. Several things should be noticed about this figure. First, since the distribution is composed of sample means, it is a sampling distribution. Second, since each sample had 4 cases, $N_{\bar{X}} = 4$. Third, the number of means that have been plotted is not given and is assumed to be infinite (he actually plotted "tens of thousands"). Fourth, as in the case with the sampling distribution for $N_{\bar{X}} = 1$, the mean of the sampling distribution equals the population mean as we know it from the other investigations: $\mu_{\bar{X}} = \mu$. This equality holds true for any population and its sampling distribution, regardless of $N_{\bar{X}}$. Fifth, comparing this value of $\sigma_{\bar{X}}$ with that obtained by the second researcher, we see that the variability of the sampling distribution—the magnitude of its standard error—*decreased* as the size of the samples on which the means were computed *increased* from one to four. This relationship is demonstrated in the following formula, which allows $\sigma_{\bar{X}}$ to be mathematically computed from $\sigma$ and $N_{\bar{X}}$ without actually going to the trouble of determining the standard deviation of a distribution of tens of thousands of sample means:

$$\sigma_{\bar{X}} = \frac{\sigma}{\sqrt{N_{\bar{X}}}} \tag{8.1}$$

where $N_{\bar{X}}$ is the size of each sample (a constant for any particular sampling distribution).

FIG. 8.3 SLIGHTLY SKEWED SAMPLING DISTRIBUTION OF MEAN RUNNING TIME; $N_{\bar{X}} = 4$.

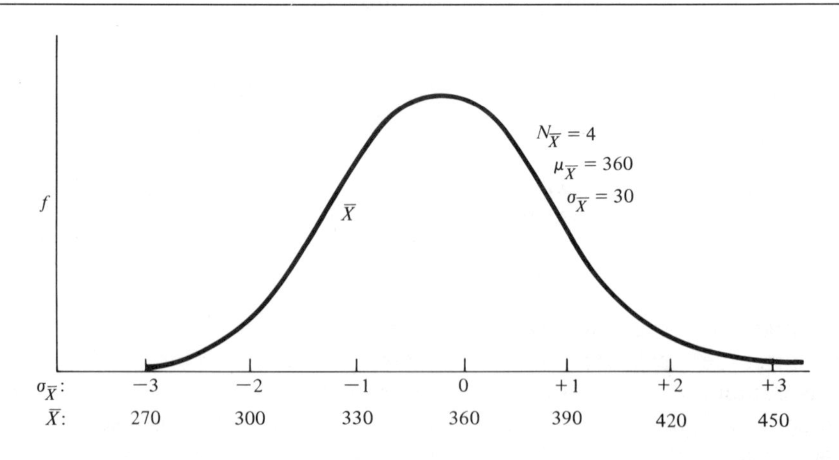

Thus, when $N_{\bar{X}} = 1$, as in the first sampling distribution,

$$\sigma_{\bar{X}} = \frac{60}{\sqrt{1}} = 60$$

And, when $N_{\bar{X}} = 4$, as in the second sampling distribution,

$$\sigma_{\bar{X}} = \frac{60}{\sqrt{4}} = \frac{60}{2} = 30$$

These are the same values that were determined through laborious data collection by the second and third researchers. If the means of tens of thousands of samples of size 100 from this population were plotted, the standard error (standard deviation) of the resulting sampling distribution would be $60/\sqrt{100} = 6$. For $N_{\bar{X}} = 250$, $\sigma_{\bar{X}}$ would be further reduced to 3.8. Thus, we have empirical evidence that the standard error of a sampling distribution ($\sigma_{\bar{X}}$) can be as large as the standard deviation of the underlying population $X$ distribution ($\sigma_{\bar{X}} = \sigma$) when $N_{\bar{X}} = 1$, and will approach zero in value ($\sigma_{\bar{X}} \rightarrow 0.0$) as $N_{\bar{X}}$ approaches infinity. In all instances, however, remember that the means of these different sampling distributions would be identical and equal to the population mean: $\mu_{\bar{X}} = \mu$.

The last point of interest in Figure 8.3 is the shape of the sampling distribution for $N_{\bar{X}} = 4$, which appears to less skewed than the population of raw $X$ values. If the sampling distribution for $N_{\bar{X}} = 100$ were also to be plotted, it would appear to be even less skewed and more normal in shape than the distribution for $N_{\bar{X}} = 4$. The shapes of sampling distributions do, in fact, become more nearly *normal* as $N_{\bar{X}}$ increases, regardless of the shape of the underlying distribution of population $X$ values.

Several of the points covered above are demonstrations of the *central limit theorem,* an interpretation of which may be stated as follows:

> If a population distribution of $X$ values (it need not be normal) has a mean $\mu$ and a standard deviation $\sigma$, a distribution of random sample means of size $N_{\bar{X}}$ drawn from this population has mean $\mu_{\bar{X}} = \mu$ and standard deviation $\sigma_{\bar{X}} = \sigma/\sqrt{N_{\bar{X}}}$ and approaches a normal distribution as the sample size $N_{\bar{X}}$ increases.

The parametric procedures which follow share the assumption that underlying population data distributions are normal in shape. This restriction would limit their application severely if it pertained primarily to population distributions of raw $X$ values, for the actual shapes of such distributions are rarely known. Fortunately, the assumption is focused on the sampling distribution of the statistic being used. According to the central limit theorem we can expect a sampling distribution of means to approach normality when $N_{\bar{X}}$ is large even though the actual population distribution of $X$ values is not normal. The sampling distribution of many statistics follows a similar pattern. For example, a sampling distribution of the chi-square statistic is severely skewed with small degrees of freedom, but approaches normality as the degrees of freedom become large. Use of parametric procedures is not limited, therefore, to situations in which the researcher is confident that the phenomenon being measured produces a purely normal distribution. As a conservative rule of thumb, the researcher should be able to apply parametric procedures to any set of data on which he would not be reluctant to use $\bar{X}$ and $S$ as descriptive statistics.

One rarely has parameter values to work with when making statistical applications. Instead, a researcher usually collects data on one sample of size $N$ from each population of interest and the resulting statistics serve as estimates of the desired, but unavailable, parameters. Obvious exceptions would include analyses of incomes by the Internal Revenue Service or studies undertaken with national examinations, such as the ACT and SAT college entrance tests, where the test publishers have the capability of generating normative information and parameters on populations of candidates. In nearly any other situation the student is likely to encounter in the behavioral and natural sciences, the mean of the sampling distribution, $\mu_{\bar{X}}$, would be estimated by the sample mean, $\bar{X}$, and $\sigma_{\bar{X}}$ would be approximated by $s_{\bar{X}}$, an estimate of the standard deviation (standard error) of the sampling distribution of means that is based purely on sample data:

$$s_{\bar{X}} = \frac{s}{\sqrt{N_{\bar{X}}}} \tag{8.2}$$

If the third investigator had concluded his study after drawing the first sample of $N_{\bar{X}} = 4$, his estimate of $\sigma_{\bar{X}}$ would have been $s_{\bar{X}} = 65.3/\sqrt{4} = 65.3/2 = 32.7$ and the remainder of his parameter estimates would be as shown below:

| KNOWN SAMPLE STATISTICS | | UNKNOWN POPULATION (X VALUE) PARAMETERS | | UNKNOWN SAMPLING DISTRIBUTION ($\bar{X}$) PARAMETERS |
|---|---|---|---|---|
| $\bar{X}$ = **355.8** | **estimate of** | $\mu$ | **and** | $\mu_{\bar{X}}$ |
| $s$ = **65.3** | **estimate of** | $\sigma$ | | |
| $s^2$ = **4264.1** | **estimate of** | $\sigma^2$ | | |
| $s_{\bar{X}}$ = **32.7** | **estimate of** | | | $\sigma_{\bar{X}}$ |

Despite the inaccuracy of these observed estimates, $\bar{X}$ and $s^2$ are in fact *unbiased* estimates of their corresponding values in the population and the bias inherent in the other estimates is minimal. Although the chances are small that a statistic computed on a single sample will exactly equal the corresponding parameter, an unbiased estimate will be neither systematically too large nor too small. Thus, if unbiased statistics were computed on a large number of samples randomly drawn from a single defined population, about one-half would be larger than their corresponding parameters and one-half smaller. It is precisely these sampling errors, or differences between statistic and parameter, which produce the variability within sampling distributions. If all the samples drawn from a particular population were to yield identical means, the standard deviation (standard error) of the sampling distribution of means would be zero; the $\bar{X}$'s would all fall on a single vertical line above the point $\mu_{\bar{X}}$. The statistic $\bar{X}$ would be termed perfectly accurate, precise, or reliable. On the other hand, if the samples yielded a diverse set of means, the sampling distribution would display a large amount of variability and the computed size of its standard error would provide an indication of the relative imprecision or unreliability of the statistic.

It may help the student to think of sampling distributions as hypothetical constructs which merely provide a method for estimating the amount of fluctuation which would be observed among means or some other statistic computed from samples of a certain size. They are seldom, if ever, actually constructed with real data by researchers. A sampling distribution may be based on any statistic; for example, median, percentage, standard deviation, and so on. Hence, any statistic has a standard error which can be computed from known population data or estimated from available sample statistics. Sampling distributions of other statistics will be discussed later as needed.

## ESTIMATION

In the previous section we were reminded that the sample mean, $\bar{X}$, is an unbiased estimate of the population mean, $\mu$. It is thus possible to draw a random sample from a defined population of interest, compute $\bar{X}$, and use this value as a single *point estimate* of $\mu$. The reader is well aware by now that sample means fluctuate; different $\bar{X}$ values will be obtained from different samples drawn from the same population. Nevertheless, a researcher typically draws only one sample from a population and computes one value of $\bar{X}$. It stands to reason that this single $\bar{X}$ has little chance of exactly equaling $\mu$, or, for that matter, $\bar{X}$'s that might subsequently be computed from other similar samples. Calculation of the value of $s_{\bar{X}}$ will give an indication of the precision of this single point estimate by providing an index of the variability of these sample means. Even more meaning would be conveyed by some presentation of the accuracy of the estimate stated directly in terms of the basic scale of measurement, or raw $X$ values; that is, indicating that the estimate will differ from the true $\mu$ by no more than a certain number of $X$ units. As an analogy, consider the type of consumer information supplied with certain products; for example, a manufacturing firm may state that a rangefinder which it markets has an accuracy of ±6 inches at a distance of 50 feet. In other words, the company is acknowledging that samples of its product fluctuate in precision, providing only an estimate of the true distance. But the company is also demonstrating confidence that the particular rangefinder we might buy will come within 6 inches (49′6″–50′6″) of measuring a true distance of 50 feet. Thus, if we can tolerate this degree of imprecision the rangefinder should satisfy our needs as a measurement tool. This additional information regarding the accuracy of the instrument's estimates of distance would be very useful in our buying decision. The same point may be made with respect to estimates of population means; the inclusion of information relative to the accuracy or goodness of a given estimate is valuable in determining whether or not that degree of imprecision is tolerable for the user's particular purpose.

The sampling distribution concept may be employed in conjunction with the known properties of the normal curve to provide such *interval estimates* of precision for statistics, as well as the probability of observing sample statistics of various magnitudes. A discussion of these estimation procedures for means follows.

## STATEMENTS ABOUT SAMPLE MEANS: PARAMETERS KNOWN

We will begin by assuming that both $\sigma$ and $\mu$ for a population being studied are either known or their values hypothesized to be 12 and 68, respectively. Parameter estimation procedures, per se, are unnecessary here. However, an investigator has drawn a sample of 100 cases from this population and has found that $\bar{X} = 71.6$. Because this difference between the sample and population mean is rather large, he is interested in determining how frequently samples of 100 drawn from this population would yield means which differ 3.6 points or more from the population mean. He may be concerned, for example, that some unexpected source of bias was present in the procedure he followed in drawing this sample, or that an error was made in measuring the $X$ values or computing $\bar{X}$.

The mean of the sampling distribution is, by definition, the population mean: $\mu_{\bar{X}} = \mu$. For this example, then, $\mu_{\bar{X}} = 68.0$. Since $\sigma$ is known, the standard deviation (standard error) is calculated by formula (8.1):

$$\sigma_{\bar{X}} = \frac{\sigma}{\sqrt{N_{\bar{X}}}} = \frac{12.0}{\sqrt{100}} = \frac{12}{10} = 1.2$$

The hypothetical, normally distributed sampling distribution for $N_{\bar{X}} = 100$ can now be depicted, as in Figure 8.4. The observed mean of this sample, 71.6, lies exactly +3 standard deviation units above the known value of $\mu_{\bar{X}}$, 68.0. The formula for converting $X$ values to $z$ values in normally distributed populations was given in Chapter 5: $z = (X - \mu)/\sigma$.

**FIG. 8.4 SAMPLING DISTRIBUTION OF MEANS; $N_{\bar{X}} = 100$.**

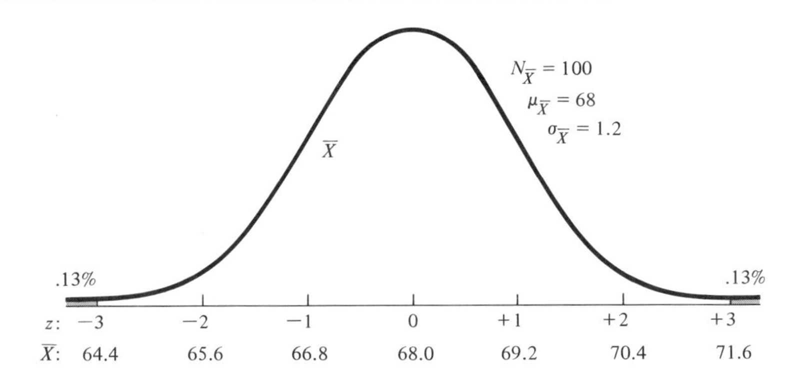


Since we assume the sampling distribution to also be normally distributed, the same formula, put into the appropriate notation, can be used for determining the $z$ value equivalent of observed sample means in sampling distributions:

$$z = \frac{\bar{X} - \mu_{\bar{X}}}{\sigma_{\bar{X}}} \tag{8.3}$$

For these data, then,

$$z = \frac{71.6 - 68.0}{1.2} = \frac{3.6}{1.2} = +3.0$$

Referring to Appendix C, we find that the proportion of cases in a normal distribution greater than or equal to a $z$ value of $+3.0$ is .0013, or .13 percent. It follows that another .13 percent of the cases are less than or equal to a $z$ value of $-3.00$. Therefore, sample means differing by 3.6 or more raw $X$ values from $\mu_{\bar{X}}$ would be obtained only about 3 times in a thousand by chance alone ($.0013 + .0013 = .0026$) with samples of size 100. Such odds would cause most researchers to review their calculations and sampling procedure to insure that some error or bias was not responsible for the extreme sample mean obtained.

With the sampling distribution constructed, it would now be possible to make any number of probability statements about the means of other samples which might be drawn from this population. For example, one could state that the probability of drawing a sample which has a mean greater than 68.0, the population mean, is .50; or that 95.44 percent of the means computed on random samples of size 100 drawn from this population would fall between $-2z$ and $+2z$, corresponding to an interval of 65.6–70.4 on the scale of $X$ values.

What percentage of samples of size 100 drawn from the underlying population of $X$ values would have means between 67.8 and 68.5? This answer is not so obvious. We must first convert these given means to $z$

values in the sampling distribution with formula (8.3) and then consult the normal table to obtain the proportion of entries (means) between the obtained $z$'s:

The larger (upper) $z$ corresponding to 68.5 is: $$z_u = \frac{68.5 - 68.0}{1.2} = +.42$$

The smaller (lower) $z$ corresponding to 67.8 is: $$z_l = \frac{67.8 - 68.0}{1.2} = -.17$$

The proportion of the area between $z_u$ and the sampling distribution mean ($\mu_{\bar{X}}$) is found to be .1628. The proportion between $z_l$ and $\mu_{\bar{X}}$ is .0675. These sum to .2303. Thus, 23 percent of the means of random samples of size 100 drawn from this population would fall between the values of 67.8 and 68.5. We could also state that the probability is .23 that any particular sample mean will have a value between 67.8 and 68.5. Notice that these exercises are merely repetitions of those in Chapter 5. The only difference is that we are dealing with means rather than $X$ values and normally distributed sampling distributions of means rather than normal distributions of raw $X$ values.

## CONFIDENCE INTERVALS FOR $\mu$ BASED ON SAMPLE DATA: PARAMETERS UNKNOWN

When $\sigma$ is known the researcher can usually assume that the theoretical sampling distribution of means is normal and can refer to the normal distribution table for making probability statements. According to the central limit theorem, this holds true even for most instances in which underlying populations of $X$ values are not normally distributed, provided that $N_{\bar{X}}$ is large. In practice, $\sigma$ is generally unknown and $\sigma_{\bar{X}}$ is estimated by $s_{\bar{X}}$. For these situations the shape of the sampling distribution is better approximated by distributions of the $t$ statistic (Appendix F). As $N_{\bar{X}}$ increases, $t$ distributions become more and more normal in shape, corresponding to normality exactly when $N_{\bar{X}} = \infty$ and approximating this shape closely enough for most practical purposes somewhere around $N_{\bar{X}} = 60$. Therefore, when a researcher's sample is large and $\sigma$ is unknown, he may elect to use the normal probability table to compute points and areas in the sampling distribution without introducing appreciable error. If this sample is small and $\sigma$ is unknown, tables of the $t$ distribution should always be used. The student may wish to adopt this arbitrary rule of thumb:

> If $\sigma$ is unknown and $N_{\bar{X}} \leq 60$, use tables of $t$ distributions for making inferences about parameters. For larger values of $N_{\bar{X}}$ the normal distribution table may (also) be used.

### Large Samples

Suppose that a single sample of 150 cases is drawn from a defined population with unknown parameters. The sample statistics are:

$N = 150$, $\bar{X} = 54.7$, and $s = 1.5$. The researcher wishes to make probability statements in terms of raw $X$ values concerning the precision of this sample estimate of $\mu$. The standard error of the sampling distribution, $\sigma_{\bar{X}}$, is estimated by $s_{\bar{X}}$ in the absence of a known $\sigma$:

$$s_{\bar{X}} = \frac{s}{\sqrt{N_{\bar{X}}}} = \frac{1.5}{\sqrt{150}} = \frac{1.5}{12.25} = .12$$

The hypothetical sampling distribution based entirely on sample estimates can now be sketched (Figure 8.5). Because $N_{\bar{X}}$ is large (150), the shape is assumed to be normal and the $z$ scale can be used along the abscissa.

The value of 54.7 is merely a point estimate of $\mu_{\bar{X}}$. The accuracy of this or other estimates of parameters can be stated in the form of a *confidence interval*. That is, we have $g$ percent confidence that the true population parameter lies between some computed lower and upper limits on the underlying scale of measurement. These limits for confidence interval statements about population means are derived through the following formula when $N$ is $> 60$:

$$g \text{ percent confidence interval} = \bar{X} \pm (z)(s_{\bar{X}}) \qquad (8.4)$$

The value of $g$ is most frequently set at 95 percent or 99 percent. If this researcher wishes to state with 95 percent confidence that a given interval of $X$ values will include the unknown population mean, $g$ is thus set at .95. Referring to the normal distribution table, he would determine the upper and lower $z$ values that mark off the middle 95 percent of the area in the sampling distribution and then convert these $z$'s to $\bar{X}$'s to define the limits of the confidence interval. According to Appendix C, 95 percent of the entries fall between $\pm 1.96z$ and $\mu_{\bar{X}}$, as illustrated in Figure 8.5. Formula (8.4) may now be applied to construct a confidence interval in terms of $X$ values:

$$\begin{aligned} 95 \text{ percent confidence interval} &= 54.70 \pm 1.96(.12) \\ &= 54.70 \pm .24 \\ &= 54.46\text{–}54.94 \end{aligned}$$

The researcher can state with 95 percent confidence that the $X$ value interval of 54.46–54.94 includes the value of the population mean, $\mu$. If the 99 percent confidence band for $\mu$ were desired, the normal table would be entered to find those $z$ values marking off the middle 99 percent of the sample means in the distribution. One-half of 1 percent (.005) would be left in each tail. Locating this value in column (4) of Appendix C gives corresponding $z$'s of $\pm 2.58$. Note that column (2) indicates a proportion of .495 cases from $z = \pm 2.58$ to the mean, and could also have been used.

$$\begin{aligned} 99 \text{ percent confidence interval} &= 54.70 \pm 2.58(.12) \\ &= 54.70 \pm .31 \\ &= 54.39\text{–}55.01 \end{aligned}$$

FIG. 8.5 HYPOTHETICAL SAMPLING DISTRIBUTION OF MEANS BASED ON LARGE SAMPLE ESTIMATES.

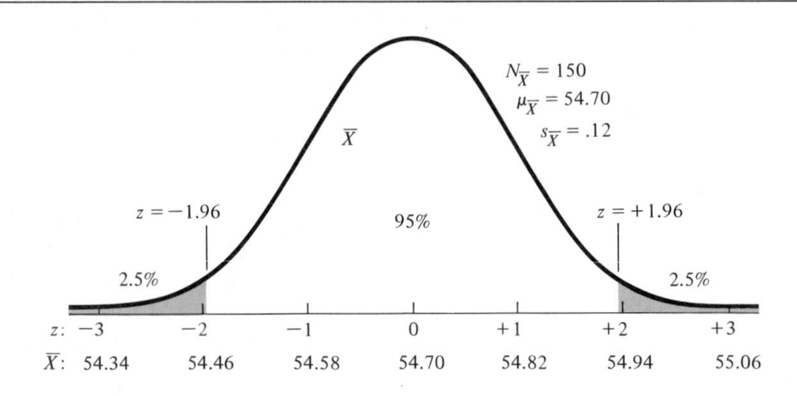


In words, the researcher is 99 percent confident that the raw data interval of 54.39–55.01 includes the value of $\mu$.

When the level of confidence that the interval will include $\mu$ with a given set of data increases, the width of the confidence interval must also increase. As the interval width approaches infinity, the confidence level approaches 1.00. At the opposite extreme, when the width of the interval approaches zero, confidence that the interval contains $\mu$ also approaches zero. This interval width is determined in the above formula by the size of $z$ (as set by the confidence level chosen) and the standard error ($s_{\overline{X}}$). The smaller the standard error, then, the less the interval width and the more precise is the estimate. Since $s_{\overline{X}}$ is partially a function of sample size, it should be apparent that large samples will generally provide more reliable, or precise, statistics.

Small Samples Tables for the $t$ distribution more adequately represent the theoretical sampling distributions when $s_{\overline{X}}$ has been used to estimate $\sigma_{\overline{X}}$ and should always be referenced in such situations when the sample size is small. The $t$ statistic is defined in the same way as $z$, except that $s_{\overline{X}}$ is substituted into the denominator:

$$t = \frac{\overline{X} - \mu_{\overline{X}}}{s_{\overline{X}}} \tag{8.5}$$

Technically, then, $t$ is actually being used whenever $\sigma$ is unknown, even though the researcher chooses to refer to the normal distribution tables. Procedures and computations learned so far for $z$ apply equally for $t$; only the shape of the sampling distribution will change.

Figure 8.6 provides an indication of the shape of $t$ distributions with small sample sizes. Notice that this $t$ distribution for $N_{\overline{X}} = 4$ is more peaked (leptokurtic) than the normal distribution and that the tails are lifted higher off the baseline.

**FIG. 8.6 SAMPLING DISTRIBUTION ($t$) OF MEANS FOR $N_{\bar{X}} = 4$. THE $t$ VALUES CORRESPONDING TO A 99 PERCENT CONFIDENCE INTERVAL ARE SHOWN.**

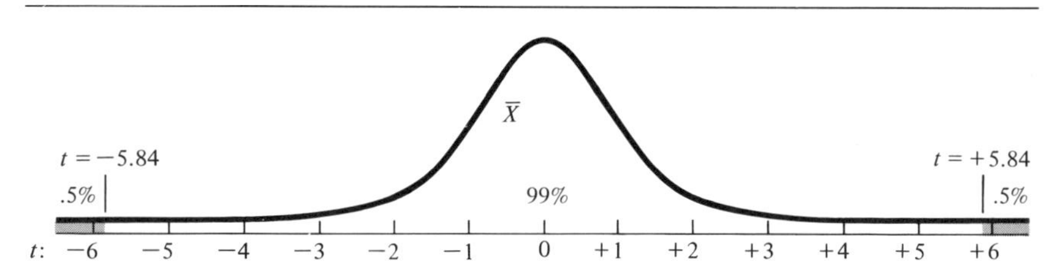


While $N$ (or $N_{\bar{X}}$) are not considered when entering the normal distribution table, a separate $t$ table exists for each sample size. The $t$ table (see Appendix F) is entered according to a *degrees of freedom* (*df*) column, where $df$ is defined as $N_{\bar{X}} - 1$ in confidence interval work. Degrees of freedom are computed in different ways for various statistical procedures. The beginning statistics student need not be overly concerned with the rationale involved; merely that he compute $df$ correctly for each application. The columns of the $t$ table under the nondirectional headings define the total proportion of the area in both tails and correspond to $\alpha$ levels, while the entries within the table are the $t$ values. Thus, when $df = \infty$, .90 of the area, or 90 percent of the entries in the distribution, lie between the $t$ values of $\pm 1.645$, the same as for $z$. Notice that all the $t$ entries for $df = \infty$ equal $z$; or, $t = z$ when $df = \infty$.

Now find the $t$ values that mark off a total of 1 percent of the cases in the tails when the sample size, $N_{\bar{X}}$, equals 4 ($df = N_{\bar{X}} - 1 = 3$). The $t$ entry is $\pm 5.841$. These $t$ values correspond to $g = 99$ percent ($\alpha = 1.00 - g = 1.00 - .99 = .01$) and have been marked off in Figure 8.6. Equivalent normal distribution $z$ values which mark off a total of 1 percent in the tails are $\pm 2.58$. Thus, whenever the sample is less than infinite in size, $t \neq z$ and some degree of error is introduced by using the normal distribution as an approximation to the appropriate $t$ distributions. In the previous section it was recommended that the student adopt an arbitrary rule of thumb that calls for using $t$ whenever $\sigma$ (and thus $\sigma_{\bar{X}}$) is unknown and $N_{\bar{X}}$ is 60 or less. In some cases, as when $df = 35$, interpolation will be required when using the $t$ table given in this text. Even so, the error is reduced somewhat as compared to using the normal table.

When $\sigma$ is unknown and $N_{\bar{X}} \leq 60$, the following formula is used for computing confidence intervals:

$$g \text{ percent confidence interval} = \bar{X} \pm (t)(s_{\bar{X}}) \qquad (8.6)$$

Suppose $N = 16$, $\bar{X} = 22.00$, and $s = 3.00$. The best available estimate of $\mu$ is 22.00. An indication of precision is desired in terms of the basic scale of measurement, $X$. To enter the confidence interval formula, $s_{\bar{X}}$ is required. For these data,

$$s_{\bar{X}} = \frac{s}{\sqrt{N_{\bar{X}}}} = \frac{3.00}{\sqrt{16}} = \frac{3.00}{4} = .75$$

The confidence level, $g$, is arbitrarily set at 99 percent. To find the appropriate value of $t$, look under the nondirectional .01 column in the $t$ table for the intersection with $df = 15$. The $t$ values which mark off the middle 99 percent of a $t$ distribution with $df = 15$ are shown to be $\pm 2.947$. These values are entered into formula (8.6):

$$\begin{aligned} \text{99 percent confidence interval} &= 22 \pm (2.947)(.75) \\ &= 22 \pm 2.21 \\ &= 19.79\text{–}24.21 \end{aligned}$$

It may be stated with 99 percent confidence that $\mu$, the mean of the underlying population of $X$ values, is included in the interval 19.79–24.21.

To illustrate an investigation which requires the use of estimation procedures, let us suppose that a researcher for a mass transit company is given the task of determining the average length of time it takes individuals to commute from their homes in a particular large suburb to their jobs in the heart of the city. The population is thus defined as a specific group of commuters, the actual number ($n$) of which is in the thousands. These employees are identified through downtown business files and a sample of 50 is randomly drawn. Each of the 50 is contacted and asked to record his driving time to work on some particular day: $X$ = a person's driving time. The mean and standard deviation are then computed for these sample data: $\bar{X} = 43.00$; $s = 9.00$. The point estimate of $\mu$ is 43.00. The researcher anticipates that he will be questioned as to how much confidence he has in this estimate of mean driving time for the entire population of workers. A confidence interval will provide an answer to this question. The estimate of the standard error of the sampling distribution, $s_{\bar{X}}$, is found to be $9/\sqrt{50} = 1.27$. The $df$ value is $50 - 1 = 49$ and the confidence level is set at 95 percent. Interpolation is required in the $t$ table between $df = 40$ and $df = 60$, resulting in a $t$ value of 2.01. Formula (8.6) may now be applied:

$$\begin{aligned} \text{95 percent confidence interval} &= 43.00 \pm (2.01)(1.27) \\ &= 43.00 \pm 2.55 \text{ minutes} \\ &= 40.45\text{–}45.55 \text{ minutes} \end{aligned}$$

The researcher is able to give his best estimate of $\mu$, 43 minutes, and state that he is 95 percent confident that the interval of 40.45–45.55 minutes includes the true mean driving time for the total defined population.

If this investigator had wanted to have even more confidence that the interval provided to his employer included the actual value of $\mu$, he might have chosen to set $g$ at some larger value. For $g$ = 99.9 percent, the nondirectional .001 column of the $t$ table would be entered to find $t$ values for formula (8.6):

$$\begin{aligned} 99.9 \text{ percent confidence interval} &= 43.00 \pm (3.51)(1.27) \\ &= 43.00 \pm 4.46 \text{ minutes} \\ &= 38.54\text{–}47.46 \text{ minutes} \end{aligned}$$

A great deal of confidence can be placed in this interval, but at the expense of a relatively large interval width (the range of the interval is 8.93 minutes).

## EXERCISES

1. What is the difference between sampling distributions and data distributions for samples and populations? How do their notations differ?
2. Define unbiased estimates. Which statistics mentioned in this chapter are unbiased estimates of parameters?
3. If a statistic is perfectly accurate or reliable, what is the shape of its sampling distribution?
4. How do sampling distributions change as $N_{\overline{X}}$ increases?
5. Contrast a point estimate with an interval estimate.

In working the following problems, sketch the sampling distributions and identify the points and areas of interest.

6. Random samples have been drawn from normally distributed populations with known parameters:
   a. $\mu = 25$, $\sigma = 4$, $\overline{X} = 24$. What is the probability of drawing a sample mean which deviates this much ($\pm 1$ unit) or more from the population mean by chance alone when $N = 64$?
   b. $\mu = 139.7$, $\sigma = 18.4$, $\overline{X} = 141.2$. What is the probability of drawing a sample mean which deviates this much or more from the population mean by chance alone when $N = 10$?
7. For a normal population with $\mu = 85.1$ and $\sigma = 8.3$, what is the probability that a sample of $N = 100$ randomly drawn from this population will have a mean between the values of 84 and 86?
8. What percentage of samples of size 20, randomly drawn from a normal population with $\mu = 59.0$ and $\sigma = 3.0$, will have means between 59.6 and 61.1?
9. A sample of size 85 is randomly drawn from a population thought to be normally distributed: $\overline{X} = 281.3$, $s = 44.8$.
   a. What are the point estimates for $\mu$, $\sigma$, $\mu_{\overline{X}}$, and $\sigma_{\overline{X}}$?
   b. Compute the 95 percent confidence interval for $\mu$.
10. A sample of 200 cases is randomly drawn from a population with an *unknown* distribution shape: $\overline{X} = 7.2$, $s = .55$.

a. Why would you not be reluctant to compute a confidence interval for this population mean, even though the distribution shape is unknown?
b. Compute the 99.9 percent confidence interval for $\mu$ and give a verbal interpretation of the result.

11. A sample of size 15 is randomly drawn from a population thought to be normally distributed: $\bar{X} = 5.4$, $s = 1.2$. Compute the 99 percent confidence interval for $\mu$ and give a verbal interpretation of the result.
12. A sample is randomly drawn from a population with an *unknown* shape: $N = 6$, $\bar{X} = 132.7$, $s = 2.8$.
    a. Compute the 90 percent confidence interval for $\mu$.
    b. Why would it be advisable to interpret this result very cautiously?

# 9

# TESTING HYPOTHESES CONCERNING PARAMETERS

The chi-square tests of homogeneity and goodness-of-fit enable a researcher to make inferences from samples about entire population distributions that have been expressed as frequencies in arbitrary categories. Significant observed values of $\chi^2$, however, may be caused by differences in population shapes, variances, central tendencies, or any combination of these factors. Chi-square tests do not permit discrimination among these possible causes. In many studies a researcher will be primarily interested in testing hypotheses concerning specific parameters; for example, means, variances, proportions, and so on. He may wish to decide, for instance, if the mean of an unknown population could have some postulated value, or if the means of two unknown populations could be equal. The tests discussed in this chapter may be used to facilitate a variety of decisions about specific parameters. Particularly emphasized are widely applicable $z$ and $t$ tests for population means and proportions.

Formulas used for calculating the $z$ and $t$ test statistics differ only with respect to how the standard error terms in their denominators are determined. Whenever the researcher has access to $\sigma$, the standard deviation of the underlying population data distribution, the standard error term may be directly computed and $z$ is the appropriate test statistic. In the more typical case, where the standard error term must be estimated from sample data, $t$ is the appropriate test statistic. Thus, $z$ tests are used when $\sigma$'s are known and $t$ tests are used when $\sigma$'s are unknown.

Theoretically, $z$ and $t$ tests can be applied only when the underlying population data distributions are normal. We will find, however, that this assumption of normality is not nearly as restrictive as it might seem. A second assumption of these tests is that the samples are truly random: Every member of a population must have an equal chance of being selected and every possible sample of size $N$ must have an equal chance of being used in the analysis.

## TESTING HYPOTHESES ABOUT MEANS: INDEPENDENT SAMPLES

The procedures to be outlined in this section are appropriate for making inferences about differences between population means when the samples that provide the research data have been randomly and independently drawn from populations. *Independent* samples have no relationship to one another. More specifically, there is no systematic way that a member of one independent sample can be paired with a member of another independent sample. If such a pairing is possible, as when the spouse of each member of one sample constitutes the membership of a second sample, the samples are said to be *dependent*. Tests using dependent sample data with paired observations are discussed in a later section.

Students should find a great deal of similarity between these parametric one-sample tests and the chi-square goodness-of-fit tests described in the previous chapter. Parametric tests of differences between population parameters are likewise quite similar to chi-square tests of homogeneity. These relationships, plus familiarity with many of the concepts which will be used, should ease the burden of comprehending this new material.

### ONE-SAMPLE $z$ TESTS, $\sigma$ KNOWN

As its name implies, only one sample is drawn for the one-sample $z$ test. It is most commonly used to compare the estimated mean of a population of interest with the mean of some known population. Formula (8.3) for computing the $z$ score equivalents of mean values in a sampling distribution of means has been defined as

$$z = \frac{\bar{X} - \mu_{\bar{X}}}{\sigma_{\bar{X}}}$$

The observed $z$ gives the relative position of $\bar{X}$ in a sampling distribution of means constructed with a known value of $\mu_{\bar{X}}$. This formula also serves as the basis for the one-sample $z$ test. $\mu_{\bar{X}}$ and $\sigma_{\bar{X}}$ are obtained from some known comparison population, while $\bar{X}$, which serves as the estimate of the mean of the unknown population of interest, is obtained from the single sample. The observed value of the test statistic, $z$, is used in testing hypotheses concerning the relative magnitudes of the mean of the unknown population of interest and the mean of this known population. When the null hypothesis of no difference between the two population means is true, the $z$ statistic is normally distributed about this common mean value.

The normality assumption mentioned above is of limited practical concern when the application of this particular test is being contemplated, for even if the population data distribution is not normal, the sampling distribution of the means will approach normality in accordance with the central limit theorem as $N$ increases. If the population shape is merely symmetrical and unimodal, for example, a sample size as small as 10 should result in a sampling distribution that approximates a normal curve. A sample size of 30 will compensate for most population distributions that are fairly skewed. When the deviation from normality is thought to be severe, however, larger sample sizes are required.

Nondirectional Tests The nondirectional test form, with which the student is familiar through the discussion of chi-square tests, will be used in this introductory illustration of the one-sample $z$ test.

Suppose that a reading skills specialist wishes to compare the average reading comprehension of adults in a particular county, an unknown population of interest, with that of the general U.S. population. A standardized test of reading comprehension is used to measure this criterion variable. According to the publisher's manual, the parameters for the U.S. population of adults (the comparison population) on this test are $\mu = 100.0$ and $\sigma = 20.0$. This distribution is normal in shape. The investigator has decided that he will draw a random sample of 16 adults from the population of adults in this county. It is now possible to determine the parameters of the *sampling distribution of means* for samples of size 16 that have been randomly drawn from the known U.S. population.

$$N_{\bar{X}} = N = 16$$

$$\mu_{\bar{X}} - \mu = 100.0$$

$$\sigma_{\bar{X}} = \frac{\sigma}{\sqrt{N_{\bar{X}}}} = \frac{20}{\sqrt{16}} = \frac{20}{4} = 5.0$$

For illustrative purposes, this sampling distribution is depicted in Figure 9.1. Even before any sample data are collected, a study of this sampling distribution will indicate, for example, that means between the values of 95 ($z = -1$) and 105 ($z = +1$) would be expected by chance from about two-thirds of the samples of size 16 that might be drawn from the U.S. population. Similarly, means greater than 115 ($z = +3$) or less than 85 ($z = -3$) would be extremely rare; the normal table indicates that they have a probability of .0013 + .0013, or that they would occur only 26 times in 10,000 samples.

**FIG. 9.1 SAMPLING DISTRIBUTION FOR DATA IN TABLE 9.1.**

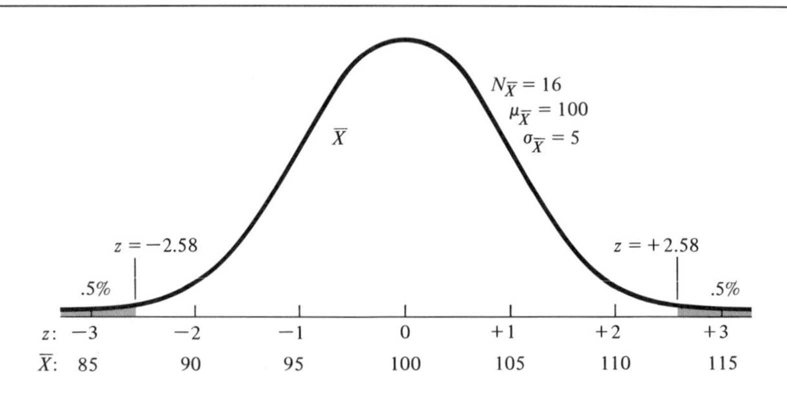


When using nondirectional $z$ and $t$ tests, we must work with both the negative and positive tails of sampling distributions, whereas with $\chi^2$ we were only concerned with one tail. A negative and a positive critical value of the test statistic must be found, based on the $\alpha$ level specified. Each critical value cuts off an area in its respective tail which is equal to $\alpha/2$. Suppose that the person investigating the reading comprehension problem feels that the Type I error is relatively crucial and decides on an $\alpha$ level of .01. Referring to Appendix C, it is determined that $\alpha/2 = .5$ percent of the entries (means) in the distribution will have $z$ values greater than +2.58 and .5 percent will have $z$ values less than −2.58. Thus, ±2.58 defines the critical values of $z$ for a nondirectional test at the .01 level. The areas in the tails of the distribution beyond these critical values are known as the *regions of rejection*, with their combined size being defined as $\alpha$. These regions are shown as shaded areas in Figure 9.1. If the mean of the sample that is yet to be drawn in this study falls in either of these regions, the researcher will reject the hypothesis that the means of the known comparison population and the specific population of interest are the same. The chances of obtaining a sample mean value this extreme in relation to the known population value would be too small for the researcher to ascribe merely to sampling fluctuations.

The sample of adults is now selected by this researcher and the mean reading comprehension score computed: $\overline{X} = 103.7$. The $z$-score equivalent is determined using formula (8.3):

$$z = \frac{\overline{X} - \mu_{\overline{X}}}{\sigma_{\overline{X}}} = \frac{103.7 - 100.0}{5.0} = \frac{3.7}{5.0} = +.74$$

Even without computing this $z$ value, it is obvious from Figure 9.1 that a mean of +103.7 does not fall in either region of rejection and the results are nonsignificant. Sample means deviating from $\mu_{\overline{X}}$ by 3.7 points or more would occur by chance rather frequently in a sampling distribution with a standard error of 5.0. Consequently, the hypothesis that the mean of the population in this county is the same as the mean for the general U.S. population cannot be rejected with these data.

With much of the logic which will underlie our one-sample $z$ tests out of the way, let us outline this same problem as a formal test of an hypothesis (see example, Table 9.1). The nondirectional hypothesis statements may be abbreviated considerably as compared to those for chi-square tests through the use of symbols that have previously been defined. If the population of interest is called population A, its unknown

**TABLE 9.1 EXAMPLE OF A NONDIRECTIONAL, ONE-SAMPLE $z$ TEST**

POPULATION A **Adult residents of a particular county during a defined period of time**
**(Parameters of the equivalent U.S. population are known)**

SAMPLE **16 individuals randomly chosen from population A**

CRITERION VARIABLE **Reading comprehension test scores**

DECISION MODEL $H_0$: $\mu_A = \mu$
$H_1$: $\mu_A \neq \mu$
$\alpha = .01$
*DR:* **Reject $H_0$ if $z_{obs}$ is $< -2.58$ or $> +2.58$; otherwise, do not reject**

| OBSERVED SAMPLE STATISTICS | KNOWN POPULATION PARAMETERS | KNOWN SAMPLING DISTRIBUTION PARAMETERS |
|---|---|---|
| $N_A = 16$ | | $N_{\bar{X}} = 16$ |
| $\bar{X}_A = 103.7$ | $\mu = 100.0$ | $\mu_{\bar{X}} = 100.0$ |
| | $\sigma = 20.0$ | $\sigma_{\bar{X}} = 5.0$ |

$$z_{obs} = \frac{\bar{X}_A - \mu_{\bar{X}}}{\sigma_{\bar{X}}} = \frac{103.7 - 100.0}{5.0} = \frac{3.7}{5.0} = +.74$$

DECISION **Since $-2.58 < +.74 < +2.58$, do not reject $H_0$**

INTERPRETATION **Evidence that the mean reading comprehension for adults in the county of interest differs from that in the general U.S. population of adults was not obtained**

(*Note:* The hypotheses are not limited to the specific form in this example. They could also have been written as $H_0$: $\mu_A = 100.0$ and $H_1$: $\mu_A \neq 100.0$.)

mean can be labeled $\mu_A$. $\mu$ can be used to denote the mean of the known U.S. population. The decision rules for nondirectional $z$ and $t$ tests are written so as to define the regions of rejection in both tails, $< -2.58$ and $> +2.58$. Similarly, the decision is written in such a way as to show the relationship of $z_{obs}$ to these critical values. The statement $-2.58 < +.74 < +2.58$ is a convenient way of symbolically abbreviating the observation that the obtained $z$ of $+.74$ falls between these critical values.

Notice that this result is not interpreted as a clear indication that the means of the unknown and the known populations are the same, for a Type II error may have been committed. It could be true, for instance, that $\mu_A$ is *actually equal* to the sample point estimate of 103.7 and the small sample size which was used made this $z$ test insensitive to such a minor difference between population means. The researcher who must make a working decision with the available data is thus placed in a position of assuming that the means are approximately equal until such time as evidence to the contrary may be presented in future studies. Type II errors will be discussed in more detail in the last section of this chapter.

Directional Tests One-sample $z$ tests may also be set up as directional tests; the assumptions of random sampling and normality are the same. Recall from Chapter 6 that in a directional test the researcher has reason to predict a specific outcome and that this outcome is stated in the alternate hypothesis. $H_1$ is extremely important in directional tests, as it establishes the way in which the decision rule is written, the sequence in which the means are entered into the formula for the observed test statistic, and the tail of the sampling distribution in which the region of rejection is located.

This discussion of one-sample, directional $z$ tests will be based on the example formally outlined in Table 9.2. A medical researcher has designed an experiment to study the effectiveness of a new medication (Y) developed to lower the diastolic blood pressure in males with mild hypertension. He administers the drug to a sample of such patients and compares their blood pressure with a set of parameters that some physicians employ to define mild hypertension: $\mu = 95.0$, $\sigma = 2.0$ (parameters of the known comparison population). He strongly suspects that this population data distribution is positively skewed, but is confident that the sampling distribution of means will approach normality in accordance with the central limit theorem if a large sample is used. Thus, a one-sample $z$ test seems appropriate. Because he is interested only in determining if the medication lowers the average diastolic pressure, a directional test form can be adopted. $H_1$ is written to state the predicted outcome, $\mu_A < \mu$ (it could also have been written as $\mu_A < 95.0$). $H_0$ then includes both a statement of no difference and the remaining possible outcome, $\mu_A \geqslant \mu$. The $\alpha$ level is set at .001 to guard against what is defined as the more crucial Type I error.

**TABLE 9.2 EXAMPLE OF A DIRECTIONAL, ONE-SAMPLE $z$ TEST**

| | |
|---|---|
| POPULATION A | Male patients in a particular city during 1977 with mild hypertension, treated with medication Y (Parameter values for the above population without treatment are assumed through medical definition of the disorder) |
| SAMPLE A | 125 patients randomly chosen from population A |
| CRITERION VARIABLE | Diastolic blood pressure |
| DECISION MODEL | $H_0$: $\mu_A \geqslant \mu$<br>$H_1$: $\mu_A < \mu$<br>$\alpha = .001$<br>*DR:* Reject $H_0$ if $z_{obs} < -3.1$; otherwise, do not reject |

| OBSERVED SAMPLE STATISTICS | KNOWN POPULATION PARAMETERS | KNOWN SAMPLING DISTRIBUTION PARAMETERS |
|---|---|---|
| $N_A = 125$ | | $N_{\overline{X}} = 125$ |
| $\overline{X}_A = 93.8$ | $\mu = 95.0$ | $\mu_{\overline{X}} = 95.0$ |
| | $\sigma = 2.0$ | $\sigma_{\overline{X}} = .18$ |

$$z_{obs} = \frac{\overline{X}_A - \mu_{\overline{X}}}{\sigma_{\overline{X}}} = \frac{93.8 - 95.0}{.18} = \frac{-1.2}{.18} = -6.67$$

| | |
|---|---|
| DECISION | Since $-6.67 < -3.1$, reject $H_0$ |
| INTERPRETATION | The mean diastolic blood pressure of the treated population was significantly less than that for the untreated population. However, the difference is insufficient for the drug to have practical value as a hypertensive medication |

The sampling distribution of means for this experiment is illustrated in Figure 9.2. $N_{\overline{X}}$ is provided by the sample size, $N = 125$. $\mu_{\overline{X}}$ is equal to the defined population mean, 95.0. $\sigma_{\overline{X}}$ is determined from the defined population standard deviation and $N_{\overline{X}}$:

$$\sigma_{\overline{X}} = \frac{\sigma}{\sqrt{N_{\overline{X}}}} = \frac{2.0}{\sqrt{125}} = \frac{2.0}{11.18} = .18$$

FIG. 9.2 SAMPLING DISTRIBUTION FOR DATA IN TABLE 9.2.

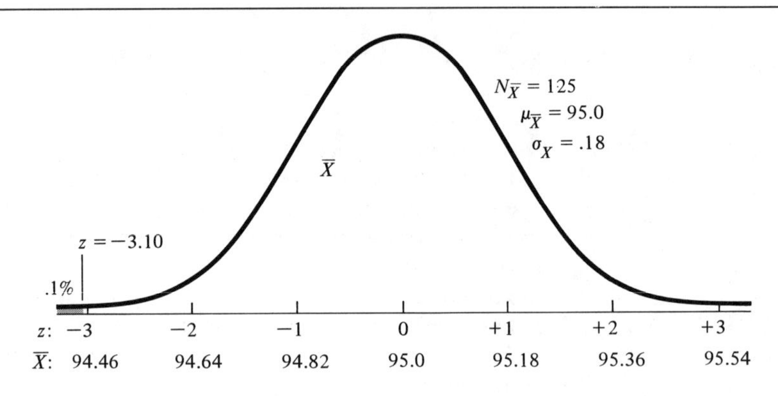


Directional $z$ and $t$ tests use only *one* tail of the sampling distribution. The inequality sign of $H_1$ points to the tail containing the single region of rejection. In this case, the inequality points to the left, so the region of rejection is in the negative tail. The size of this region is equal to $\alpha$, or .001. Reference to Appendix C shows that the critical value of $z$ which cuts off .1 percent of the means in the negative tail is −3.1. The student should note that, all else being equal, it should be easier to reject $H_0$ with a directional than with a nondirectional test.

Decision rules for directional $z$ and $t$ tests are similar to those for $\chi^2$ tests; only one critical value is given. The inequality sign must be consistent with the inequality sign in $H_1$. In this case, then, we would reject any observed $z$ value that is *less* than the critical value of −3.1.

The researcher proceeds to draw the sample and collect data on the criterion variable. $\overline{X}_A$ is found to be 93.8. In computing $z_{\text{obs}}$, the sequence of the mean values in the numerator must be the same as the order in which they appear in $H_1$. In this example, the estimated mean for the population of interest comes first and the defined population mean second. The observed value of $z$, −6.67, is smaller than the critical value, −3.1, and falls in the region of rejection. Thus, means of 93.8 or smaller would occur by random sampling less than 1 in 1000 times. With a probability this small the researcher rejects the hypothesis that the sample could have been drawn from the defined population. In other words, the directional test permits the specific inference that the diastolic blood pressure of the treated population is significantly less than the mean for a population of mildly hypertensive males. From a practical standpoint, however, the results are discouraging, for most physicians would still classify the average patient who received the medication as being mildly hypertensive. Here, then, we have an example of a difference which is *statistically,* but not *practically,* significant. The medication appears to help, but not enough to be considered a truly successful treatment for mild hypertension.

Before leaving this example, let us consider the effects on the procedures if the sequence of $\mu_A$ and $\mu$ were to be reversed in the alternate hypothesis. Obviously, $H_1$ could just as easily have been stated as $\mu > \mu_A$; the original statement that $\mu_A$ is less than $\mu$ is still retained. $H_0$ would then be written as $\mu \leq \mu_A$ and the decision rule would state: "Reject $H_0$ if $z_{obs} > +3.1$." Since the inequality points to the right-hand, positive tail, the critical value given in the decision rule will be positive. The means in the formula for $z_{obs}$ must now be reversed to be consistent with their sequence in $H_1$, with the defined population mean appearing first: $\mu_{\bar{X}} - \bar{X}_A$. This would result in a positive value of $z_{obs}$ for these data: +6.67. Since +6.67 is greater than +3.1, we would reject $H_0$ in accordance with the decision rule. The decision is the same as it was originally and either way of writing $H_1$ is appropriate as long as the researcher is careful to be consistent at each succeeding step. As any inconsistency could lead to an erroneous research decision, the student should make sure that he understands the above discussion. In summary, the inequality sign and order of the means in $H_1$ establishes the model to be used for each succeeding step in a directional test.

## ONE-SAMPLE $t$ TESTS, $\sigma$ UNKNOWN

We will find that $t$ tests between means differ from $z$ tests only with respect to the formula that is used for calculating the observed test statistic ($t$) and the tables that are referenced for finding critical values. In Chapter 8 it was suggested that $t$ be used for computing confidence intervals whenever the population standard deviation, $\sigma$, is unknown and that tables for the $t$ distribution be used to find critical values for $t$ whenever the sample size, $N$, is 60 or less. The same arbitrary rule of thumb applies equally well to one-sample test situations. When $\sigma$ is unknown but $N$ is greater than 60, the student will still use the $t$ formula to compute the test statistic but has the option of referencing tables of the normal distribution to find the critical value(s).

Recall, also, that $\sigma_{\bar{X}}$ is estimated by $s_{\bar{X}}$ when $\sigma$ is unknown and that formula (8.5) for computing $t$ was given as

$$t = \frac{\bar{X} - \mu_{\bar{X}}}{s_{\bar{X}}}$$

When the null hypothesis is true, this statistic has a $t$ distribution with $N - 1$ degrees of freedom. This relationship is exact only when the population data distribution is normal in shape. However, several empirical sampling studies have shown that nondirectional tests are robust with regard to this assumption. If the population distribution is at least approximately unimodal and symmetrical, the relationship is sufficiently close even for small samples and $N$'s of 30 or more will usually compensate for fairly large deviations from normality. As a rough guide, then, the same rules for sample size apply to one-sample, nondirectional $t$ tests as for $z$ tests. This assumption is more critical for the directional test.

Nondirectional Tests As with nondirectional $z$ tests, nondirectional $t$ tests are used whenever the researcher is unable to specify the direction of the probable difference. Consider the following example, which is outlined in Table 9.3.

Federal statutes requiring documented proof of advertising claims have caused most manufacturers to carefully field-test their products. As an illustration we will cite a fictitious automotive tire manufacturer who

**TABLE 9.3 EXAMPLE OF A NONDIRECTIONAL, ONE-SAMPLE $t$ TEST**

POPULATION A **Production LM series tires, size FR78-15 (The mean of a comparison population is defined)**

SAMPLE **12 members (tires) randomly chosen from population A**

CRITERION VARIABLE **Test track tread life in miles**

DECISION MODEL $H_0$: $\mu_A = 45{,}000$
$H_1$: $\mu_A \neq 45{,}000$
$\alpha = .10$
*DR:* **Reject $H_0$ if $t_{obs}$ is $< -1.80$ or $> +1.80$; otherwise, do not reject**

| OBSERVED SAMPLE STATISTICS | KNOWN POPULATION PARAMETERS | KNOWN SAMPLING DISTRIBUTION PARAMETERS |
|---|---|---|
| $N_A = 12$ | | $N_{\bar{X}} = 12$ |
| $\bar{X}_A = 43{,}957.0$ | $\mu = 45{,}000$ | $\mu_{\bar{X}} = 45{,}000$ |
| $s_A = 1752.84$ | | |
| $s_{\bar{X}} = \dfrac{s_A}{\sqrt{N_{\bar{X}}}} = 506.0$ | | |

$$t_{obs} = \frac{\bar{X}_A - \mu_{\bar{X}}}{s_{\bar{X}}} = \frac{43{,}957.0 - 45{,}000.0}{506.0} = \frac{-1043.0}{506.0} = -2.06$$

DECISION **Since $-2.06 < -1.80$, reject $H_0$**

INTERPRETATION **The population of LM tires does not have an average tread life of 45,000 miles.**

has developed a new long-mileage tire line designated as LM. The advertising department wishes to make the claim that the average tread life of the LM line is 45,000 miles. Although prototypes have undergone many laboratory tests, no actual field data are available and a study is formulated to determine if this mileage claim can be substantiated.

A specific value against which the LM tire population mean will be compared is defined by the proposed claim, $\mu = 45{,}000$. (It could also have been stated $\mu_A = \mu$, for in effect an estimated mean is to be compared with a known mean.) $H_1$ is written as $\mu \neq 45{,}000$, signifying a nondirectional test. The $\alpha$ level is set at .10 to guard against the potentially embarrassing and costly Type II error of not rejecting $H_0$ when it is in fact false.

Because of the expense involved in this field-testing operation, a sample of only 12 tires can be included in the project. The $N$ of 12 results in degrees of freedom of $12 - 1 = 11$ for this test. The critical value for a nondirectional $t$ at the .10 $\alpha$ level with $df = 11$ is found in Appendix F to be $\pm 1.80$. The sampling distribution depicted in Figure 9.3 shows the two regions of rejection, one in each tail beyond the $t$ values of $\pm 1.80$. These regions are also expressed by the two inequalities in the decision rule. If the observed value of $t$ is either $< -1.80$ or $> +1.80$, it falls in a region of rejection.

Test track data for the 12 sample tires show a mean tread life of 43,957 miles. The $t$ value corresponding to this mean is $-2.06$. As $t_{obs}$ is less than the critical value of $-1.80$, $H_0$ must be rejected at the .10 level. On the basis of these data, the company has little choice but to abandon its plan to use the targeted advertising claim. Notice that the potentially embarrassing Type II error could not have been committed in this case, as the decision was to reject $H_0$. Type II errors can be made only when $H_0$ is *not* rejected.

**FIG. 9.3 SAMPLING DISTRIBUTION FOR DATA IN TABLE 9.3.**

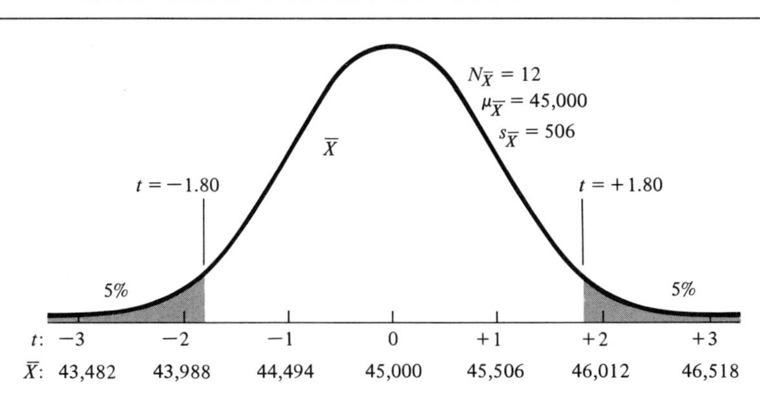

Directional Tests When the direction of the possible difference between the mean for an unknown population and some specified value can be predicted, a one-sample, directional $t$ test may be appropriate. The researcher should be aware, however, that the normality assumption for the underlying population data distribution is more crucial for this test than for any test covered thus far. Common sense should be sufficient to demonstrate why this is so. First, the single estimate of $s_{\bar{X}}$ may or may not be a good approximation of $\sigma_{\bar{X}}$ for the sampling distribution. Secondly, the critical value of $t$ is based on the shape of a true $t$ distribution. If the actual sampling distribution of observed values of $t$ is skewed, for example, the effective size of the single region of rejection will be either larger or smaller than the area specified by $\alpha$, due to the uncharacteristic shape of the tail. These factors may work separately or in combination to produce erroneous decisions. If the population shape is completely unknown or if it is thought to be nonnormal, only a large sample should be used when applying this test. Unless some justification can be found for assuming that the $X$ values in the population are approximately normally distributed, a minimum sample size of 30 should probably be planned.

To illustrate the use of this test, let us suppose that a team of chemists has developed a coating that is designed to protect steel from salt water corrosion for an average period of more than two years. A study as outlined in Table 9.4 is formulated to determine if the objective of an average of more than 730 days protection in salt water has been met with the coating. Because a value for the comparison population mean and a direction for the outcome is specified, a directional test is called for and $H_1$ is stated as $\mu_A > 730.00$. Previous research with coatings lead the investigators to predict that the population data distribution is somewhat skewed. Consequently, a large sample of 61 pieces of coated steel will be evaluated. To guard against the crucial Type I error, $\alpha$ is set at a very low level, .005. The large sample option of using the standard normal table for determining critical values is exercised. Reference to Appendix C gives a $z$ value of 2.58 for a directional test at the .005 level. The inequality in the decision rule indicates that this value will be positive and that the single region of rejection will be in the right-hand tail of the sampling distribution, as shown in Figure 9.4.

The chemists are able to reject $H_0$. Their objective has seemingly been met. However, let us analyze the effects of using the normal approximation to the $t$ distribution in this particular study. The critical value of $t$ for a directional test at the .005 level with $df = 61 - 1 = 60$ is $+2.66$. This is found by using the *directional* $\alpha$ level heading in the $t$ distribution tables. If this value (which is likely to be more appropriate than the $z$ value used) were to be entered in the decision rule, the results of the investigation would have been nonsignificant: $+2.64 \not> +2.66$. Situations in which the observed value of $t$ falls between the critical value of $z$ and $t$ will rarely occur, but the student must be aware that the possibility of making an erroneous decision through use of the normal approximation to $t$ does exist unless $N$ is extremely large.

**TABLE 9.4 EXAMPLE OF A DIRECTIONAL, ONE-SAMPLE $t$ TEST**

POPULATION A: **Steel pieces coated with a new protective compound (The mean of a comparison population is defined)**

SAMPLE: **61 pieces of steel with random shapes and sizes**

CRITERION VARIABLE: **Days of immersion in salt water without detectable traces of corrosion**

DECISION MODEL: $H_0$: $\mu_A \leq 730.00$

$H_1$: $\mu_A > 730.00$

$\alpha = .005$

*DR:* **Reject $H_0$ if $t_{\text{obs}} > +2.58$; otherwise, do not reject**

(*Note:* The large-sample option of finding the critical value in the normal table has been exercised.)

| OBSERVED SAMPLE STATISTICS | KNOWN POPULATION PARAMETERS | KNOWN SAMPLING DISTRIBUTION PARAMETERS |
|---|---|---|
| $N_A = 61$ | | $N_{\overline{X}} = 61$ |
| $\overline{X}_A = 753.45$ | $\mu = 730.00$ | $\mu_{\overline{X}} = 730.00$ |
| $s_A = 69.35$ | | |
| $s_{\overline{X}} = 8.88$ | | |

$$t_{\text{obs}} = \frac{\overline{X}_A - \mu_{\overline{X}}}{s_{\overline{X}}} = \frac{753.45 - 730.00}{8.88} = \frac{23.45}{8.88} = +2.64$$

DECISION: **Since $+2.64 > +2.58$, reject $H_0$**

INTERPRETATION: **The sample was drawn from a population with a mean greater than 730 days. The coating meets the prescribed standards**

(*Note:* Equivalent hypothesis statements to those above are $H_0$: $\mu_A \leq \mu$ and $H_1$: $\mu_A > \mu$.)

FIG. 9.4 SAMPLING DISTRIBUTION FOR DATA IN TABLE 9.4.

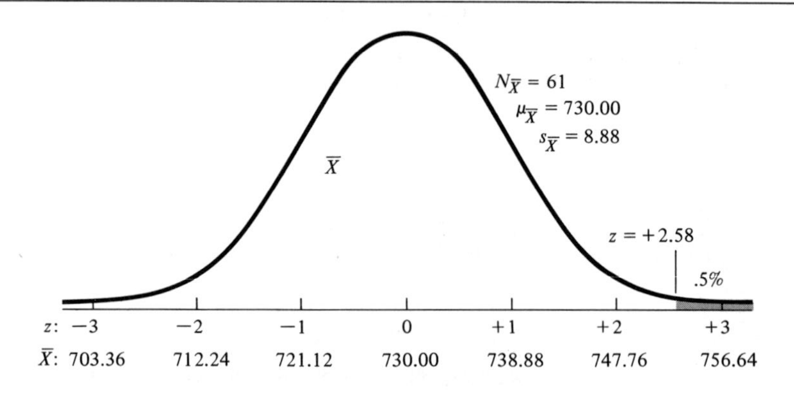


## TWO-SAMPLE TESTS BETWEEN POPULATION MEANS

The $z$ and $t$ tests to be described in this section are appropriate for testing hypotheses concerning differences between two unknown population means using sample data drawn from each population. They are thus similar to chi-square tests of homogeneity, which compare entire populations rather than specific parameters of those populations. Observed values of $t$ or $z$ for two-sample tests are obtained by finding the ratio of the difference between the means to the standard error of the difference (the standard deviation of the sampling distribution of differences).

To understand two-sample tests, we must first consider the model for the sampling distribution of differences between means. Instead of each entry in the sampling distribution being a sample mean, $\overline{X}$, as was the case for one-sample tests, we find that each entry is now the *difference* between two sample means, or $\overline{X}_1 - \overline{X}_2$; $\overline{X}_1$ having been computed on a sample drawn from population 1 and $\overline{X}_2$ having been computed on a sample independently drawn from population 2. In Figure 9.5 two population distributions of raw $X$ values are depicted. The number of entries ($X$'s) is infinite in each case ($n_1 = \infty, n_2 = \infty$). To construct an empirical sampling distribution of differences between means, tens of thousands of samples of the same size would be drawn with replacement from population 1 and the mean computed for each. The same procedure would then be independently followed for population 2, although the $N$ used throughout need not be the same as the size used for population 1 samples. The difference between all possible pairs of samples in the form $\overline{X}_1 - \overline{X}_2$ would now be computed. These difference values would be used as entries in constructing the sampling distribution of differences. To provide a highly abbreviated example of how this would be done, three samples have been drawn from population 1 and three more have been independently drawn from population 2 in Figure 9.5. The nine possible differences between pairs of population 1 and population 2 means have then been computed.

(Since we assume that the number of such entries is infinite, we would never attempt to even approximate the construction of a sampling distribution by hand.) The means of both populations are equal to 50.0. When we add up the nine differences between pairs of means, the algebraic sum is equal to zero; $\Sigma(\bar{X}_1 - \bar{X}_2) = 0$. By definition, when the means of the two parent populations are equal, the sum of all possible differences between pairs of sample means will be zero and the mean of the sampling distribution of differences will be zero; $\mu_{\bar{X}_1-\bar{X}_2} = 0$. If the means of the parent populations differ by some value, $V$, the sum of all possible differences between pairs of sample means will be equal to that value and the mean of the sampling distribution will also equal $V$. Therefore, if $\mu_1 - \mu_2 = 38$; $\mu_{\bar{X}_1-\bar{X}_2} = 38$.

**FIG. 9.5 SAMPLING DISTRIBUTION OF DIFFERENCES BETWEEN MEANS.**

Distributions of Raw $X$ Values

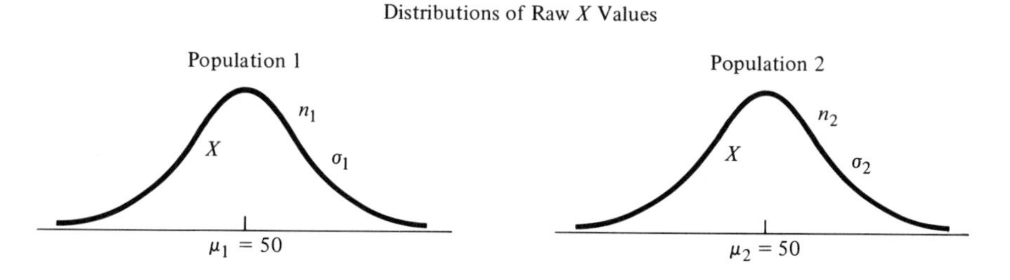


| SAMPLES FROM POPULATION 1: $N = 20$ | SAMPLES FROM POPULATION 2: $N = 25$ |
|---|---|
| $\bar{X}_{1_1} = 47$ | $\bar{X}_{2_1} = 51$ |
| $\bar{X}_{1_2} = 55$ | $\bar{X}_{2_2} = 46$ |
| $\bar{X}_{1_i} = 48$ | $\bar{X}_{2_i} = 53$ |

Differences between means in the form $\bar{X}_1 - \bar{X}_2$ for those samples shown are:

| | | |
|---|---|---|
| $\bar{X}_{1_1} - \bar{X}_{2_1} = 47 - 51 = -4$ | $\bar{X}_{1_2} - \bar{X}_{2_1} = 55 - 51 = +4$ | $\bar{X}_{1_i} - \bar{X}_{2_1} = 48 - 51 = -3$ |
| $\bar{X}_{1_1} - \bar{X}_{2_2} = 47 - 46 = +1$ | $\bar{X}_{1_2} - \bar{X}_{2_2} = 55 - 46 = +9$ | $\bar{X}_{1_i} - \bar{X}_{2_2} = 48 - 46 = +2$ |
| $\bar{X}_{1_1} - \bar{X}_{2_i} = 47 - 53 = -6$ | $\bar{X}_{1_2} - \bar{X}_{2_i} = 55 - 53 = +2$ | $\bar{X}_{1_i} - \bar{X}_{2_i} = 48 - 53 = -5$ |

$$\Sigma(\bar{X}_1 - \bar{X}_2) = 18 - 18 = 0$$

Sampling distribution of differences between sets of sample means, $\bar{X}_1 - \bar{X}_2$, when $\mu_1 = \mu_2$

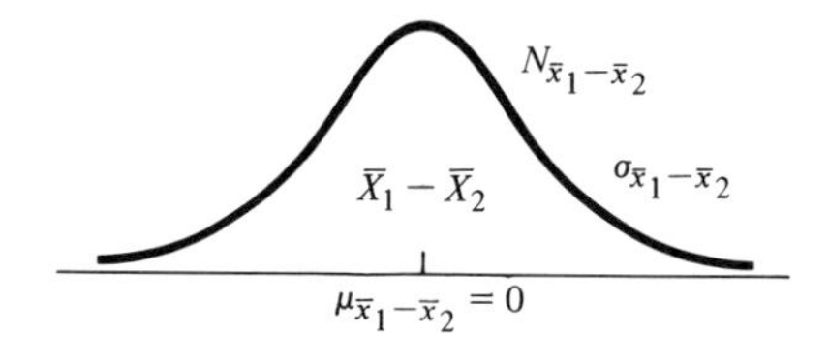

In summary, the mean of the sampling distribution of differences, which we will symbolize as $\mu_{\bar{X}_1-\bar{X}_2}$, will be zero if $\mu_1 = \mu_2$. If $\mu_1$ is greater than $\mu_2$, $\mu_{\bar{X}_1-\bar{X}_2} > 0$. If $\mu_1$ is smaller than $\mu_2$, $\mu_{\bar{X}_1-\bar{X}_2} < 0$. If we could assume a given shape for this sampling distribution of differences and describe it in a mathematical table, we would then have a basis for making inferences from sample mean differences to population mean differences. Fortunately, the shape of sampling distributions of differences between means can usually be described by the familiar $z$ and $t$ distributions, just as in the one-sample test case. Again, we find that the sampling distribution of differences between means becomes more and more normal in shape as the sample sizes increase, even if the parent data distributions are not normal. As before, the decision to use $z$ or $t$ depends on whether $\sigma_1$ and $\sigma_2$ are known.

In practice, only one sample need be drawn from population 1 and one sample from population 2. The difference between them, $\bar{X}_1 - \bar{X}_2$, provides a point estimate of $\mu_{\bar{X}_1-\bar{X}_2}$. If this value is considerably greater or less than zero we may be able to reject the hypotheses that it is a likely entry in a hypothetical sampling distribution that has a mean of zero; in turn, rejecting the null hypothesis that $\mu_1 = \mu_2$.

The only other information we must have before two-sample tests can be applied is a method of computing the standard deviation (standard error) of a sampling distribution composed of differences between sample means. If $\sigma_1$ and $\sigma_2$ are *known*, this standard error is computed as follows:

$$\sigma_{\bar{X}_1-\bar{X}_2} = \sqrt{\sigma_{\bar{X}_1}{}^2 + \sigma_{\bar{X}_2}{}^2} \tag{9.1}$$

where

$\sigma_{\bar{X}_1-\bar{X}_2}$ is the standard error of the differences
$\sigma_{\bar{X}_1}$ is the standard error of the means for population 1
$\sigma_{\bar{X}_2}$ is the standard error of the means for population 2

When the population $\sigma$'s are *unknown* and a $t$ test is appropriate, sample estimates of $s_{\bar{X}}$ are substituted in the formula for each $\sigma_{\bar{X}}$. The result is now called a *pooled variance* estimate of the standard error of the differences, symbolized as $s_{\bar{X}_1-\bar{X}_2}$. The general computational formula for this estimate is

$$s_{\bar{X}_1-\bar{X}_2} = \sqrt{\frac{\Sigma x_1{}^2 + \Sigma x_2{}^2}{N_1 + N_2 - 2}\left(\frac{1}{N_1} + \frac{1}{N_2}\right)} \tag{9.2}$$

The term in the denominator of formula (9.2), $N_1 + N_2 - 2$, defines the degrees of freedom for all two-sample $t$ tests conducted on independent samples. When the sample sizes are equal, $N_1 = N_2 = N$, formula (9.2) reduces to

$$s_{\bar{X}_1-\bar{X}_2} = \sqrt{\frac{\Sigma x_1{}^2 + \Sigma x_2{}^2}{N(N - 1)}} \tag{9.3}$$

Two-Sample $z$ Tests, $\sigma$'s Known Research situations in which population standard deviations are known but their means are unknown, calling for a $z$ test, occur very infrequently. If the student should find that a two-sample $z$ test is appropriate for a particular problem, formula (9.1) would be used to compute the standard error and the critical value(s) would be located in the normal distribution table. All other procedures for the $z$ test are equivalent to those which will be described for two-sample $t$ tests.

Two-Sample $t$ Tests, $\sigma$'s Unknown The basic formula for the two-sample $t$ test is

$$t = \frac{\bar{X}_1 - \bar{X}_2}{s_{\bar{X}_1 - \bar{X}_2}} \tag{9.4}$$

The generalized computational formula is

$$t = \frac{\bar{X}_1 - \bar{X}_2}{\sqrt{\dfrac{\Sigma x_1^2 + \Sigma x_2^2}{N_1 + N_2 - 2}\left(\dfrac{1}{N_1} + \dfrac{1}{N_2}\right)}} \tag{9.5}$$

When $N_1 = N_2$, formula (9.3) may be substituted into the denominator of (9.5).

For this test statistic to approximate the hypothetical $t$ distributions, the following assumptions must be met:

1. The samples are random and independently drawn
2. The underlying population data distributions are normal in shape
3. The variances of populations 1 and 2 are equal (frequently called the *homogeneity of variance* assumption)

Assumption 1 has been explained in previous sections. Assumption 2 is not critical, particularly if the $N$'s are large. If one can reasonably expect that the data distributions of populations 1 and 2 approach normality, sample $N$'s as small as 10 will usually suffice. When the shapes are completely unknown, as is usually the case, the cautious investigator would probably feel more comfortable with an $N$ of 20 to 30 for each sample. Even larger sample sizes should be planned if distributions are thought to differ substantially from the normal shape. The two-sample $t$ test is not so robust with regard to assumption 3. In older textbooks a test of this assumption was often suggested. However, by far the simplest and most effective method of meeting this assumption is to use sample sizes that are *equal,* or very nearly so.

When the degrees of freedom for a two-sample $t$ test are 60 or less, critical values should be found in the table of $t$ distributions. The normal table may be used, if desired, for problems with larger degrees of freedom. Because the student is now familiar with the role of sampling distributions, they need not be depicted for subsequent examples in this chapter.

Nondirectional Tests An example of a two-sample, nondirectional $t$ test is outlined in Table 9.5. A speech pathologist has developed a standardized technique for the measurement of speech defects which is designed for rough screening purposes in secondary schools. It must be determined if the technique is sensitive to sex differences before normative data can be established. Since no population information of any kind is available, large samples of equal size are randomly and independently drawn from each population of interest.

The hypothesis to be tested is that the means of the populations are equal, $H_0$: $\mu_1 = \mu_2$. Because the direction of any possible difference cannot be predicted, the nondirectional hypothesis form is appropriate. The

**TABLE 9.5 EXAMPLE OF A NONDIRECTIONAL, TWO-SAMPLE $t$ TEST**

| | |
|---|---|
| POPULATIONS | **Secondary school students in a particular state, fall 1976**<br>**1: Males**<br>**2: Females** |
| SAMPLES | **1: 45 members randomly drawn from population 1**<br>**2: 45 members randomly drawn from population 2** |
| CRITERION VARIABLE | **Composite index of speech defects** |
| DECISION MODEL | $H_0$: $\mu_1 = \mu_2$<br>$H_1$: $\mu_1 \neq \mu_2$<br>$\alpha = .05$<br>*DR:* **Reject $H_0$ if $t_{obs}$ is $< -1.96$ or $> +1.96$; otherwise, do not reject** |

OBSERVED SAMPLE STATISTICS

| | |
|---|---|
| $N_1 = 45$ | $N_2 = 45$ |
| $\bar{X}_1 = 63.8$ | $\bar{X}_2 = 62.7$ |
| $\Sigma x_1^2 = 678.0$ | $\Sigma x_2^2 = 536.0$ |

$$t_{obs} = \frac{\bar{X}_1 - \bar{X}_2}{\sqrt{\dfrac{\Sigma x_1^2 + \Sigma x_2^2}{N(N-1)}}} = \frac{63.8 - 62.7}{\sqrt{\dfrac{678.0 + 536.0}{45(45-1)}}} = \frac{1.1}{\sqrt{.61}} = \frac{1.1}{.78} = +1.41$$

| | |
|---|---|
| DECISION | **Since $-1.96 < +1.41 < +1.96$, do not reject $H_0$** |
| INTERPRETATION | **Evidence for population sex differences in average speech defect scores was not obtained** |

$\alpha$ level is set at .05. Critical values for a nondirectional $t$ at the .05 level with $df = N_1 + N_2 - 2 = 88$ are found to be approximately $\pm 1.99$ in Appendix F. Since the $df$ value is greater than 60, however, the researcher chooses the option of using critical values from the normal distribution table in the decision rule: $\pm 1.96$.

Means and $\Sigma x^2$ values are computed for each sample. No parameter values are shown, for the only parameter available is the size of the samples used in computing the difference values in the sampling distribution. The test statistic, a difference divided by the standard error of that difference, is computed using formula (9.3) for equal sample sizes in the denominator. The general formula for $t$, (9.5), would give the same result within rounding error.

Since $t_{obs}$ is neither less than $-1.96$ nor greater than $+1.96$, $H_0$ cannot be rejected. The average speech defect scores for girls in the population was not found to be significantly different from that for the population of boys. Although $H_0$ has not been proven to be correct, it would seem safe at this time for the designer of the screening technique to establish and use a single set of normative data for both sexes.

Directional Tests The two-sample, directional $t$ test is applicable whenever $\sigma$'s are unknown and the researcher is able to predict that the mean of a particular population will be larger than the other. This test is not as sensitive to underlying population shapes as its one-sample counterpart. Therefore, the suggested sample sizes given for nondirectional $t$ tests in the previous section apply equally well here.

To illustrate these tests, a social-psychological study of creative problem-solving behavior in small groups is cited (see Table 9.6). Specifically, the investigation has been devised to test the theory that creative solutions to problems are offered more frequently by individuals in leaderless, unstructured group atmospheres than by individuals in groups which have a formal structure and leaders imposed by a researcher. Notice in this example that the researcher first drew a random sample from the general population of interest, then randomly assigned its members to groups representing subpopulations differing only with respect to type of work atmosphere. This procedure produces a *randomized group* design. All differences between sample members (sex, motivation level, problem-solving ability, etc.) are now randomly distributed throughout the two groups. If significant mean differences are observed, they can be attributed directly to the influence of type of work atmosphere. Without such initial randomization of members, the researcher would not know if significant results were due to differences in group atmosphere or to other extraneous, uncontrolled factors.

Previous studies have indicated that distributions of the criterion variable (number of creative solutions per person during defined time

**TABLE 9.6 EXAMPLE OF A DIRECTIONAL, TWO-SAMPLE $t$ TEST**

| | |
|---|---|
| POPULATIONS | **Sophomore students at a western university, fall 1976** |
| | **1: Working in small, unstructured group atmosphere** |
| | **2: Working in small, structured group atmosphere** |
| SAMPLES | **1: 10 students randomly chosen from population 1** |
| | **2: 11 students randomly chosen from population 2** |
| CRITERION VARIABLE | **Number of creative solutions per person in 2-hour session** |
| DECISION MODEL | $H_0: \mu_1 \leq \mu_2$ |
| | $H_1: \mu_1 > \mu_2$ |
| | $\alpha = .025$ |
| | *DR:* **Reject $H_0$ if $t_{obs} > +2.09$; otherwise, do not reject** |

OBSERVED SAMPLE STATISTICS

| | |
|---|---|
| $N_1 = 10$ | $N_2 = 11$ |
| $\bar{X}_1 = 8.7$ | $\bar{X}_2 = 6.3$ |
| $\Sigma x_1^2 = 38.0$ | $\Sigma x_2^2 = 44.0$ |

$$t_{obs} = \frac{\bar{X}_1 - \bar{X}_2}{\sqrt{\dfrac{\Sigma x_1^2 + \Sigma x_2^2}{N_1 + N_2 - 2}\left(\dfrac{1}{N_1} + \dfrac{1}{N_2}\right)}}$$

$$= \frac{8.7 - 6.3}{\sqrt{\dfrac{38 + 44}{10 + 11 - 2}\left(\dfrac{1}{10} + \dfrac{1}{11}\right)}} = \frac{2.4}{\sqrt{\dfrac{82}{19}(.10 + .09)}}$$

$$= \frac{2.4}{\sqrt{(4.32)(.19)}} = \frac{2.4}{\sqrt{.82}} = \frac{2.4}{.91}$$

$$= +2.64$$

| | |
|---|---|
| DECISION | **Since $+2.64 > +2.09$, reject $H_0$** |
| INTERPRETATION | **The number of creative solutions per person is greater in the population of students in unstructured groups** |

periods) tend to approach normality for similar populations. Because a specific outcome is expected, the sample $N$'s will be very nearly equal, and the $\sigma$'s are unknown, a two-sample directional $t$ test appears to be appropriate and its assumptions justified. The researcher wants to be rather sure of himself in rejecting $H_0$, so $\alpha$ is set at a fairly low level; .025. This decision, in combination with the small $N$'s, make the risk of the less crucial Type II error rather substantial.

From the inequality sign in $H_1$ it is known that the region of rejection will be in the right-hand, positive tail of the sampling distribution, beyond the $t$ value of $+2.09$. In going through the example, the student should again observe the consistency which is essential in directional tests. The inequality is the same in the decision rule and the decision statement, and the means are listed in the same sequence in both the hypotheses and the $t$ formula. The use of formula (9.5) results in an observed $t$ that is greater than the critical value for a directional $t$ with $N_1 + N_2 - 2 = 19$ degrees of freedom at the .025 level. $H_0$ is rejected. Therefore, a Type II error could not have been committed. The data support the researcher's theory concerning creativity in different group atmospheres.

Tests When $\mu_1$ and $\mu_2$ Are Not Assumed Equal In all the examples given so far with two-sample tests, it has been assumed under $H_0$ that the population means are equal; $H_0$: $\mu_1 = \mu_2$. An optional way of stating this is $H_0$: $\mu_1 - \mu_2 = 0$. These $z$ and $t$ tests can also be set up to determine if the difference between the population means corresponds to some value other than zero. For example, $H_0$: $\mu_1 - \mu_2 = 7.2$, $H_1$: $\mu_1 - \mu_2 \neq 7.2$. The mean of the sampling distribution of differences would be $\mu_{\bar{X}_1 - \bar{X}_2} = 7.2$ when this $H_0$ is true. The numerator of the formula for computing the observed value of the test statistic must be modified to take this probable difference between means into account, as shown below for $t$:

$$t = \frac{(\bar{X}_1 - \bar{X}_2) - (\mu_1 - \mu_2)}{s_{\bar{X}_1 - \bar{X}_2}} \tag{9.6}$$

The specific value of the difference hypothesized in $H_0$, $\mu_1 - \mu_2$, is subtracted from $\bar{X}_1 - \bar{X}_2$. For the above example, the numerator would be $(\bar{X}_1 - \bar{X}_2) - 7.2$. Except for these minor adjustments in the hypothesis statements and the test statistic formula, the procedures followed are identical to those used in testing the hypothesis that two population means are equal.

## TESTING HYPOTHESES ABOUT MEANS: DEPENDENT SAMPLES

When data in different samples may be paired, as when one-half of selected sets of twins or matched partners compose one sample and the second sample is composed of the remaining half, the samples and the populations they represent are said to be *dependent*. For example, sets of siblings may be chosen for an experiment on the effects of loss of sleep, where one member of each set is assigned to a sample allowed normal sleep and the siblings of these individuals are assigned to another sample in which less sleep is permitted. Or, in a study of the effects of freezing on battery life, two identical hand calculators might be selected from each of 10 manufacturers. One calculator of each make might be assigned to a sample tested under freezing temperatures and its twin assigned to a sample tested under normal room temperatures.

Repeated-measures experiments, in which sample members are measured twice on the same criterion variable, also generate dependent sample data. For instance, a study of differential effects of stressful vs. nonstressful conditions on typing speed may be designed so that the speed of each typist is measured under both conditions and compared.

In these and in other examples of dependent sample studies, it is possible to *match* systematically members of the two samples and compare their performance on a common criterion variable. The samples represent populations which differ with respect to some treatment or condition of interest to the investigator. Matching is not feasible with independent samples randomly drawn from populations, for no systematic method exists for pairing members of the two samples.

Population standard deviations are seldom known in studies of dependent population means. Therefore, only the $t$ test will be described. The assumptions include random sampling and normality of underlying population data distributions. Sample sizes as small as $N = 10$ (10 sets of partners or 10 members observed twice) may be used if the underlying distributions are known to approach normality. As with other $t$ or $z$ tests, larger sample sizes are called for if no such information is available.

Procedurally, calculation of the test statistic is accomplished by finding $\Sigma X$ and $\bar{X}$ for each sample or group of observations. The difference ($D$) between each matched pair and the square of each $D$ value are then found and entered in separate columns. Both columns are summed to obtain $\Sigma D$ and $\Sigma D^2$. The mean of $D$, symbolized $\bar{D}$, should equal the difference between $\bar{X}_1$ and $\bar{X}_2$ if the calculations are correct:

$$\bar{D} = \frac{\Sigma D}{N} = \bar{X}_1 - \bar{X}_2 \tag{9.7}$$

The $t$ statistic for a test of two dependent population means is given by

$$t = \frac{\bar{X}_1 - \bar{X}_2}{\sqrt{\dfrac{\Sigma D^2 - (\Sigma D)^2/N}{N(N-1)}}} \tag{9.8}$$

$\bar{D}$ could also be used as the numerator, but the term shown is more consistent with preceding discussions. The denominator of formula (9.8) provides an estimate of the standard error of the sampling distribution of differences between the means.

### REPEATED MEASURES

The primary benefit of using dependent samples is the reduction of experimental error. As an illustration of this with a repeated-measures design, suppose that a consumer testing agency is interested in evaluating the effectiveness of an alleged fuel-saving device for automobiles. The researchers could draw a random sample of cars without the device and compare average fuel consumption with the average for another independent sample on which the device has been installed. Unless the samples are very large, however, it should be apparent that chance differences in the fuel consumption of various sizes and makes of cars in the two samples are likely to be very large relative to the somewhat small differences, if any, that most "add-on" gadgets could make. Statistical tests which might be employed using independent samples would probably not have sufficient power to detect a minor, but real, improvement or decrement in average fuel consumption. To circumvent this problem of population heterogeneity, a random sample of cars could be drawn and gas consumption tested under controlled conditions. The devices could then be installed in these same cars and gas consumption again measured under identical conditions. Any change in efficiency should now be due to influence of the device, not chance sampling fluctuations or other uncontrolled factors. In a majority of instances, such use of dependent samples will result in a reduction of the standard error of the sampling distribution of differences between means, effectively increasing the magnitude of the observed test statistic and making the test more sensitive to small actual population differences.

Hypothetical data for this study are presented in Table 9.7. Assume that $H_0$: $\mu_1 = \mu_2$, $H_1$: $\mu_1 \neq \mu_2$, and the $\alpha$ level had been set at .05. The degrees of freedom for dependent sample $t$ tests is $N - 1$. With $df = 5$ and $\alpha = .05$, the critical values of $t$ for a nondirectional test are $\pm 2.57$. Even though the standard error was minimized with the repeated-measures design, $t_{\text{obs}}$ falls between these values and the null hypothesis of no difference between fuel consumption with and without the device cannot be rejected at the .05 level. The probability of a Type II error is rather high with such unrealistically small samples, however.

### MATCHED PAIRS

In certain research settings it is inadvisable or even impossible to use repeated-measures designs, for the first condition or treatment may influence measures obtained under subsequent conditions, often through practice effects. A repeated-measures design would be inappropriate to compare two methods of teaching first-aid skills, for example, if either

TABLE 9.7 DATA FOR A REPEATED-MEASURES $t$ TEST

| CAR | DEVICE | NO DEVICE | $D$ | $D^2$ |
|---|---|---|---|---|
| 1 | 22 | 23 | −1 | 1 |
| 2 | 14 | 12 | 2 | 4 |
| 3 | 17 | 16 | 1 | 1 |
| 4 | 28 | 28 | 0 | 0 |
| 5 | 21 | 19 | 2 | 4 |
| 6 | 19 | 18 | 1 | 1 |
| $N = 6$ | $\Sigma X_1 = 121$<br>$\bar{X}_1 = 20.167$ | $\Sigma X_2 = 116$<br>$\bar{X}_2 = 19.333$ | $\Sigma D = 5$<br>$\bar{D} = .834$ | $\Sigma D^2 = 11$ |

$$t_{\text{obs}} = \frac{\bar{X}_1 - \bar{X}_2}{\sqrt{\dfrac{\Sigma D^2 - (\Sigma D)^2/N}{N(N-1)}}}$$

$$= \frac{20.167 - 19.333}{\sqrt{\dfrac{11 - 5^2/6}{6(6-1)}}} = \frac{.83}{\sqrt{\dfrac{11 - 4.17}{30}}} = \frac{.83}{\sqrt{.228}} = \frac{.83}{.48}$$

$$= +1.73$$

method 1 or method 2 were always taught first. In these situations a matched-pairs design should be considered. The primary objective in matching is to insure that the two members of each pair are as alike as possible on all relevant dimensions. Familial relationship was the basis for matching pairs in the two examples described in the introductory paragraph to this topic. Extent of prior first-aid training or knowledge would be one of several appropriate matching variables which might be used for the above instructional methods comparison example. Other commonly applied bases for matching people include age, socioeconomic status, intelligence, educational level, and so forth.

A study using matched pairs of individuals and a directional test will now be formally outlined. A toothpaste manufacturer wishes to demonstrate that a new chemical additive, FC, results in significantly fewer cavities among children. To reduce sampling error, it is decided that the additive will be tested on a sample of identical twins, with one of each pair using toothpaste containing the additive and the other using the same toothpaste without this compound (see Table 9.8). A random sample of twins is drawn from the population of interest and then randomly split into these two experimental groups, which now represent two distinct

**TABLE 9.8 EXAMPLE OF A DIRECTIONAL, MATCHED-PAIRS $t$ TEST**

POPULATIONS **Identical twins, ages 8–10, in city Y; January 1974 to January 1975**
**1: Twin half using toothpaste without FC**
**2: Twin half using toothpaste with FC**

SAMPLE **10 sets of twins randomly drawn from population**

CRITERION VARIABLE **Number of new cavities during one-year period**

DECISION MODEL $H_0: \mu_1 \leq \mu_2$
$H_1: \mu_1 > \mu_2$
$\alpha = .05$
*DR:* **Reject $H_0$ if $t_{obs} > +1.83$; otherwise, do not reject**

**Observed Data**

| PAIR | WITHOUT | WITH | D | $D^2$ |
|---|---|---|---|---|
| *a* | 4 | 4 | 0 | 0 |
| *b* | 0 | 1 | −1 | 1 |
| *c* | 2 | 0 | 2 | 4 |
| *d* | 5 | 3 | 2 | 4 |
| *e* | 3 | 4 | −1 | 1 |
| *f* | 4 | 2 | 2 | 4 |
| *g* | 5 | 1 | 4 | 16 |
| *h* | 1 | 1 | 0 | 0 |
| *i* | 4 | 0 | 4 | 16 |
| *j* | 3 | 1 | 2 | 4 |
| $N = 10$ | $\bar{X}_1 = 3.1$ | $\bar{X}_2 = 1.7$ | $\Sigma D = 14$ | $\Sigma D^2 = 50$ |

$$t_{obs} = \frac{\bar{X}_1 - \bar{X}_2}{\sqrt{\dfrac{\Sigma D^2 - (\Sigma D)^2/N}{N(N-1)}}}$$

$$= \frac{3.1 - 1.7}{\sqrt{\dfrac{50 - 14^2/10}{10(10-1)}}} = \frac{1.4}{\sqrt{\dfrac{50 - 196/10}{90}}} = \frac{1.4}{\sqrt{\dfrac{30.4}{90}}} = \frac{1.4}{\sqrt{.338}} = \frac{1.4}{.58}$$

$$= +2.41$$

DECISION **Since $+2.41 > +1.83$, reject $H_0$**

INTERPRETATION **Additive FC results in fewer average cavities (see text)** populations. As only a positive difference in favor of the additive is in keeping with the firm's objective, a directional test is specified. The crucial error is defined by the researcher as a decision that the additive is not beneficial when it actually is beneficial in the population. However, the researcher cannot control this error by stating it in the alternate hypothesis (thus making it a Type I error), because the specific outcome of fewer cavities with additive use is the only decision of interest to him. Therefore, he chooses an $\alpha$ level of .05 to help reduce the probability of making the crucial Type II error. A large sample size should also be used, both to minimize Type II error and to compensate for underlying distributions that are undoubtedly skewed. The small sample of $N = 10$ is shown for this study only to simplify the presentation.

The inequality of $H_1$ points to a region of rejection in the right-hand tail of the sampling distribution, resulting in a positive critical value: $t = +1.83$ with $10 - 1$ degrees of freedom. Notice the consistency in this directional test of the inequality signs as used throughout, as well as the sequence of the means; $\mu_1$ or $\bar{X}_1$ is always listed before $\mu_2$ or $\bar{X}_2$. The dental researcher is able to reject $H_0$ at the .05 level with these tightly controlled dependent sample data, but now faces the task of interpreting the extent to which a random sample of twins in this city represent all children in the population of potential FC additive users.

Sandler's *A* Test As a substitute for the method of using a $t$ test with dependent sample data, Sandler has derived a simple statistic, $A$, which is defined as

$$A = \frac{\Sigma D^2}{(\Sigma D)^2}$$

Using the data in Table 9.8, we obtain

$$A = \frac{50}{(14)^2} = \frac{50}{196} = .26 \qquad (9.9)$$

$A$ is then interpreted using Appendix M, which is entered with $N - 1$ degrees of freedom. Notice that an observed value of $A$ is significant if it is *equal to or less than* the tabled critical value. This relationship holds true for both directional and nondirectional tests. Since .26 is less than .28, the results from Table 9.8 are again significant at the .05 level. Sandler's $A$ and the $t$ test for dependent samples are mathematically equivalent. The same decisions will be made with either technique.

## TESTING HYPOTHESES ABOUT PROPORTIONS

Tests may be conducted on population proportions ($p$) with either $z$ tests or $\chi^2$ tests if certain conditions can be met. Suppose it is desired to compare an obtained sample proportion, $P$, with a known or assumed population proportion, $p$; where $P$ and $p$ define the relative frequency of a particular event or phenomenon. For example, $p = .30$ may indicate that 30 out of every 100 people in some population have college degrees, or that 30 percent of the products manufactured by some company are defective. A $P$ value of .923 indicates that some defined event or phenomenon is observed among 92.3 percent of the sample members; for example, 92.3 percent of a sample of light bulbs is found to give over 800 hours of continuous service.

Distributions constructed of sample $P$ values are examples of *binomial* sampling distributions. The central tendency of such a distribution will be the population $p$ value, just as the central tendency of sampling distributions of means is $\mu$. With large samples the shape of the binomial sampling distribution is approximated by the normal distribution, and inferential problems may be solved using $z$ tests or $\chi^2$ tests (the same decision will be made with either test). These $z$ and $\chi^2$ approximations are fairly good when $Np > 5$ and $Nq > 5$; where $q = 1 - p$. The approximations are very good when $Np$ and $Nq$ are both $>10$.

## HYPOTHESES CONCERNING SINGLE PROPORTIONS

Suppose a researcher wishes to determine if a sample with a certain $P$ value could have been drawn from a particular population with a known or postulated $p$. This problem should sound familiar, as it is a type of goodness-of-fit test. Let us reconsider the coin-tossing example of Chapter 7 (Table 7.1.). If we define $P$ as the proportion of observed heads in the sample and $p$ as the proportion of expected heads in the population, then

$$P = \frac{530}{1000} = .53$$

and

$$p = \frac{500}{1000} = .50$$

Now, we can simply interpret this study as one which has been conducted to determine if this sample $P$ value of .53 could have occurred by chance (at the .01 level) in a population with $p = .50$. Conditions for an approximation to the binomial have been met, as $Np = (1000)(.50) = 500$ and $Nq = (1000)(.50) = 500$; both values being much greater than 10. Since $H_0$ was not rejected with the chi-square test shown, we may conclude that the difference between $P = .53$ and $p = .50$ could be due to chance sampling fluctuation and is not significant at the .01 level.

Still another approach to this problem is offered by the normal approximation to the binomial. The test statistic, $z$, is defined as

$$z = \frac{P - p}{\sqrt{\frac{pq}{N}}} \tag{9.10}$$

Where $N$ is the total number of observations in the sample.

For the data in Table 7.1,

$$z_{\text{obs}} = \frac{.53 - .50}{\sqrt{\frac{(.50)(.50)}{1000}}}$$

$$= \frac{.03}{\sqrt{\frac{.25}{1000}}} = \frac{.03}{\sqrt{.00025}} = \frac{.03}{.0158}$$

$$= +1.90$$

The critical value for a nondirectional $z$ test when $\alpha = .01$ is $\pm 2.58$. Once again, $H_0$ cannot be rejected. This is as it must be, for $\chi^2 = z^2$ when $df = 1$ for the chi-square problem. Squaring $z_{\text{obs}}$ gives $(1.90)^2 = 3.61$, within rounding error of the observed value of $\chi^2$, 3.60. Besides being relatively easy to compute, application of the $z$ approximation to the binomial gives the researcher the option of being able to design directional $z$-tests on proportions (procedures for directional chi-square tests are not discussed in this book).

## TESTS OF DIFFERENCES BETWEEN PROPORTIONS

Tests of the difference between two independent population proportions, $p_1 - p_2$, may also be approached as chi-square or $z$ problems. Approximations to the binomial will be sufficiently accurate for most problems if $NP$ and $NQ$ are $>5$ for both samples. The approximations improve as these values increase.

Referring once again to Chapter 7 for an example, we will interpret the problem in Table 7.3 as one concerning the difference between population proportions. $p_A$ may be defined as the proportion of seedlings with a growth rate of 5 inches or more in population A. $p_B$ is then the proportion of seedlings with a growth rate of 5 inches or more in population B. The hypothesis statements given in Table 7.3 could be stated equivalently as $H_0$: $p_A = p_B$ and $H_1$: $p_A \neq p_B$ for a test of proportions. The significant result of the $\chi^2$ test as shown could be interpreted without any procedural changes as an indication that the proportion of seedlings with growth rates of 5 inches or more are not the same in the two populations. Thus, differences between population proportions may be tested with chi-square tests of homogeneity.

Application of the $z$ approximation to the binomial will result in the same decision when this formula for the test statistic is used

$$z = \frac{P_1 - P_2}{\sqrt{P'Q'\left(\frac{1}{N_1} + \frac{1}{N_2}\right)}} \tag{9.11}$$

where $P'$ is defined as

$$P' = \frac{N_1P_1 + N_2P_2}{N_1 + N_2} \tag{9.12}$$

and

$$Q' = 1 - P' \tag{9.13}$$

Notice that the numerator of $P'$ is actually the sum of the sample frequencies, $f_1 + f_2$. To analyze the seedling data, let us redefine population A as 1 and B as 2. First, $P_1$ and $P_2$ are found for the samples:

$$P_1 = \frac{87}{96} = .906 \qquad P_2 = \frac{71}{94} = .755$$

Next, $P'$ and $Q'$ are computed using formulas (9.12) and (9.13):

$$\begin{aligned} P' &= \frac{(96)(.906) + (94)(.755)}{96 + 94} = \frac{87 + 71}{190} = \frac{158}{190} \\ &= .832 \\ Q' &= 1 - .832 \\ &= .168 \end{aligned}$$

Finally, the observed value of $z$ is determined with formula (9.11).

$$\begin{aligned} z_{\text{obs}} &= \frac{.906 - .755}{\sqrt{(.832)(.168)(1/96 + 1/94)}} \\ &= \frac{.151}{\sqrt{(.140)(.021)}} = \frac{.151}{\sqrt{.003}} = \frac{.151}{.054} \\ &= +2.78 \end{aligned}$$

This observed value is greater than the critical value of +1.96 and would be rejected once again at $\alpha = .05$. Notice that $z^2 = \chi^2$, within rounding error: $z^2 = (2.78)^2 = 7.73$; $\chi^2 = 7.73$. To arrive at the precise solution, the maximum number of significant digits must remain in the calculator for each intermediate answer (only a portion are shown above).

If the forester conducting this study had chosen to set up the directional alternate hypothesis that $p_A > p_B$, he would have been able to infer from these data that the proportion of 5 inch or taller seedlings was greater in the Oregon population.

## TESTING HYPOTHESES ABOUT VARIANCES

Occasionally a researcher will want to make inferences about population variances. We will consider here only one form which such inferences might take; specificially, *directional* hypotheses concerning differences between the variances of two independent populations. To illustrate, we will describe a study which has as its objective the reduction of population variance.

A professor is concerned about the large degree of variability among students in his course. By directing his lectures and assignments to the "average" student, those with greater than average ability show signs of boredom and those with less ability cannot keep up. At the beginning of a term he decides to study a possible solution to this problem and randomly assigns half the students registered for the course to one class section and the other half to another section. Section 1 will constitute the control sample with regular instruction. Remedial instruction will be offered as needed to students in section 2, the experimental sample. He hypothesizes that section 2 will be more homogeneous (less variable) in final test scores than section 1. The directional hypotheses are stated as follows:

$$H_0: \sigma_1^2 \leqslant \sigma_2^2$$
$$H_1: \sigma_1^2 > \sigma_2^2$$

The $\alpha$ level is set at .05. The test statistic which is used to evaluate differences between variances in a problem of this type is $F$, the ratio between two sample variances. The alternate hypothesis determines how the formula for an observed value of $F$ will be structured, for the sample variance *hypothesized* to be *larger* will appear in the numerator of the ratio. The general formula for $F$ is

$$F = \frac{s_i^2}{s_j^2} \tag{9.14}$$

So defined, this statistic will have a sampling distribution which is approximated by a defined $F$ distribution when $\sigma_i^2 = \sigma_j^2$. Its theoretical range is zero to infinity. Since $s_1^2$ is hypothesized to be the larger variance in $H_1$, it will appear in the numerator of $F$ whether or not it actually is larger.

The professor's data are shown below:

$$N_1 = 22 \qquad N_2 = 25$$
$$\bar{X}_1 = 63.8 \qquad \bar{X}_2 = 63.2$$
$$s_1^2 = 35.31 \qquad s_2^2 = 15.67$$

$$F_{\text{obs}} = \frac{s_1^2}{s_2^2} = \frac{35.31}{15.67} = 2.25$$

This observed value of $F$ will be compared with a critical value for a specified $\alpha$ level in the tables of the $F$ distribution (see Appendix E). To look up this critical value, the $\alpha$ level and two separate degrees of freedom are required: one for the variance in the numerator ($N_i - 1$) and a second for the variance in the denominator ($N_j - 1$). For these data, the *df* values are $N_1 - 1 = 21$ for the numerator and $N_2 - 1 = 24$ for the denominator. The critical value is found to equal 2.03. Because the observed value of $F$ is greater than the critical value, a decision can thus be made to reject $H_0$ in favor of the alternate hypothesis.

Use of this $F$ test for variances is based on the assumption that the sample data have been randomly and independently drawn from normally distributed population distributions. If the normality assumption cannot be justified, the test should be applied only when very large samples are available.

## TYPE II ERRORS, POWER, AND *N*

Many applied researchers either fail to understand or underestimate the importance of Type II errors and the concept of power with respect to hypothesis testing. While some consideration has been given to the relationship between Type II errors and $\alpha$ levels in the examples outlined in this and previous chapters, any detailed discussion of these topics has purposely been postponed until the student has acquired a basic understanding of more fundamental issues. We will begin by reviewing the definitions of Type I and Type II errors and adding a definition of power:

Type I error: Rejection of $H_0$ when this hypothesis is actually true. The probability of committing this error is $\alpha$.

Type II error: Failure to reject $H_0$ when this hypothesis is actually false. The probability of committing this error is $\beta$.

Power: Correctly rejecting $H_0$ when this hypothesis is actually false. The probability of this decision is defined as $1 - \beta$.

*Power,* then, is an index of test sensitivity. It describes the efficiency of a statistical test in detecting that $H_0$ is an incorrect statement. The numerical value of power can approach the limits of 0 (lowest power) to 1 (highest power). Ideally, a test should minimize the probability of making either error and maximize the probability of correctly rejecting $H_0$. The magnitudes of $\alpha$ and $\beta$ should be small; and power, as defined by $1 - \beta$, should be large. Power and Type II error are thus directly related; techniques which *reduce* $\beta$ will *increase* power.

An exact value of a difference must be stated in $H_1$ before $\beta$ and $1 - \beta$ can be determined. In other words, *exact* hypothesis statements are necessary. As only *inexact* hypotheses have been used in previous examples, a new research problem is needed to form the basis for this presentation.

Suppose that an institutional researcher in a university has developed a rating scale to measure student evaluations of teaching. The following norms (parameters) for the scale were generated several years previously for the entire population of professors at a number of large institutions: $\mu = 100$; $\sigma = 10$. The data distribution was found to be normal in shape. Since the purpose of the scale was to provide professors with information which they could use to improve their teaching skills, the researcher is now curious as to whether the rating instrument has served its purpose. He somewhat arbitrarily defines improvement as a minimum increase of 6 points in the population mean. Nothing less than this has any practical meaning to him, even though some smaller increase might actually be statistically significant. Unfortunately, funds for the project have been depleted and he cannot afford to collect current population parameters. These values must be estimated with data collected from one small random sample.

An $\alpha$ level of .01 is set for the study. By specifying an $N$ of 25, he can now proceed to determine $\beta$ for a directional $z$ test and power with respect to correctly rejecting $H_0$. No new observed data are needed for these calculations. He computes $\sigma_{\bar{X}} = \sigma/\sqrt{N} = 10/\sqrt{25}$ and uses this standard error of 2.0 in drawing curves (a) and (b) in Figure 9.6. The solid curve (a) represents the sampling distribution of means when the current population mean still equals $\mu = 100.0$ (no change). The dotted curve (b) represents a sampling distribution with a mean of 106.0, depicting the minimum meaningful increase of 6 points. We will denote this postulated mean as $\mu'$. The critical value for a directional $z$ test with $\alpha = .01$ is $+2.33$. This corresponds to a mean value on curve (a) of $(\sigma_{\bar{X}})(z_{.01}) + \mu = (2)(2.33) + 100 = 104.67$. The region of rejection beyond 104.67 is darkened in curve (a).

He first assumes that curve (a) represents the *true* state of affairs in the current population; the mean evaluations of teaching have not changed. If the observed current sample mean is $\leq 104.67$, the null hypothesis of no difference (an exact hypothesis that $\mu = 100$) will not be rejected. A correct inference will be made. But what if the observed sample mean should turn out to be $>104.67$? The null hypothesis would be falsely rejected (a Type I error would be committed) with these sample data. The risk of making this error—the probability of getting a mean this large or larger by chance alone when $\mu = 100$—has been predetermined by the $\alpha$ level, .01.

Now, the researcher draws curves (c) and (d) and assumes that distribution (d) represents the *true* state of affairs in the current population: $\mu' = 106.0$. A minimum positive change of 6 points has actually occurred. The region of rejection is again shown beyond the point 104.67 in curve (c). If the observed sample mean is now $>104.67$, the null hypothesis of no difference ($\mu = 100$) will be correctly rejected. No error will be made.

FIG. 9.6 SAMPLING DISTRIBUTION AREAS CORRESPONDING TO $\alpha$, $\beta$, AND POWER.

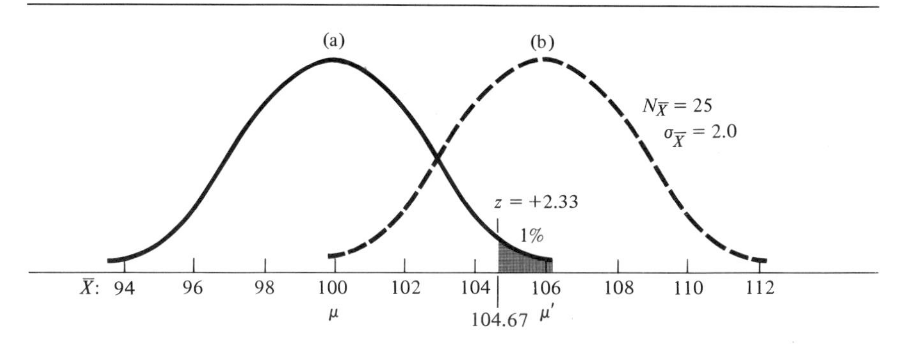


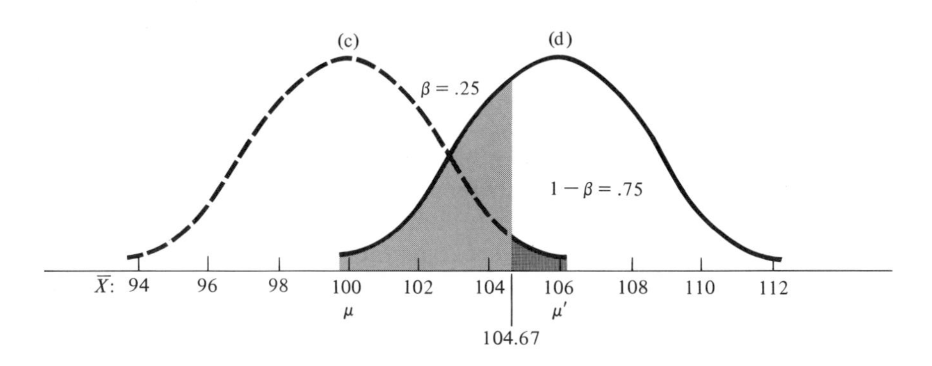


The ability of the test to produce this decision when $H_0$ is false is called its power. If, however, the sample mean is $\leq 104.67$, the researcher will fail to reject the incorrect null hypothesis and a Type II error will have been committed.

The probability of the Type II error, $\beta$, is shown by the shaded area of curve (d). Any time an observed sample mean falls in this area, a Type II error will be produced. The size of the area is easily determined by finding the $z$ equivalent of a mean value of 104.67 for curve (d):

$$z = \frac{\overline{X} - \mu'}{\sigma_{\overline{X}}} = \frac{104.67 - 106.00}{2} = \frac{-1.33}{2} = -.67$$

Reference to the standard normal table will show that this smaller, shaded area of distribution (d) is equivalent to .25, or 25 percent. $\beta$, the probability of a Type II error, is thus .25. The unshaded portion of (d) must now equal $1 - \beta$, or .75. The probability that the test will be sensitive enough to reject $H_0$ when it is false is .75; a statement of the test's power under the defined circumstances. In plain words, there is one chance in four that $H_0$ will not be rejected when it should be (Type II error). Conversely, there are three chances in four that $H_0$ will be rejected when it should be (power). The researcher may feel somewhat comfortable with these values or he may feel very uncomfortable, depending in large part on the consequences of making a Type II error. Related discussions in previous examples should enable the student to now understand that his reaction will depend on the actual problem at hand and the consequences of the various errors.

How may this researcher, or any researcher, go about decreasing $\beta$ and increasing the power of a statistical test? Four practical solutions are presented below:

1. Increase the size of $\alpha$. If the institutional researcher were to set $\alpha$ at .10 instead of .01, the critical value on curve (c) would equal +1.28, or $\bar{X} = 102.56$. $\beta$ would now equal .04 and the power would be .96. This solution is practicable only when the higher risk associated with Type I error can be tolerated.
2. Increase the sample size, $N$. Suppose that the researcher could afford to increase the projected sample size from 25 to 100. This would decrease the standard error, as $\sigma_{\bar{X}} = \sigma/\sqrt{N}$. The standard error term in the example would now equal 1.0 instead of 2.0. By mentally redrawing curves (a) and (b) using the same means but with $\sigma_{\bar{X}} = 1.00$, you will see that the two distributions will be almost totally separated from one another. With the overlap nearly eliminated, $\beta$ is now very small and $1 - \beta$ very large. While this solution is quite effective, an increase in sample sizes is not always feasible.
3. Use an experimental design which reduces standard error. While this suggestion can apply to many techniques, an excellent illustration is offered by application of the matched pairs and repeated-measures designs described in the previous section. Power is typically increased with these designs as compared to using independent samples, again through reduction of curve overlap given a fixed difference between distribution means. Use of a repeated-measures design would be a viable option in the example we are studying, for professors' new ratings could be compared with the ratings these same individuals received several years earlier when data were gathered on the entire population.
4. Use directional, rather than nondirectional, tests. This technique, already employed in the example, will decrease $\beta$ provided that the hypothesized region of rejection is in the direction of the true population mean. Look again at Figure 9.6. Notice that the vertical line above the critical value for a directional test on curve (c) cuts off less area on curve (d) than would the right-hand critical value alone for a nondirectional test, given the same $\alpha$ level. Directional test forms thus reduce the area defining $\beta$ in the same

way as an increase in $\alpha$ reduces $\beta$, by increasing the distance between the critical value and $\mu'$. However, a word of warning is in order. If the researcher should happen to guess wrong in setting up the directional hypothesis—if the true population mean is actually in the opposite direction—a severe penalty is paid through the loss of virtually all power.

It should be apparent that there is no such thing as *the* power or *the* $\beta$ error of statistical tests. Power and $\beta$ depend first of all on the type of test. While only $z$ tests have been discussed, the power of $t$ tests will be nearly the same if $N$ is large and the concept applies to all other types of tests as well. These values also depend on $\alpha$, $N$, test directionality, and the postulated value of the true mean; the greater the difference between $\mu$ and $\mu'$, the greater the power.

## EXERCISES

For the following problems, assume the population data distributions tend to be normally distributed.

1. $\mu = 60$ and $\sigma = 10$ for a defined population. A random sample of 22 members thought to belong to this population has a mean of 65 and a standard deviation of 9.
   a. Conduct and interpret a one-sample, nondirectional $z$ test for $\alpha = .05$.
   b. Conduct and interpret a one-sample $z$ test at the .01 level when $H_1$ states that the unknown population mean is larger.
2. The manual for a standardized management training game provides national normative information on manager scores: $\mu = 168.5$, $\sigma = 17.4$. The average (mean) score for 11 managers at company A is 159.2. Formally outline a study, set up like the examples in this chapter, to determine if the company A average differs significantly from the national average. Be sure to state your decision and interpret the results.
   a. Use a nondirectional test at the .10 level.
   b. Use a directional test ($H_1$: $\mu_A < \mu$) at the .001 level.
   c. Sketch the sampling distribution for the above directional test, identifying the points and areas of interest.
3. Given that $\mu = 18.8$, $\bar{X} = 19.1$, $s = 1.8$, and $N = 80$, conduct a one-sample, nondirectional test at the .05 level and sketch the sampling distribution used.
4. Given that $\mu = 258$, $\bar{X} = 242$, $s = 40$, $N = 28$, conduct an appropriate one-sample, directional test at the .005 level.
5. Assume that the mean height of 10-year-old males in America is 55 inches. An investigator has selected a sample (A) of 19 males of this

age identified as being raised on a protein-deficient diet and obtains these statistics on height: $\bar{X} = 52$ inches, $s = 4$ inches.

a. Formally outline and interpret a nondirectional test at the .05 level.
b. Sketch the sampling distribution for the above test.
c. State the hypotheses for a directional test and make a decision at the .05 level.

6. Statistics for two independent samples are: $N_1 = 9$, $\bar{X}_1 = 85$, $\Sigma x_1^2 = 1800$; $N_2 = 10$, $\bar{X}_2 = 64$, $\Sigma x_2^2 = 2916$. Conduct a nondirectional test between population means at the .10 level.
7. Statistics for two independent samples are: $N_1 = 45$, $\bar{X}_1 = 12.3$, $\Sigma x_1^2 = 344.96$; $N_2 = 45$, $\bar{X}_2 = 13.1$, $\Sigma x_2^2 = 275.0$. Conduct a directional test ($H_1$: $\mu_1 < \mu_2$) at the .025 level.
8. A parachute manufacturer is testing new, inexpensive webbing as a possible replacement for the expensive webbing now used in making its chutes. Six random pieces of the new webbing and six random pieces of the old webbing are tested for breaking strength. The results are shown below in pounds. The manufacturer is interested only in determining if the new webbing is stronger. Formally outline this study and interpret the results. Justify the $\alpha$ level you choose. (You may code the data, if desired, by subtracting a constant; for example, 10,000. The value of $t_{\text{obs}}$ will be the same.)

| X: OLD | X: NEW |
|---|---|
| **10,423** | **10,554** |
| **10,161** | **10,417** |
| **10,324** | **10,391** |
| **10,442** | **10,539** |
| **10,253** | **10,228** |
| **10,298** | **10,353** |

9. Statistics for two independent samples are : $N_1 = 26$, $\bar{X}_1 = 539.4$; $N_2 = 26$, $\bar{X}_2 = 528.3$. The standard error of the difference between means, $s_{\bar{X}_1-\bar{X}_2}$, equals 3.1. Conduct a nondirectional test of the null hypothesis that $\mu_1 - \mu_2 = 4$ at the .001 level, using formula (9.6). Notice that $\mu_1$ and $\mu_2$ are not assumed to be equal in this problem.
10. The following statistics are obtained from two dependent samples (matched-pairs): $N = 35$, $\bar{X}_1 = 96$, $\bar{X}_2 = 93$, $\Sigma D = 105$, $\Sigma D^2 = 335$. Conduct a nondirectional test between population means at the .05 level.

a. Use a $t$ test.
b. Use Sandler's $A$.

11. Random air samples are measured for particulate matter within a 5-mile radius of a factory before and after installation of pollution-control devices. Readings at 10 random locations are given below for this repeated-measures problem.

| LOCATION | (1) BEFORE | (2) AFTER |
|---|---|---|
| a | 16 | 14 |
| b | 7 | 10 |
| c | 34 | 28 |
| d | 29 | 29 |
| e | 18 | 19 |
| f | 23 | 18 |
| g | 9 | 8 |
| h | 14 | 16 |
| i | 31 | 24 |
| j | 25 | 25 |

a. Formally outline and conduct a directional test ($H_1$: $\mu_1 > \mu_2$) between the before and after means at the .025 level.
b. Apply Sandler's test to the above data and compare your result with that obtained in part a.

12. The value of $P$ for a sample of 200 is found to be .60. Use both $\chi^2$ and the $z$ approximation to test the hypothesis that this sample was drawn from a population with $p = .75$. Set $\alpha$ at .01. (Be sure to convert to frequencies before computing $\chi^2$.)
13. Given the following data for two independent samples: $P_1 = .42$, $N_1 = 36$; $P_2 = .58$, $N_2 = 40$. Compute $P'$ and $Q'$ and conduct a non-directional test between population proportions at the .10 level.
14. Given the following frequency data on the relative effectiveness of a particular therapy for independent samples of males and females:

| | EFFECTIVE | INEFFECTIVE |
|---|---|---|
| MALES (1) | 18 | 82 |
| FEMALES (2) | 26 | 74 |

Conduct a $z$ approximation test to determine if the proportion of males for whom the therapy was judged effective is less than for females ($H_1$: $p_1 < p_2$). Make your decision at the .01 level.

15. Given the following data on two independent samples: $N_1 = 21$, $s_1 = 6.8$; $N_2 = 21$, $s_2 = 11.3$. Conduct a test at the .05 level to determine if

the variance for population 2 is larger than the variance for population 1.

16. An investigator wishes to determine the power of a proposed directional $z$ test with these defined or computed parameters: $\mu = 50$, $\mu' = 53$, $\sigma = 8.0$, $N = 64$, $\alpha = .05$.
    a. Compute the value of $\beta$ and power.
    b. If an $\alpha$ level of .001 were to be used, what would be the test's power?

# 10

# SINGLE-FACTOR ANALYSIS OF VARIANCE

The analysis of variance (*ANOVA*) is one of the most highly developed statistical techniques available to the applied researcher. Despite the implication in its title, the elementary *ANOVA* designs presented in this book are *not* used to test hypotheses about population variances, but rather to test hypotheses about population means. We will see that this task is accomplished through the procedure of dividing sample variances into components, some of which describe the extent of observed differences between sample means and others of which describe the extent of the expected differences between these means when $H_0$ is assumed to be true. Thus, *ANOVA* consists of analyzing variances in order to make inferences about population means.

*ANOVA* techniques exist for the analysis of data in extremely complex research designs, as the student might already have discovered by looking through an advanced text or browsing through the statistical program library of a university computing center. Since computations in large problems can be very laborious with hand or desk calculators, the student should familiarize himself with the procedures for using the *ANOVA* computer programs that are available at his institution. The following presentation is merely an attempt to develop a fundamental understanding of the basic logic on which the *ANOVA* relies so that appropriate designs can be selected and the results accurately interpreted. Only a few of *ANOVA*'s simplest applications will be demonstrated here.

If the techniques presented so far in the text were applied to a situation involving the comparison of means from three or more populations, the researcher would have two possibilities: (1) use of the chi-square test of homogeneity, or (2) use of repeated $z$ or $t$ tests between each possible pair of means. Possibility (1) is a viable option, but the $\chi^2$ test would compare simultaneously the shapes, variances, and central tendencies of the data distributions. There would be no way to determine if significant differences were due to any one or all of these characteristics. Possibility (2) would not be acceptable with the $z$ and $t$ techniques as presented in the last chapter. The first problem which the researcher would face involves the number of comparisons which must be made. If three means, A, B, and C, were to be compared, three separate $z$ or $t$ tests would be necessary: A vs. B, A vs. C, and B vs. C. If 10 means were to be compared, a little calculation will demonstrate that 45 separate comparisons must be conducted. The second problem involves the repeated analysis of the same sample data, which violates the assumptions used to establish probabilities. The actual probability of $\alpha$ and $\beta$ errors associated with each individual $t$ or $z$ test would no longer be known. Lastly, consider the more obvious effects of making a larger number of tests, say, 100 $t$ tests at the .05 level. By chance alone, (.05)(100), or 5 of these tests would be termed significant. Application of *ANOVA* techniques not only provides solutions to these problems, but in addition allows the researcher to explore such phenomena as overall effects, trends, interaction effects, and a variety of other topics.

## RATIONALE OF *ANOVA*

Suppose that 200 random samples of size 10 have been independently drawn from a defined population and we wish to obtain an estimate of the population variance, $\sigma^2$, with these data. It stands to reason that the average of the 200 sample variances should provide a closer approximation of $\sigma^2$ than one would expect to get by arbitrarily using the variance of a single sample of size 10 as an estimate. A pooled variance estimate like that employed in connection with $t$ tests is therefore in order. If the sample $N$'s are fairly large, or if many samples have been drawn (as in this example), this pooled variance estimate should be very accurate.

There is also another way in which $\sigma^2$ might be estimated with these samples. We know that a sampling distribution of means consists of an infinite number of sample means. Although we do not have an infinite number of such means here, we could actually plot the 200 means that are available and thereby approximate the theoretical sampling distribution for $N_{\bar{X}} = 10$. This is really no different than collecting a sample of 200 raw $X$ values and using this sample data distribution to approximate the underlying population data distribution of $X$ values; something the student has done many times in previous chapters. The only difference is that the entries in the distribution are now $\bar{X}$'s rather than $X$'s.

After the sampling distribution of means for samples of size 10 is approximated with observed sample means, we could compute indices of its central tendency and variability. The variance so computed ($s_{\bar{X}}^2$) will be an *unbiased* estimate of the variance of the theoretical sampling distribution of means ($\sigma_{\bar{X}}^2$). Now, recall that the relationship between $\sigma$ and $\sigma_{\bar{X}}$, or $s$ and $s_{\bar{X}}$, is

$$\sigma_{\bar{X}} = \frac{\sigma}{\sqrt{N}}$$

$$s_{\bar{X}} = \frac{s}{\sqrt{N}}$$

Solving for $\sigma$ or $s$, we get

$$\sigma = \sqrt{N}\,\sigma_{\bar{X}}$$

$$s = \sqrt{N}\,s_{\bar{X}}$$

Squaring both sides to eliminate the radical provides formulas for the variance:

$$\sigma^2 = N\,\sigma_{\bar{X}}^2$$

$$s^2 = N\,s_{\bar{X}}^2$$

Since $N$ and $s_{\bar{X}}^2$ are known in our sample, we now have a somewhat circuitous, but different, approach to determining $s^2$ and estimating $\sigma^2$. By using the distribution of the 200 sample means to approximate the standard error of the sampling distribution of means and multiplying this estimate by $N$, the result provides a second unbiased estimate of $\sigma^2$ which is completely *independent* of our original pooled variance estimate of $\sigma^2$. This independence derives from the fact that we did not consider differences between the individual sample means in calculation of the first pooled variance estimate and we did not consider the magnitudes of individual sample variances in calculating the second estimate via the standard error of the means.

Since both these estimates of $\sigma^2$ are unbiased, they should be nearly equal when all the samples being studied *have* in fact been randomly drawn from the same population. In other words, the ratio of the two variances should be *approximately* 1.0:

$$\frac{\sigma^2 \text{ estimated from sample means}}{\sigma^2 \text{ estimated from sample variances}} \approx 1.0$$

What would be the result when the sample means have actually been drawn from several populations with different means? The differences between the observed sample means would now be greater and the sampling distribution constructed with these means would display more variability. Its standard error would be larger than before, so the estimate of $\sigma^2$ via $Ns_{\overline{X}}^2$ would have a larger value. But what effect would this have on the denominator; the pooled variance estimate of $\sigma^2$? As long as we can assume that these several parent populations have equal variances there would be no reason for the individual sample variances to change. The pooled variance estimate should still be about the same size as it was previously when the samples were all drawn from the same population. The ratio of the two variance estimates would now be greater than 1.0:

$$\frac{\sigma^2 \text{ estimated from sample means}}{\sigma^2 \text{ estimated from sample variances}} > 1.0$$

All that remains before hypotheses may be tested is to use this ratio to define a test statistic and to construct theoretical tables which define its distribution under prescribed conditions. This statistic is $F$, already familiar to us from the discussion in Chapter 9 on testing hypotheses about variances.

To facilitate this discussion, let us now add some definitions:

$s_b^2$ = the variance estimate of $\sigma^2$ based on differences *between* sample means (the numerator in the ratio presented above)
$s_w^2$ = the variance estimate of $\sigma^2$ based on the pooled variance *within* samples (the denominator in the ratio presented above)

$F$ is defined as the ratio of these two variance estimates:

$$F = \frac{s_b^2}{s_w^2} \tag{10.1}$$

When the sample means have in fact been drawn from the same population, this statistic has a sampling distribution which is approximated by an $F$ distribution with $df = k - 1$ in the numerator and $df = k(N - 1)$ in the denominator; where $k$ is the number of means being compared and $N$ is the sample size. The two variance estimates, $s_w^2$ and $s_b^2$, will be nearly equal in magnitude and the $F$ ratio will have a value of approximately 1.0. When the samples have been drawn from several populations with equal variances but with different means, $s_b^2 > s_w^2$ and the $F$ ratio will be $>1.0$. Another way of putting this is that $s_b^2$ describes the extent of *observed* differences between means and $s_w^2$ describes the extent of *expected* differences when $H_0$ is true. We still must face the matter of sampling fluctuations, of course. A second set of 200 samples drawn from the same population(s) will probably not give us exactly the same estimates of $\sigma_w^2$ and $\sigma_b^2$. We must allow for this sampling fluctuation when $F_{\text{obs}}$ is interpreted by deciding how much greater than 1.0 the $F_{\text{obs}}$ value must be before it may be termed significant. As with previous tests, this is done by setting an $\alpha$ level and finding an associated critical value of $F$ against which $F_{obs}$ is compared. Sampling fluctuations are likely to be greater when the sample $N$'s are small or when only a few samples have been drawn ($k$ is small). Therefore, we need two $df$ values for $F$; one associated with the number of sample means used in computing $s_b^2$, $df_b$, and one associated with the sample sizes involved in the computation of $s_w^2$, $df_w$. Magnitudes of observed and critical $F$'s will have a theoretical range of zero to infinity.

## AN *ANOVA* DESIGN WITH TWO INDEPENDENT SAMPLES

In practice, 200 means are not necessary for computing $s_b^2$. It can be done with as few as two sample means, although the probable accuracy of the estimate will suffer unless the $N$'s are large. Since the homogeneity of variance assumption is the same for both $F$ and $t$ tests, we thus have the option of using the $F$ test as an alternative to the two-sample $t$ test. To relate these techniques and hopefully provide the student with more of an intuitive feel for the *ANOVA* $F$ test, both have been used to analyze the same hypotheses on the male and female sample data given in Table 10.1.

No difficulty should be encountered in examining the $t$ test portions of this table, but many of the procedures and formulas used in association with *ANOVA* are new. Consider first the decision rule for the $F$ test ($DR_F$). Even though the hypothesis statements are nondirectional, $DR_F$ appears as if it belongs to a directional test. However, as was the case with $\chi^2$, we will consider the *ANOVA* only as a nondirectional test (when 3 or more populations are being compared, in fact, only nondirectional tests are permissible with the procedures outlined). Because both $\chi^2_{obs}$ and $F_{obs}$ are always positive in value, we need merely to be concerned with one (positive) tail of these distributions. Consequently, a single critical value of $F$ is sufficient to test a nondirectional hypothesis. In determining this value, the degrees of freedom associated with the numerator and the denominator of the $F$ ratio are defined respectively as

$$df_b = k - 1 \tag{10.2}$$

where $k$ is the number of means being compared (the number of samples), and

$$df_w = k(N - 1) \tag{10.3}$$

where $N$ is the size of each sample (equal $N$'s are assumed).

The degrees of freedom for the numerator of $F$ in the example is $df_b = k - 1 = 2 - 1 = 1$. The $df$ for the denominator is $df_w = k(N - 1) = 2(9) = 18$. Referring to Appendix E, we find that the critical value of $F$ at the .05 level with 1 degree of freedom in the numerator and 18 degrees of freedom in the denominator is 4.41.

The total number of cases in the problem, $N_T$, is the sum of the sample $N$'s; 10 + 10 = 20. The *total* (or *grand*) mean is the average of all observations, or

$$\bar{X}_T = \frac{\Sigma X_T}{N_T} \tag{10.4}$$

where $\Sigma X_T$ is the sum of the measures across all samples.

You will recall that $\Sigma X^2$ is also known as the *variation,* or $SS$. The general formula for a variance can thus be given as

$$s^2 = \frac{\Sigma x^2}{N - 1} = \frac{SS}{N - 1}$$

Since $N - 1$ is actually a $df$ term, we can generalize from this formula to $s_b^2$ and $s_w^2$.

$$s_b^2 = \frac{\Sigma x_b^2}{df_b} = \frac{SS_b}{df_b} \tag{10.5}$$

$$s_w^2 = \frac{\Sigma x_w^2}{df_w} = \frac{SS_w}{df_w} \tag{10.6}$$

**TABLE 10.1 ABBREVIATED EXAMPLE OF NONDIRECTIONAL *t* AND *F* TESTS BETWEEN TWO MEANS USING THE SAME DATA**

DECISION MODEL $H_0: \mu_1 = \mu_2$

$H_1: \mu_1 \neq \mu_2$

$\alpha = .05$

$DR_t$: **reject** $H_0$ **if** $t_{obs} < -2.10$ **or** $> +2.10$**; otherwise, do not reject**

$DR_F$: **Reject** $H_0$ **if** $F_{obs} > 4.41$**; otherwise do not reject**

**OBSERVED *X* VALUES AND STATISTICS**

| SAMPLE 1 (MALES) | SAMPLE 2 (FEMALES) |
|---|---|
| *X*: 11, 15, 9, 13, 11, 12, 8, 16, 8, 14 | *X*: 15, 13, 18, 19, 17, 16, 14, 12, 15, 16 |

$N_1 = 10$ $\Sigma X_1 = 117$

$\bar{X}_1 = 11.7$ $\Sigma X_1^2 = 1441$

$\Sigma x_1^2 = 72.1$

$N_2 = 10$ $\Sigma X_2 = 155$

$\bar{X}_2 = 15.5$ $\Sigma X_2^2 = 2445$

$\Sigma x_2^2 = 42.5$

TOTALS

$N_T = N_1 + N_2 = 20$

$\Sigma X_T = \Sigma X_1 + \Sigma X_2 = 272$ $\bar{X}_T = \frac{\Sigma X_T}{N_T} = 13.6$

$\Sigma X_T^2 = \Sigma X_1^2 + \Sigma X_2^2 = 3996$

(*Table continues*)

**Table 10.1 (*Continued*)**

*t* TEST; $N_1 = N_2$

$$t_{\text{obs}} = \frac{\bar{X}_1 - \bar{X}_2}{\sqrt{\dfrac{\Sigma x_1^2 + \Sigma x_2^2}{N(N-1)}}} = \frac{11.7 - 15.2}{\sqrt{\dfrac{72.1 + 42.5}{10(10-1)}}} = \frac{-3.8}{\sqrt{\dfrac{114.6}{90}}} = \frac{-3.8}{\sqrt{1.2733}} = \frac{-3.8}{1.1284}$$
$$= -3.386$$

DECISION **Since $-3.368 < -2.10$, Reject $H_0$**

*F* TEST

$$SS_b = \frac{\Sigma(\Sigma X_k)^2}{N} - \frac{(\Sigma X_T)^2}{N_T}$$
$$= \frac{(\Sigma X_1)^2 + (\Sigma X_2)^2}{N} - \frac{(\Sigma X_T)^2}{N_T} = \frac{(117)^2 + (155)^2}{10} - \frac{(272)^2}{20}$$
$$= \frac{37{,}714}{10} - \frac{73{,}984}{20} = 3771.4 - 3699.2$$
$$= 72.2$$

$$SS_w = \Sigma X_T^2 - \frac{\Sigma(\Sigma X_k)^2}{N} = \Sigma X_T^2 - \frac{(\Sigma X_1)^2 + (\Sigma X_2)^2}{N} = 3886 - 3771.4$$
$$= 114.6$$

(*Note:* The second term in $SS_w$ was previously computed in $SS_b$.)

$$F_{\text{obs}} = \frac{SS_b/df_b}{SS_w/df_w} = \frac{72.2/1.0}{114.6/18} = \frac{72.2}{6.37}$$
$$= 11.34$$

DECISION **Since $11.34 > 4.41$, Reject $H_0$**

*F* may be defined in these terms as

$$F = \frac{SS_b/df_b}{SS_w/df_w} \tag{10.7}$$

Basic and computational formulas can now be presented for the calculation of $SS_b$ and $SS_w$, respectively:

$$SS_b = N\Sigma(\bar{X}_k - \bar{X}_T)^2 \tag{10.8}$$

$$= \frac{\Sigma(\Sigma X_k)^2}{N} - \frac{(\Sigma X_T)^2}{N_T} \tag{10.9}$$

where $\bar{X}_k$ is the mean of sample (group) $k$ and $\Sigma X_k$ is the sum of the measures in sample $k$

$$SS_w = \Sigma(X_k - \bar{X}_k)^2 \tag{10.10}$$

$$= \Sigma X_T^2 - \frac{\Sigma(\Sigma X_k)^2}{N} \tag{10.11}$$

where $X_k$ is a measure in sample $k$ and $\Sigma X_T^2$ is the sum of all squared measures.

$SS_b$ is thus the sum of the squared deviations of each sample mean from the total mean multiplied by the sample size, $N$. $SS_w$ is the sum of the squared deviations of each $X$ value about its respective mean, $\bar{X}_k$. These two sums-of-squares provide independent estimates of the population variation when the samples have been drawn from a single population (or multiple populations having equal means). When divided by their respective degrees of freedom, unbiased and independent estimates of population variance are obtained. While the computational formulas require fewer calculations, the large intermediate terms generated with some problems may tax the internal capacities of inexpensive pocket calculators.

Detailed calculations of $SS_b$ and $SS_w$ via the computational formulas are provided for the observed data in Table 10.1. If basic formula (10.8) were to be used for determining $SS_b$, the total mean, 13.6, would be subtracted from each of the two sample means. These two differences would be squared and their sum multiplied by $N = 10$. To use basic formula (10.10) in determining $SS_w$, the mean of each sample is subtracted from every measure in that sample and the differences squared. The 20 squared differences are then summed.

Variance due to sample mean differences ($s_b^2$) is found to be considerably larger than the pooled within-sample variance ($s_w^2$) for the male and female samples in Table 10.1, raising the suspicion that these samples may have been drawn from different populations; that is, $\mu_1 \neq \mu_2$. To test the hypothesis of no difference, $F_{\text{obs}}$ is determined for this problem by dividing $s_b^2$ by $s_w^2$ and the result is then compared with the critical value of $F$ at the .05 level with $df = 1$ for the numerator and $df = 18$ for the denominator. Since $11.34 > 4.41$, $H_0$ of no difference between population means can again be rejected at the .05 level.

When the difference between two population means is being tested ($df_b = 1$ for the $F$ test) use of either $t$ or $F$ will *always* result in the same decision and $t^2_{\text{obs}}$ will equal $F_{\text{obs}}$. In this example, $t^2_{\text{obs}} = (-3.368)^2 = 11.34$, the value of $F_{\text{obs}}$. The *ANOVA* and the $t$ test may thus be used interchangeably for the specific situation in which two independent population means are being compared. With the formulas given, however, $t$ would usually be the choice of a researcher using only a small calculator as a computational aid.

## TERMINOLOGY IN *ANOVA*

Special sets of symbols, terms, and conventions have developed around the *ANOVA*. Some of these will be presented here to facilitate explanation of the elementary designs to be covered and to provide the student with a basic set of knowledge which can be built upon in subsequent courses and readings.

One of the most frequently used symbols in the *ANOVA*, *MS* (mean square), was introduced in Chapter 4 as being synonomous with variance. Mean square may thus be defined in the following ways:

$$MS = s^2 = \frac{\Sigma x^2}{df} = \frac{SS}{df} \tag{10.12}$$

*F* can now be correctly defined either as the ratio of two variances or the ratio of two mean squares. Formula (10.13) defines the basic *F* ratio in terms of all the notations used in this and previous chapters:

$$F = \frac{MS_b}{MS_w} = \frac{s_b{}^2}{s_w{}^2} = \frac{SS_b/df_b}{SS_w/df_w} \tag{10.13}$$

A *factor* in the *ANOVA* is a variable, the effects of which are being studied. Factors also serve as the basis for classifying data into categories. In contrast to criterion variables, factors are under the control of, and in some cases manipulated by, the investigator. The factor in the example outlined in Table 10.1 was sex. Such *ANOVA* studies with only one factor are variously called *single-factor, single-classification,* or *one-way* analyses. Categories of a factor into which data are classified are called *levels*. The factor of sex had two levels, male and female.

A universe of levels exist for any factor. Although the universe of levels for sex consists of only two categories, the universe of levels for a factor defined as "AKC breed of dog" consists of more than 100 categories and the universe of levels for a factor defined as "electric-shock intensity" is nearly infinite. When a researcher purposely and nonrandomly selects certain levels of a factor for study, his *ANOVA* design corresponds to a *fixed-effects* model, or *Model I*. If, on the other hand, levels are randomly selected for study from the universe of a factor's levels, the design corresponds to a *random-effects* model, or *Model II*. All the examples to be used in this chapter will employ the fixed-effects model. Results of studies which correspond to the fixed-effects model are interpreted only with respect to the specific levels actually analyzed; generalizations to the entire universe of a factor's levels are not permissible. Upon finding that an electric-shock factor is significant, for example, the researcher can draw inferences only to the levels (intensity) of shock actually administered, much as the results of a chi-square test are interpretable only in terms of the categories of the contingency table. If generalization to the universe of all possible factor levels is desired, a random-effects model should be employed. Discussions of this model will be found in any advanced experimental design text.

Sampling procedures can be rather involved in the more complex *ANOVA* designs. In certain of the elementary designs, on the other hand, sampling is simply a matter of randomly drawing members from the populations underlying each level of the factor(s) being investigated. For example, separate samples were drawn directly from the two populations of males and females (the two levels of the sex factor) in the problem cited previously in this chapter. Even with elementary designs, however, it is frequently the case that populations underlying factor levels are not actually available for sampling purposes. Consider an evaluation of the effects of three experimental instructional methods (levels of a factor) on learning (the criterion variable). Populations of students taught via the three methods are not yet in existence, so the researcher must randomly draw a single sample of students from the general defined population and then subdivide it by randomly assigning its members to one of the three instructional methods, or levels (an example of a randomized group design). It is then assumed that each subsample represents the theoretical population of students taught by a particular method. Since the subsamples are being treated differently, they may be referred to as *treatment groups* and the methods may be called *treatments,* as in the following diagram:

**FACTOR: INSTRUCTIONAL METHOD**

| **First Method** | **Second Method** | **Third Method** |
|---|---|---|
| LEVEL 1<br>(TREATMENT 1) | LEVEL 2<br>(TREATMENT 2) | LEVEL 3<br>(TREATMENT 3) |
| **Treatment Group 1** | **Treatment Group 2** | **Treatment Group 3** |

Thus, each level of a factor defines a treatment that a treatment group, or groups, will receive. It may be somewhat difficult for the student to think of the levels of some nominal factors, such as sex, as being treatments. The convention of referring to levels as treatments is so ingrained among researchers, however, that this notation would not be deemed unusual.

Observed differences among the treatment groups on the criterion variable are referred to as *treatment effects,* assumed to be due to the differential treatments accorded the groups. If the treatment effects for a factor are found to be significant, the researcher may infer that population differences exist. In terms of the above fixed-effects study of instructional methods, significant results would imply that populations of students taught by methods A, B, and C differ with respect to average learning as a function of the particular method (treatment) administered.

## ASSUMPTIONS IN THE FIXED-EFFECTS *ANOVA*

The assumptions of fixed-effects *ANOVA* techniques are basically the same as those for $t$ and $z$ tests. First, each sample must be *randomly* selected from a defined population. This assumption is specifically concerned with the randomness of treatment groups in the *ANOVA*. If underlying levels of populations exist and are available for sampling purposes, the members of each treatment group must be randomly selected from the corresponding population. If the underlying populations are unavailable, the assumption is met by drawing a random sample from the general population of interest and then randomly assigning members to treatment groups.

A second assumption is that the treatment groups must be *independent*. The student will recall from Chapter 9 that samples (treatment groups) are said to be independent when there is no systematic way of matching or pairing members of one group with members of any other group. This is a particularly important assumption with the fixed-effects model and violation can easily precipitate erroneous decisions. Consequently, the methods presented here are not appropriate for matched or repeated-measures designs.

Third, the population data distributions underlying each treatment group are assumed to be *normal* in shape. Tests in the *ANOVA* are rather *robust* with regard to this assumption. Even if the population distributions should depart from normality, the probability levels of the $F$ statistic will be reasonably accurate if the treatment group $N$'s are fairly large (an extension of the central limit theorem).

Finally, *homogeneity of variance* is assumed with respect to the population data distributions underlying the treatment groups. Statisticians formerly recommended that special tests of this assumption be conducted before an *ANOVA* was initiated. Now it has been determined that this equal variance assumption presents no serious risk in the vast majority of situations as long as the treatment groups are *equal* in size. Just as with the $t$ test, then, the researcher can usually satisfy these last two assumptions by planning to use equal treatment group sizes and by increasing these sizes in relationship to the amount of his doubt or lack of knowledge concerning the normality of population distributions.[1]

[1] The importance of understanding the implications of equal vs. unequal treatment group sizes in the more complex *ANOVA* designs cannot be overemphasized. The student should become familiar with these considerations before attempting to use a particular design that is not discussed in this text (see Winer, 1971).

## PARTITIONING THE SUMS OF SQUARES

In the first example, the between sum-of-squares, $SS_b$, and the within sum-of-squares, $SS_w$, were defined and used as the computational bases for estimating population variance. If the observed treatment group means should all be equal in a single-factor *ANOVA* problem, $SS_b$ would be zero and the total variation in the problem would be accounted for by $SS_w$. If, on the other hand, the treatment group means should differ but the $X$ values within any given group were all equal, $SS_w$ would be zero and the total variation would be accounted for by $SS_b$. Typical problems will display both within and between variation and neither $SS$ term above will account for the *total variation,* defined as the sum of the squared deviation of each $X$ value in the study about the total mean, $\bar{X}_T$:

$$SS_T = \Sigma(X - \bar{X}_T)^2 \tag{10.14}$$

where $X$ is each measure in the treatment groups.

The equivalent computational formula for the total variation is

$$SS_T = \Sigma X_T{}^2 - \frac{(\Sigma X_T)^2}{N_T} \tag{10.15}$$

Again, it should be noted that these computational formulas may overtax the capacities of inexpensive hand calculators when $N$'s and/or $X$ values are large.

*ANOVA* procedures serve to *partition* the total sum of squares into component *sources* of variation. In a single-factor *ANOVA* with independent treatment groups, the sum of components $SS_b$ and $SS_w$ will equal $SS_T$:

$$SS_T = SS_b + SS_w \tag{10.16}$$

A large number of components may be generated through partitioning $SS_T$ in complex *ANOVA* designs.

The total degrees of freedom associated with $SS_T$ is defined as

$$df_T = N_T - 1 \tag{10.17}$$

The total degrees of freedom may also be partitioned into components associated with each source of variation. For the same elementary design as described above,

$$df_T = df_b + df_w \tag{10.18}$$

It should be stressed that this partitioning into additive components is appropriate only for variation, not for variance. The mean square components are not additive. Thus,

$$MS_T \neq MS_b + MS_w$$

## SINGLE-FACTOR *ANOVA* DESIGNS WITH THREE OR MORE LEVELS

Presentations in this and subsequent sections will again be limited to the *fixed-effects model* with *independent* treatment groups of *equal size*. The procedures outlined in the first example can rather easily be generalized to single-factor problems with any number of levels. Although not all the *ANOVA* notation, conventions, and formulas to be used have been discussed, those which remain can perhaps be introduced most efficiently within the context of a formal numerical example.

Suppose an archaeologist in charge of research at four excavation sites in Central America wishes to determine if these locations were humanly inhabited simultaneously, or during different time periods. Each site contains a population of artifacts, some of which may have previously been removed and some of which will probably remain to be discovered by future groups of scientists. Using a prescribed excavation strategy, the present scientific team has uncovered what may be considered as a random sample of artifacts from each site. Artifact age, the criterion variable, has been estimated for each member of these samples through carbon dating and other techniques. By comparing the average age of artifacts at each site, the archaeologist plans to draw inferences about the relative ages of the populations of artifacts and consequently to the periods of human habitation (see Table 10.2).

This study has one factor (excavation site location) with four levels (sites). The treatment groups can be sampled directly from these sites. Since these particular sites were deliberately chosen for study from a universe of sites in the area, a fixed-effects model applies. Remember that these levels may be referred to as treatments under *ANOVA* notation and that any differences in average artifact age by location may be termed treatment effects.

Assumptions for the use of *ANOVA* procedures with this study are seemingly met. The excavation sampling strategy was designed to obtain a random sample of artifacts from each site. The treatment groups are probably independent, for the excavations are in physically separate locations. Past experience of the archaeologist indicates that data distributions of artifact age at similar sites tend to be normal in shape. Lastly, equal-size treatment groups will be used, thereby minimizing the risk of violating the homogeneity of variance assumption.

The sample size chosen for the actual study would, in all likelihood, consist of a fairly large number of artifacts. To simplify this illustration, however, we will arbitrarily use very small treatment groups of $N = 5$. With four levels of the single factor, the number of treatment groups is thus $k = 4$. The total number of artifacts studied would then be $N_T = Nk = (5)(4) = 20$.

**TABLE 10.2 EXAMPLE OF A SINGLE-FACTOR *ANOVA* APPLICATION WITH FOUR LEVELS**

FACTOR **Excavation site location**
LEVELS **Four separate sites**
POPULATION **Artifacts in 4 defined sites**
SAMPLES (4) **20 artifacts; 5 randomly chosen from each site**
CRITERION VARIABLE **Artifact age in years**
DECISION MODEL $H_0: \mu_1 = \mu_2 = \mu_3 = \mu_4$
$H_1$: not $H_0$
$\alpha = .01$
*DR:* Reject $H_0$ if $F_{obs} > 5.29$; otherwise, do not reject

**OBSERVED DATA**

**Excavation Site Location**
**Levels (Sites)**

| 1 | 2 | 3 | 4 | TOTALS |
|---|---|---|---|---|
| *X:* 1300, 1260, 1340, 1220, 1370 | *X:* 1010, 1150, 1070, 970, 1080 | *X:* 1070, 1050, 1130, 1240, 1100 | *X:* 1250, 1140, 1330, 1180, 1290 | |
| $N_1 = 5$ | $N_2 = 5$ | $N_3 = 5$ | $N_4 = 5$ | $N_T = 20$ |
| $\Sigma X_1 = 6490$ | $\Sigma X_2 = 5280$ | $\Sigma X_3 = 5590$ | $\Sigma X_4 = 6190$ | $\Sigma X_T = 23550$ |
| $\bar{X}_1 = 1298.0$ | $\bar{X}_2 = 1056.0$ | $\bar{X}_3 = 1118.0$ | $\bar{X}_4 = 1238.0$ | $\bar{X}_T = 1177.5$ |
| $\Sigma X_1^2 = 8438500$ | $\Sigma X_2^2 = 5594800$ | $\Sigma X_3^2 = 6271900$ | $\Sigma X_4^2 = 7687500$ | $\Sigma X_T^2 = 27992700$ |

$$SS_T = \Sigma X_T^2 - \frac{(\Sigma X_T)^2}{N_T}$$
$$= 27992700 - \frac{(23550)^2}{20} = 27992700 - 27730125$$
$$= 262575$$

$$SS_b = \frac{\Sigma(\Sigma X_k)^2}{N} - \frac{(\Sigma X_T)^2}{N_T} = \frac{(\Sigma X_1)^2 + (\Sigma X_2)^2 + (\Sigma X_3)^2 + (\Sigma X_4)^2}{N} - \frac{(\Sigma X_T)^2}{N_T}$$
$$= \frac{(6490)^2 + (5280)^2 + (5590)^2 + (6190)^2}{5} - 27730125$$
$$= \frac{139562700}{5} - 27730125 = 27912540 - 27730125$$
$$= 182415$$

(*Note:* The second term, $(\Sigma X_t)^2/N_T$, was calculated in $SS_T$ above.)

$$SS_w = \Sigma X_T^2 - \frac{\Sigma(\Sigma X_k)^2}{N}$$
$$= 27992700 - 27912540$$
$$= 80160$$

(*Note:* Both terms have been calculated previously.)

(*Table continues*)

**Table 10.2 (*Continued*)**

***ANOVA* SUMMARY TABLE**

| SOURCE OF VARIATION | *SS* | *df* | *MS* | *F* |
|---|---|---|---|---|
| **Between** | **182415** | **3** | **60805** | **12.14*** |
| **Within** | **80160** | **16** | **5010** | |
| **Total** | **262575** | **19** | | |

* Significant at .01 level.

DECISION **Since 12.14 > 5.29, reject $H_0$**

INTERPRETATION **Treatment effects are significant. The average age of artifacts differs among these 4 sites (see text)**

Nondirectionality of the *ANOVA* is depicted in the hypothesis statements of the decision model. $H_0$ states that the means of the populations underlying each treatment group are equal and $H_1$ merely states that $H_0$ is not true. The $\alpha$ level is set at .01, because the archeologist is somewhat concerned with guarding against the Type I error of deciding that the means are unequal when, in fact, they are equal. Since the $F$ test will be the ratio of $MS_b$ to $MS_w$, the degrees of freedom for the $F$ test are $df_b = k - 1 = 3$ and $df_w = k(N - 1) = 16$. These parameters are now used to determine the critical value of $F$ from Appendix E for the decision rule: $F = 5.29$.

Observed $X$ values by treatment group are given in the data table. Even with these small $N$'s, the $X$ values are so large that use of computational formulas for the sums of squares will create overflow conditions in some hand calculators. A researcher finding himself in this situation could divide the data into parts and complete final calculations by hand. He might also try using the basic rather than the computational formulas, thus reducing the magnitude of the intermediate terms. Perhaps the best solution would be to code the data by subtracting a constant from each $X$ value. If a constant of 950 were subtracted from each observation in Table 10.2, for example, neither the $SS$ terms nor $F_{\text{obs}}$ would change.

The analysis begins by finding $\Sigma X$ and $\Sigma X^2$ for each treatment group ($\Sigma X^2$ by treatment group is not essential, but serves to break calculations down into more manageable pieces). Totals across all groups are then found: $N_T$, $\Sigma X_T$, and $\Sigma X_T^2$. The sums of squares terms $SS_T$, $SS_b$, and $SS_w$ may now be computed using the above values. Actually, only two of the three $SS$ terms would have to be computed, since the third may be determined by the relationship $SS_T = SS_b + SS_w$. Calculation of all three, however, allows the relationship to be used as a valuable accuracy check. Although either the basic or the computational formulas might have been used, step-by-step calculations have been completed only for the latter.

By convention, *ANOVA* results are typically summarized in a table such as the one presented. These *ANOVA* summary tables list the sources of variation; the *SS, df,* and *MS* terms associated with each; and the observed $F$ ratio(s). The student should study the table for this problem carefully to make sure that the derivations of all the terms and their interrelationships are fully understood.

The observed $F$ is much greater than the critical value of 5.29. Consequently, $H_0$ can be rejected at the prescribed $\alpha$ level of .01. Although the archaeologist can reject the hypothesis that the average age of artifacts differs among the site populations, it should be apparent that the question of simultaneous habitation has not been laid to rest. For one thing, observed ages of artifacts in the samples overlap among the sites. Only between sites 1 and 2 is no overlapping of artifact ages observed. In addition, the means for sites 1 and 4 are very similar, as are the means for sites 2 and 3. A significant overall $F$ tells us only that at *least one* comparison between treatment group means is significant. It may be possible, then, that only sites 1 (with the highest mean) and 2 (with the lowest mean) actually differ in average artifact age. To answer such questions, the researcher may wish to use a *post hoc* procedure for testing the differences between treatment group means.

## *POST HOC* TESTS AMONG MEANS

The *post hoc* procedure to be described in this section is appropriate for determining which sets of treatment group means differ when the overall $F$ for a factor has been found to be significant. This particular method was developed by Scheffé (1959) and is specially designed to allow the researcher to make selected, or every possible, comparisons between means after the data have been observed. Perhaps the simplest way to use the Scheffé method is to solve for the *smallest significant difference* (*ssd*) in an absolute sense between any two treatment group means, using the same $\alpha$ level as for the preceding overall test. The following formula may be used when the treatment group $N$'s are equal:

$$ssd = \sqrt{\frac{(2N)(k-1)(F_\alpha)(MS_w)}{N^2}} = \frac{1}{N}\sqrt{(2N)(k-1)(F_\alpha)(MS_w)} \qquad (10.19)$$

where $F_\alpha$ is the critical value used in a preceding *ANOVA F test.*

For the data in Table 10.2,

$$ssd = \sqrt{\frac{(2)(5)(4-1)(5.29)(5010)}{25}}$$

$$= \sqrt{\frac{795087}{25}}$$

$$= 178.34$$

Observed absolute differences of 178.34 or greater between any two treatment group means are thus significant at the .01 level. Referring to the observed data table in the archeological example, we find three comparisons which exceed this *ssd* value; $\bar{X}_1 - \bar{X}_2 = 242$, $\bar{X}_1 - \bar{X}_3 = 180$, and $\bar{X}_4 - \bar{X}_2 = 182$.

These *post hoc* tests have given the archeologist considerably more information with regard to average artifact age by site. The more general hypothesis of simultaneous human habitation, however, can be tentatively rejected only for site 1 vs. site 2 because of the overlap in $X$ values for other comparisons.

## DEMOGRAPHIC vs. EXPERIMENTAL FACTORS

In the preceding example, a natural relationship existed between sample members and factor categories (sites). Each excavation site defined a population from which a sample of artifacts was drawn. Any given artifact automatically belonged to a particular level of the site location factor. It was not possible, therefore, for the archaeologist to assign artifacts randomly to treatment groups. For convenience, factors which naturally categorize sample members into treatment groups will be termed *demographic factors*. When treatment groups are composed of people, demographic factors might include sex, aptitude, intelligence, socioeconomic status, educational achievement, and so on. A fixed-affects *ANOVA* design using a single demographic factor is applied to determine if the means of several existing populations differ.

Suppose that a pharmacologist wishes to evaluate differences in the effectiveness of four pain-relieving drugs for a population of adults. His *ANOVA* design has one factor (type of drug) with four levels (drug formulas). The terms treatment, treatment group, and treatment effects now begin to take on some intrinsic meaning in the context of this study. Type of drug is not a demographic factor. No natural relationship exists between its categories and the human sample members. The pharmacologist is free to assign the people to any drug formula level in a randomized group design. Since treatment groups can be randomly constructed from the members of one general sample that has been drawn without regard to the factor categories, any variability among treatment group means (treatment effects) can be attributed directly to the differential effects of the drugs under investigation. Type of medication is an example of an *experimental factor*; no natural categorization of sample members occurs. Other experimental factors in studies involving human subjects might include working conditions, level of reinforcement, teaching method, task difficulty, and so on. The distinction between these two classes of factors becomes extremely important in multiple-factor designs.

## EXERCISES

1. Compute both an *ANOVA* $F$ test and a nondirectional $t$ test for independent samples on the following data, using an $\alpha$ level of .01. Verify that $t^2 = F$ for two-sample problems.

| SAMPLE 1 | SAMPLE 2 |
|---|---|
| *X:* **20, 16** <br> **22, 19** | *X:* **23, 31** <br> **32, 28** |

2. Code the data in Table 10.2 by subtracting a constant of 950 from each of the 20 observed $X$ values.
   a. Compute $SS_T$, $SS_b$, and $SS_w$ for these coded data using both the basic *and* the computational formulas. Show your work.
   b. Complete the *ANOVA* summary table based on these new $SS$ values.
   c. What is the decision at the .01 level?
   d. Run a *post hoc* test on the coded means by determining the *smallest significant difference* with formula (10.19). Which pairs of means are significantly different from one another?
3. Three procedures are being evaluated for improving the tensile strength of a certain type of alloy steel. A random sample of steel box girders, all of the same dimensions and thickness, is randomly assigned to the procedures (treatments). The criterion variable is measured tensile strength in pounds per square inch. The following coded data are obtained:

| PROCEDURE | | |
|---|---|---|
| 1 | 2 | 3 |
| *X:* **152, 150** <br> **148, 153** | *X:* **144, 141** <br> **143, 146** | *X:* **157, 153** <br> **156, 160** |

   a. Formally outline this problem, as illustrated in Table 10.2. Use an $\alpha$ level of .05 in the decision model.

b. Construct the *ANOVA* summary table. State your decision and interpretation of the results.
c. If justified, conduct a *post hoc* test on the means.
d. List the assumptions made when applying *ANOVA* to this problem.

4. An industrial psychologist wishes to determine if supervisory skills differ among the five divisions of a large corporation. Samples of first-line supervisors are randomly drawn from each division and are administered a "personnel practices" inventory. The following scores are obtained:

| | | DIVISION | | |
|---|---|---|---|---|
| 1 | 2 | 3 | 4 | 5 |
| $X$: 6, 3<br>2, 8<br>4 | $X$: 7, 9<br>2, 1<br>3 | $X$: 5, 8<br>2, 4<br>6 | $X$: 2, 7<br>4, 5<br>3 | $X$: 9, 2<br>4, 7<br>1 |

a. Formally outline this problem, using an $\alpha$ level of .01.
b. Construct the *ANOVA* summary table. State your decision and interpret the results.
c. If justified, conduct a *post hoc* test on the means.

5. An experiment is devised to compare the effects of coaching on the scores obtained on an aptitude test for a professional school. Three levels of coaching are used: None, 4 hours, and 10 hours. A random sample of 18 applicants is drawn from the 1977 national applicant population and each member is randomly assigned to a coaching level (treatment). The results are given below.

| | COACHING | |
|---|---|---|
| NONE | 4 HOURS | 10 HOURS |
| $X$: 27, 20<br>24, 30<br>26, 22 | $X$: 25, 29<br>32, 27<br>24, 30 | $X$: 30, 33<br>26, 32<br>27, 35 |

a. Formally outline this problem, using an $\alpha$ level of .05.
b. Construct the *ANOVA* summary table. State your decision and interpret the results.
c. If justified, conduct a *post hoc* test between means.

# 11

# TWO-FACTOR ANALYSIS OF VARIANCE

Two or more factors may be jointly analyzed with the *ANOVA*, enabling the researcher to make inferences about the effects of individual factors plus the effects of combinations of factor. Such a design is variously termed a *factorial, multiple-classification,* or *multiple-factor* design. Since almost any measure is, in reality, influenced by combinations of factors, these designs permit a researcher to approximate more closely the actual conditions which exist in nature. Particularly useful is the opportunity to observe and analyze *interactions*—effects due to unique treatment combinations which are not predictable from knowledge of the individual effects of each factor alone. A simple example of interaction can be found in the manufacture of alloys from two basic elements. As the relative amounts (levels) of the two basic elements (factors) in the alloy are varied, physical properties of the alloy (the criterion variables) are changed. It is very important that jointly analyzed factors be *independent* of one another. In practice, this is usually assured if only one or none of the factors in a design is demographic and the remainder are experimental, with sample members randomly assigned to levels. The $F$ test then indicates whether or not this manipulation produces or causes variation in the criterion variable. More will be said in later sections about types of factors and the meaning of interaction.

## REPRESENTATION

Representation of two-factor designs is accomplished through tables similar to the contingency tables associated with chi-square analyses. Consider the following schematic of an *ANOVA* design with treatment groups of size 4 and two factors: a column factor, *C*, which has three levels, and a row factor, *R*, with two levels.

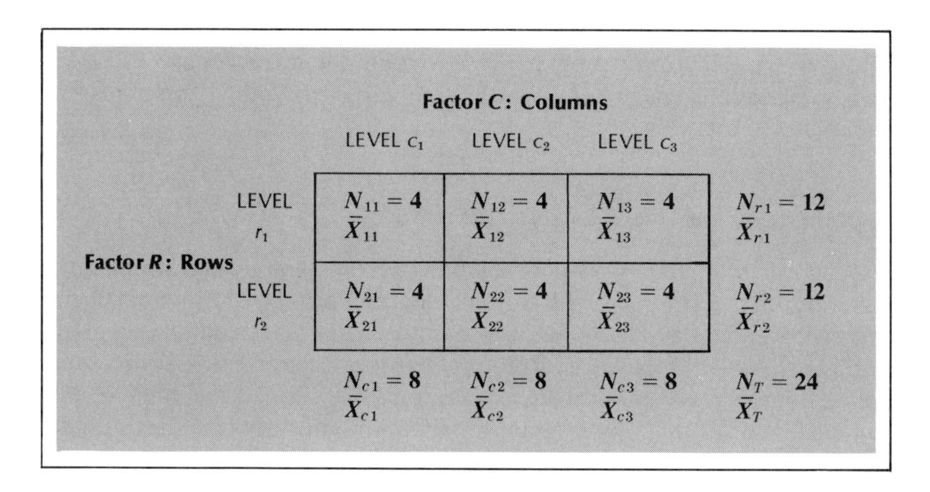

| | | **Factor C: Columns** | | | |
|---|---|---|---|---|---|
| | | LEVEL $c_1$ | LEVEL $c_2$ | LEVEL $c_3$ | |
| **Factor *R*: Rows** | LEVEL $r_1$ | $N_{11} = 4$<br>$\bar{X}_{11}$ | $N_{12} = 4$<br>$\bar{X}_{12}$ | $N_{13} = 4$<br>$\bar{X}_{13}$ | $N_{r1} = 12$<br>$\bar{X}_{r1}$ |
| | LEVEL $r_2$ | $N_{21} = 4$<br>$\bar{X}_{21}$ | $N_{22} = 4$<br>$\bar{X}_{22}$ | $N_{23} = 4$<br>$\bar{X}_{23}$ | $N_{r2} = 12$<br>$\bar{X}_{r2}$ |
| | | $N_{c1} = 8$<br>$\bar{X}_{c1}$ | $N_{c2} = 8$<br>$\bar{X}_{c2}$ | $N_{c3} = 8$<br>$\bar{X}_{c3}$ | $N_T = 24$<br>$\bar{X}_T$ |

There are two rows ($r = 2$), three columns ($c = 3$), and six cells ($k = 6$) in this table. Each cell corresponds to a treatment group, a sample from an underlying population that has received a unique treatment combination. The group in the upper right-hand corner, for example, receives the unique treatment associated with level 1 of factor *R* and level 3 of factor *C*. The mean of this treatment group may be conveniently labeled as $\bar{X}_{13}$, with the first subscript identifying the row and the second subscript the column in which the group is located. Effects of the treatment combination $r_1$ and $c_3$ will be reflected in the magnitude of the mean of this treatment group relative to the means of the other five treatment groups. Effects of an entire factor (factor *main* effects) are determined through comparisons of the marginal means. Differences between the marginal row means $\bar{X}_{r1}$ and $\bar{X}_{r2}$, then, reflect the main effect of factor *R* across all levels of factor *C*. Similarly, differences between the column marginal means $\bar{X}_{c1}$, $\bar{X}_{c2}$, and $\bar{X}_{c3}$ are analyzed to determine the main effect of factor *C*.

Notice that there are 12 measures in each row ($N_r = 12$) and 8 measures in each column ($N_c = 8$) of the table. The total number of measures is 24 ($N_T = 24$). The mean of all the measures, the total mean, is symbolized as $\bar{X}_T$. The above notation may be generalized to two-factor problems of any size.

## ASSUMPTIONS

The assumptions of two-factor *ANOVA* designs parallel those given previously for single-factor designs. For the fixed-effects model without repeated measures, the factors must be *independent* and each treatment group is assumed to be an *independent* sample that has been *randomly* drawn from the population underlying that particular treatment combination. The population data distributions which underlie each treatment group are assumed to be *normal* in shape and to have *equal variances*. As in the one-factor case, the $F$ test is robust with respect to these last two assumptions, particularly with fairly large and equal sample (treatment group) sizes.

## CONSTRUCTING TREATMENT GROUPS

Procedures followed in the construction of treatment groups are mainly a function of the classes of factors (demographic vs. experimental) being studied. Remember, also, that a researcher using the fixed-effects model will usually wish to construct treatment groups of equal size in deference to the homogeneity of variance assumption. For this reason, and because unequal sample sizes have many implications in terms of the formulas used, the results obtained, and the interpretation of outcomes, the presentation in this chapter will be restricted to the equal $N$ case.

### TWO DEMOGRAPHIC FACTORS

Problems with two demographic factors are rarely adaptable to simple *ANOVA* procedures, for if natural relationships exist between sample members and the categories of both factors, there is a strong likelihood that the factors are also related to one another, or are dependent. A study with a population of people concerning the simultaneous effects of the demographic factors of age and income level on some criterion variable, for instance, obviously violates the assumption of factor independence. Chi-square tests of independence or correlational procedures are appropriately applied to such studies.

### ONE DEMOGRAPHIC AND ONE EXPERIMENTAL FACTOR

Frequently, the researcher will find that sample members are naturally categorized on one factor in a study but may be assigned to levels of another factor. Suppose that one factor is defined as intelligence, a demographic factor, and the other factor is type of instruction given for a laboratory problem. No natural relationship exists between the subjects used in the experiment and instruction type; consequently, it is an experimental factor which the behavioral researcher may manipulate. Construction of treatment groups would proceed by drawing random samples from each intelligence level population and then randomly assigning the members of each sample to instruction type categories. In such studies the primary interest is usually in the main effects of the experimental factor; the demographic factor is included to check for possible interaction effects across demographic levels. While main effects of the experimental factors are assumed to be due to the differential influence of its levels, caution must be exercised when interpreting main effects of the demographic factor. Differences between intelligence levels, for example, may actually be caused by greater motivation to perform well among the more intelligent subjects, or some other variable that happens to be related to intelligence categories. Since the subjects could not be randomized by intelligence level, possible influences of these other variables were uncontrolled.

### TWO EXPERIMENTAL FACTORS

Only one sample is drawn from a single defined population when two experimental factors are being investigated. These sample members are assigned to the treatment groups in a completely random fashion. Any member, then, has an equal chance of being assigned to a particular treatment combination. If this behavioral researcher should decide to study the joint effects of background noise and type of instruction on task performance in the laboratory, each sample member could be randomly assigned to a treatment group corresponding to a particular level of noise and a particular type of instruction. Individual differences between the sample members would thus be completely randomized and hopefully controlled, enabling interpretation of significant treatment effects as being due to, or caused by, combinations of experimental conditions. Because levels of noise were purposely varied and all other influences were controlled through randomization, the researcher may feel reasonably safe in making the inference that a significant main effect for noise is actually *caused* by differential amounts of noise. Similarly, significant main effects for instruction are assumed to have been caused only by differences in the type of instruction given the sample members.

## PARTITIONING THE SUMS-OF-SQUARES

The sums-of-squares formulas given previously for $SS_T$ and $SS_w$ [(10.12) and (10.10)] apply equally well to one- and two-factor designs. However, $SS_b$ need not be shown in the *ANOVA* summary table or even computed (except as an accuracy check), for in two-factor designs the between sum-of-squares is *partitioned* into the variation due to the main effect of $R$ ($SS_{\text{rows}}$), the main effect of $C$ ($SS_{\text{cols}}$), and the interaction effect ($SS_{r\times c}$):

$$SS_b = SS_{\text{rows}} + SS_{\text{cols}} + SS_{r\times c} \tag{11.1}$$

Computational formulas for $SS_{\text{rows}}$ and $SS_{\text{cols}}$ are given as

$$SS_{\text{rows}} = \frac{\Sigma(\Sigma X_r)^2}{N_r} - \frac{(\Sigma X_T)^2}{N_T} \tag{11.2}$$

where $\Sigma X_r$ is the sum of the $X$ values in each row and $N_r$ is the overall sample size in each row; and

$$SS_{\text{cols}} = \frac{\Sigma(\Sigma X_c)^2}{N_c} - \frac{(\Sigma X_T)^2}{N_T} \tag{11.3}$$

where $\Sigma X_c$ is the sum of the $X$ values in each column and $N_c$ is the overall sample size in each column.

Variation due to interaction ($SS_{r \times c}$) may be found by subtraction after computing the other essential $SS$ terms:

$$SS_{r \times c} = SS_T - SS_w - SS_{\text{rows}} - SS_{\text{cols}} \tag{11.4}$$

Each of the three components of $SS_b$ may be converted to mean square values by dividing by their respective degrees of freedom. The $MS$ values so obtained are all independent estimates of population variance when the related null hypotheses are true. Thus, tests of the main effects of factors $R$, $C$, and the interaction effects can be conducted by using these $MS$ values as numerators of $F$ ratios. $MS_w$, which could be referred to as the *error* term, serves as the denominator for all these tests.

## DETERMINING THE DEGREES OF FREEDOM

The degrees of freedom associated with the total sum of squares, $df_T$, and the within sum-of-squares, $df_w$, are determined with the same formulas for one- and two-factor designs [(10.17) and (10.3)]. Degrees of freedom values associated with the between variation components of rows, columns, and interaction are obtained with the following formulas:

$$df_{\text{rows}} = r - 1 \tag{11.5}$$

$$df_{\text{cols}} = c - 1 \tag{11.6}$$

$$df_{r \times c} = (r - 1)(c - 1) \tag{11.7}$$

## TWO-FACTOR EXAMPLE

We will now consider an example of a fixed-effects *ANOVA* with two independent factors and treatment groups of equal size. Suppose that an agronomist has designed an experiment to study the effects of chemical vs. physical weed control techniques on corn production. He is particularly interested in determining if these effects differ according to various soil compositions. Review of the assumptions for a two-factor *ANOVA* convinces him that violation risks are minimal with equal sample sizes.

The descriptive portion of this study is presented in Table 11.1 Ob-

**TABLE 11.1 EXAMPLE OF A TWO-FACTOR *ANOVA* APPLICATION: DESCRIPTIVE PORTION**

FACTORS **R: Weed control technique. Levels (2): Physical and chemical**
**C: Soil type. Levels (3): Sand, clay, and loam**

POPULATION **Predominately sand, clay or loam grain-production acres in a nine-county area during the 1976 growing season**

SAMPLES **One-acre plots (24). Each sample consists of 8 plots randomly chosen from a specific soil composition population**

CRITERION VARIABLE **Yield in bushels per acre**

DECISION MODEL $H_0$: $\mu_{r1} = \mu_{r2} = \mu_{r3}$
$H_1$: **Not** $H_0$

$H_0$: $\mu_{c1} = \mu_{c2} = \mu_{c3}$
$H_1$: **Not** $H_0$

$H_0$: **Interaction effects are not present**
$H_1$: **Interaction effects are present**
$\alpha = .05$
$DR_{\text{rows}}$: **Reject** $H_0$ **if** $F_{\text{obs}} > 4.41$**; otherwise, do not reject**
$DR_{\text{cols}}$: **Reject** $H_0$ **if** $F_{\text{obs}} > 3.55$**; otherwise, do not reject**
$DR_{r \times c}$: **Reject** $H_0$ **if** $F_{\text{obs}} > 3.55$**; otherwise, do not reject**

***ANOVA* SUMMARY TABLE**

| SOURCE OF VARIATION | *SS* | *df* | *MS* | $F_{\text{obs}}$ |
|---|---|---|---|---|
| **Between** | | | | |
| **Rows (technique)** | **84.37** | **1** | **84.37** | **1.16** |
| **Columns (soils)** | **4140.75** | **2** | **2070.38** | **28.50*** |
| **Interaction** | **7.76** | **2** | **3.88** | **0.05** |
| **Within (error)** | **1307.75** | **18** | **72.65** | |
| **Total** | **5540.63** | **23** | | |

* Significant at .05 level

(Table continues)

**Table 11.1 (*Continued*)**

| | |
|---|---|
| DECISIONS | **(1) Since 1.16 $\not>$ 4.41, the main effects for rows are not significant** |
| | **(2) Since 28.50 > 3.55, the main effects for columns are significant** |
| | **(3) Since 0.05 $\not>$ 3.55, interaction effects are not significant** |
| INTERPRETATION | **(1) No significant difference in average yield was observed between the two weed control techniques for the soil types studied** |
| | **(2) Significant differences in average yield were observed by soil type across the weed control techniques. Since yield differences were expected among the populations of sand, clay, and loam soils, this finding is of little importance** |
| | **(3) Significant interaction effects were not obtained** |

served data and computations are given in Table 11.2. Factors are weed control technique (two levels) and predominent soil composition or type (three levels). Yield in average number of bushels per acre is the criterion variable. The population of soils is restricted to grain-producing acres in a nine-county area; all of which are potentially usable as sample members.

Because the population of acre plots may be categorized by predominant soil type, this factor is demographic. Weed control technique is obviously an experimental factor which can be manipulated. Construction of treatment groups is accomplished by sampling acre plots directly from the population of acres in each soil category and then randomly assigning these plots to a weed control technique. To simplify the computations, treatment groups of size 4 will be used in the example: $N = 4$, $N_T = 24$. (A larger sample would normally be recommended for this particular experiment.)

Three sets of nondirectional hypotheses are needed; one each for the main effects of factors $R$ and $C$, plus one pertaining to interaction effects. The $\alpha$ level is set at .05 for all tests to reduce the probability of Type II errors, which might be substantial with such small sample sizes.

Decision rules, symbolized as $DR_{\text{cols}}$ for factor $C$, $DR_{\text{rows}}$ for factor $R$, and $DR_{r\times c}$ for interaction, are constructed after determining the $df$ values. The $df$ value associated with the denominator of the $F$ ratios, $MS_w$, is defined as

$$df_w = k(N - 1) = 6(4 - 1) = 18$$

**TABLE 11.2 EXAMPLE OF A TWO-FACTOR *ANOVA* APPLICATION: NUMERICAL PORTION**

**OBSERVED DATA**

**Factor C: Soil Type**

| Factor R: Weed Control Techniques | | CLAY (1) | SAND (2) | LOAM (3) | ROW TOTALS |
|---|---|---|---|---|---|
| | PHYSICAL (1) | $X$: 79, 84, 101, 96<br>$\Sigma X = 360$<br>$\bar{X} = 90.0$<br>$\Sigma X^2 = 32714$ | $X$: 107, 91, 81, 103<br>$\Sigma X = 382$<br>$\bar{X} = 95.5$<br>$\Sigma X^2 = 36900$ | $X$: 120, 116, 130, 119<br>$\Sigma X = 485$<br>$\bar{X} = 121.25$<br>$\Sigma X^2 = 58917$ | $k_{r1} = 3$<br>$N_{r1} = 12$<br>$\Sigma X_{r1} = 1227$<br>$\bar{X}_{r1} = 102.25$<br>$\Sigma X_{r1}^2 = 128531$ |
| | CHEMICAL (2) | $X$: 85, 94, 95, 102<br>$\Sigma X = 376$<br>$\bar{X} = 94.0$<br>$\Sigma X^2 = 35490$ | $X$: 99, 92, 100, 111<br>$\Sigma X = 402$<br>$\bar{X} = 100.5$<br>$\Sigma X^2 = 40586$ | $X$: 124, 129, 114, 127<br>$\Sigma X = 494$<br>$\bar{X} = 123.5$<br>$\Sigma X^2 = 61142$ | $k_{r2} = 3$<br>$N_{r2} = 12$<br>$\Sigma X_{r2} = 1272$<br>$\bar{X}_{r2} = 106.00$<br>$\Sigma X_{r2}^2 = 137218$ |
| | COLUMN TOTALS | $k_{c1} = 2$<br>$N_{c1} = 8$<br>$\Sigma X_{c1} = 736$<br>$\bar{X}_{c1} = 92.00$<br>$\Sigma X_{c1}^2 = 68204$ | $k_{c2} = 2$<br>$N_{c2} = 8$<br>$\Sigma X_{c2} = 784$<br>$\bar{X}_{c2} = 98.0$<br>$\Sigma X_{c2}^2 = 77486$ | $k_{c3} = 2$<br>$N_{c3} = 8$<br>$\Sigma X_{c3} = 979$<br>$\bar{X}_{c3} = 122.40$<br>$\Sigma X_{c3}^2 = 120059$ | $k_T = 6$<br>$N_T = 24$<br>$\Sigma X_T = 2499$<br>$\bar{X}_T = 104.13$<br>$\Sigma X_T^2 = 265749$ |

$$SS_T = \Sigma X_T^2 - \frac{(\Sigma X_T)^2}{N_T}$$

$$= 265749 - \frac{(2499)^2}{24}$$

$$= 265749 - \frac{6245001}{24} = 265749 - 260208.38$$

$$= 5540.63$$

$$SS_w = \Sigma X_T^2 - \frac{\Sigma(\Sigma X_k)^2}{N}$$

$$= 265749 - \frac{(360)^2 + (382)^2 + (485)^2 + (376)^2 + (402)^2 + (494)^2}{4}$$

$$= 265749 - \frac{1057765}{4} = 265749 - 264441.25$$

$$= 1307.75$$

*(Table continues)*

**Table 11.2 (*Continued*)**

$SS_b$ COMPONENTS

$$SS_{\text{rows}} = \frac{\Sigma(\Sigma X_r)^2}{N_r} - \frac{(\Sigma X_T)^2}{N_T} = \frac{(\Sigma X_{r1})^2 + (\Sigma X_{r2})^2}{N_r} - \frac{(\Sigma X_T)^2}{N_T}$$
$$= \frac{(1227)^2 + (1272)^2}{12} - 260208.38$$
$$= \frac{3123513}{12} - 260208.38 = 260292.75 - 260208.38$$
$$= 84.37$$

$$SS_{\text{cols}} = \frac{\Sigma(\Sigma X_c)^2}{N_c} - \frac{(\Sigma X_T)^2}{N_T} = \frac{(\Sigma X_{c1})^2 + (\Sigma X_{c2})^2 + (\Sigma X_{c3})^2}{N_c} - \frac{(\Sigma X_T)^2}{N_T}$$
$$= \frac{(736)^2 + (784)^2 + (979)^2}{8} - 260208.38$$
$$= \frac{2114793}{8} - 260208.38 = 264349.13 - 260208.38$$
$$= 4140.75$$

$$SS_{r\times c} = SS_T - SS_w - SS_{\text{rows}} - SS_{\text{cols}}$$
$$= 5540.63 - 1307.75 - 84.37 - 4140.75$$
$$= 7.76$$

Degrees of freedom for the numerators of the three $F$ ratios are

$$df_{\text{rows}} = r - 1 = 2 - 1 = 1$$
$$df_{\text{cols}} = c - 1 = 3 - 1 = 2$$
$$df_{r\times c} = (r - 1)(c - 1) = (1)(2) = 2$$

These sum to the total degrees of freedom, defined as

$$df_T = N_T - 1 = 24 - 1 = 23$$

Critical values of $F$ given in the decision rules for each test are found in Appendix E using these $df$ values and the chosen $\alpha$ level of .05.

After data collection, $\Sigma X$ and $\Sigma X^2$ values are computed for each treatment group and then totaled across rows, columns, and over the entire table to facilitate use of the computational formulas for the sums of squares (Table 11.2). The cell, column, row, and total mean values have been presented for illustrative purposes only. The obtained sums of squares and $df$ terms are entered into the *ANOVA* summary table. The $F$ values shown were found by dividing the $MS$ corresponding to each effect by $MS_w$. Only the null hypothesis associated with soil type can be rejected at the .05 level. The results of the experiment thus provide no indication that average yield in the population of soil differs for chemical vs. physical weed control techniques. Similarly, differential influences of treatment *combinations* (*interaction effects*) were not apparent.

## *POST HOC* TESTS

The Scheffé *post hoc* procedures can be used to make comparisons among the marginal (level) means of significant factors in two-factor problems. In the above example, the agronomist might wish to determine if the marginal mean associated with loam soils ($\bar{X}_{c3}$) differs significantly from the clay and sand marginals ($\bar{X}_{c1}$ and $\bar{X}_{c2}$). This would be accomplished by treating the column marginal data for factor $C$ as an isolated $1 \times 3$ single-factor *ANOVA* problem. Terms in the formula for the smallest significant difference (10.19) would include $N = N_c = 8$, $k = k_c = 3$, $MS_w$ as given in Table 11.1, and a critical value of $F$ based on $k_c - 1 = 2$ and 18 degrees of freedom.

## TWO-FACTOR EXAMPLE WITH INTERACTION

To illustrate interaction, consider a psychological study of the simultaneous effects of strategy and reinforcement (the factors) on achievement with regard to a complex prescribed task (the criterion). The reinforcement factor has three levels: punishment, reward, and no reinforcement. The strategy factor also has three levels, each of which corresponds to a specific strategy which a sample member is to follow while working on the task. The example is simplified by showing only two individuals per treatment group. The $\alpha$ level is set at .01. Since populations do not exist for either set of levels (both factors are experimental), a single sample is drawn and members are randomly assigned to the nine treatment groups.

The numerical portion only of this study is presented in Table 11.3. Except for the computation of treatment group means, calculations have been kept to the minimum required for application of the computational formulas to these data.

All the $F$ tests in the analysis are found to be significant at the .01 level. Treatment effects are in evidence for both factors and interaction is present. Interpretation is no longer straightforward, for the results imply that a complex relationship exists among the factors and the criterion variable. The effects of interaction can best be depicted by constructing a simple graph of the treatment group means. The first graph (a) in Figure 11.1 has been constructed from the data in Table 11.3 by placing points corresponding to levels of factor $C$ on the abscissa and a scale for criterion variable measures on the ordinate. The treatment group means for each level of factor $R$ are now plotted separately on this graph and connected by lines. It usually does not matter which factor is listed along the abscissa; interpretation of the graph is basically the same in either case. Notice that these profiles of means by level of factor $R$ are nonparallel. Whenever interaction is *significant,* such profiles will not be parallel. The graph shows that almost all of this nonparallelism is due to level 1 of factor $C$; punishment appears to have a completely different effect on performance across the three strategies than does reward or no reinforcement. Although the kind of reinforcement makes a difference, the effects are not the same for the various strategies.

Now look at graph (b), which was constructed from the data in Table

**TABLE 11.3 TWO-FACTOR *ANOVA* WITH SIGNIFICANT INTERACTION**

**OBSERVED DATA**

**Factor *C*: Reinforcement**

| | | LEVEL 1 PUNISHMENT | LEVEL 2 REWARD | LEVEL 3 NONE | |
|---|---|---|---|---|---|
| **Factor *R*: Strategy** | LEVEL 1 STRATEGY $S_1$ | $X$: 8, 8<br>$\Sigma X = 16$<br>$\bar{X} = 8$ | $X$: 9, 7<br>$\Sigma X = 16$<br>$\bar{X} = 8$ | $X$: 6, 6<br>$\Sigma X = 12$<br>$\bar{X} = 6$ | $N_{r1} = 6$<br>$\Sigma X_{r1} = 44$ |
| | LEVEL 2 STRATEGY $S_2$ | $X$: 1, 3<br>$\Sigma X = 4$<br>$\bar{X} = 2$ | $X$: 7, 7<br>$\Sigma X = 14$<br>$\bar{X} = 7$ | $X$: 4, 6<br>$\Sigma X = 10$<br>$\bar{X} = 5$ | $N_{r2} = 6$<br>$\Sigma X_{r2} = 28$ |
| | LEVEL 3 STRATEGY $S_3$ | $X$: 9, 9<br>$\Sigma X = 18$<br>$\bar{X} = 9$ | $X$: 4, 6<br>$\Sigma X = 10$<br>$\bar{X} = 5$ | $X$: 2, 2<br>$\Sigma X = 4$<br>$\bar{X} = 2$ | $N_{r3} = 6$<br>$\Sigma X_{r3} = 32$ |
| | | $N_{c1} = 6$<br>$\Sigma X_{c1} = 38$ | $N_{c2} = 6$<br>$\Sigma X_{c2} = 40$ | $N_{c3} = 6$<br>$\Sigma X_{c3} = 26$ | $N_T = 18$<br>$\Sigma X_T = 104$<br>$\Sigma X_T^2 = 712$ |

$$SS_T = \Sigma X_T^2 - \frac{(\Sigma X_T)^2}{N_T}$$

$$= 712 - \frac{(104)^2}{18} = 712 - 600.89$$

$$= 111.11$$

$$SS_w = \Sigma X_T^2 - \frac{\Sigma(\Sigma X_k)^2}{N}$$

$$= 712 - \frac{(16)^2 + (16)^2 + \cdots + (4)^2}{2} = 712 - \frac{1408}{2} = 712 - 704$$

$$= 8.0$$

(*Table continues*)

**Table 11.3 (*Continued*)**

$SS_b$ **COMPONENTS**

$$SS_{\text{rows}} = \frac{\Sigma(\Sigma X_r)^2}{N_r} - \frac{(\Sigma X_T)^2}{N_T}$$
$$= \frac{(44)^2 + (28)^2 + (32)^2}{6} - 600.89 = \frac{3744}{6} - 600.89$$
$$= 624 - 600.89$$
$$= 23.11$$

$$SS_{\text{cols}} = \frac{\Sigma(\Sigma X_c)^2}{N_c} - \frac{(\Sigma X_T)^2}{N_T}$$
$$= \frac{(38)^2 + (40)^2 + (26)^2}{6} - 600.89 = \frac{3720}{6} - 600.89$$
$$= 620.0 - 600.89$$
$$= 19.11$$

$$SS_{r \times c} = SS_T - SS_w - SS_{\text{rows}} - SS_{\text{cols}}$$
$$= 111.11 - 8.0 - 23.11 - 19.11$$
$$= 60.89$$

***ANOVA* SUMMARY TABLE**

| SOURCE OF VARIATION | *SS* | *df* | *MS* | *F* |
|---|---|---|---|---|
| **Between** | | | | |
| **Rows (strategy)** | **23.11** | **2** | **11.56** | **13.01*** |
| **Columns (reinforcement)** | **19.11** | **2** | **9.56** | **10.76*** |
| **Interaction** | **60.89** | **4** | **15.22** | **17.12*** |
| **Within (error)** | **8.0** | **9** | **.89** | |
| **Total** | **111.11** | **17** | | |

* Significant at .01 level.

11.2. This time factor $R$ was arbitrarily placed on the abscissa. The profiles, plotted from data in which interaction was nonsignificant, tend to be parallel (within sampling error, since factor $R$ was also nonsignificant). Factor $C$, soil type, can be clearly interpreted in this situation.

The presence of interaction must be acknowledged in the interpretation of factor main effects. With respect to the data in Table 11.3 once again, it is not enough to state that the type of reinforcement received makes a difference, for the effects of reinforcement type vary according to the particular strategy that is being followed. Similarly, it must be

**FIG. 11.1 GRAPHIC PRESENTATIONS OF SIGNIFICANT (a) AND NONSIGNIFICANT (b) INTERACTION EFFECTS.**

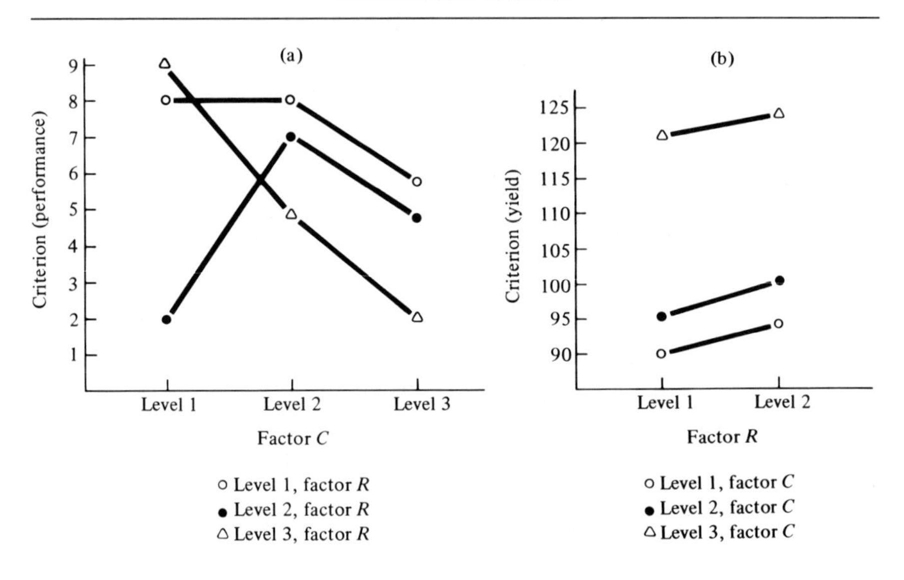


acknowledged that, while the type of strategy followed makes a difference, the effects vary according to the type of reinforcement received. Such weak interpretations may be of little help to an applied researcher looking for decision-making evidence. More specific and precise interpretations, however, are not justified with the analyses presented. For instance, it cannot be stated that strategy $S_2$ produces significantly poorer performance than strategies $S_1$ or $S_3$ when punishment is used as reinforcement, even though this possible relationship could be the most important single finding in the entire study. Although methods of conducting such *post hoc* comparisons, called tests of *simple effects* and *individual comparisons,* are available and should be applied when interaction is significant in multiple-factor *ANOVA* studies, presentation of these techniques is beyond the scope of this book. The interested student should consult an advanced experimental design text (see, in particular, Winer, 1971).

Before concluding this discussion of interaction effects, the student should be cautioned that not all significant interactions are really what they seem. Interaction is rather closely tied to the scale of measurement used for the criterion variable. A logarithmic, square root, or other type of transformation on the original criterion measures may serve to reduce or even eliminate interaction effects in some instances, thus simplifying interpretation of main effects and perhaps modifying decisions concerning the data.

## USE OF COMPUTER PROGRAMS

Widespread availability of computer programs for analyzing data with *ANOVA* techniques is somewhat of a mixed blessing. On the positive side, a computer can work problems in seconds which might take days to solve with hand or desk calculators. On the negative side, it has become increasingly easy for the untrained student or researcher to *misapply* techniques and *misinterpret* results provided on printouts. Before using any program, the researcher should first write out descriptions of the factors, populations, and samples and construct the entire decision model, just as if the problem were to be computed by hand. It must be kept in mind that the computer will indiscriminately analyze and make probability statements for any set of data which is fed into it, without regard to the appropriateness of the program. If, after reading a program description, there is any question concerning its applicability to a set of observed data, the advice of a statistical consultant should be obtained.

## EXERCISES

(*Note:* Small $N$'s have been used to simplify calculations. Larger samples would ordinarily be used in actual studies.)

1. a. Compute $SS_T$, $SS_w$, $SS_{\text{rows}}$, $SS_{\text{cols}}$, $SS_{r \times c}$, and the associated *df* values for this simple set of observed data:

| | | Factor C | |
|---|---|---|---|
| | | 1 | 2 |
| Factor R | 1 | X: 1, 3 | X: 5, 6 |
| | 2 | X: 4, 4 | X: 9, 7 |

   b. Compute $MS$ terms associated with all the above sources of variation except $SS_T$.
   c. Compute $F$ terms for the *between* components of variance.
   d. Which, if any, of the $F$ values are significant at the .01 level?
   e. What assumptions are being made in this two-factor *ANOVA*?
2. Random samples of beams cut from two varieties of wood and subjected to one of three seasoning processes are being evaluated for load-bearing strength by a lumber company. Observed data (coded) are given on the following page:

| | | Factor C: Seasoning Process | | |
|---|---|---|---|---|
| | | LEVEL 1 (TREATMENT 1) | LEVEL 2 (TREATMENT 2) | LEVEL 3 (TREATMENT 3) |
| Factor *R*: Wood Variety | LEVEL 1 | *X*: 3, 5, 4, 4, 3 | *X*: 6, 4, 5, 5, 6 | *X*: 8, 8, 9, 9, 7 |
| | LEVEL 2 | *X*: 2, 3, 3, 4, 4 | *X*: 4, 5, 4, 6, 4 | *X*: 7, 6, 8, 6, 7 |

a. Which factor is demographic and which is experimental?
b. How would the treatment groups be constructed for this study?
c. Formally outline this experiment, using the examples in this chapter as a guide. Use an $\alpha$ level of .001 in the decision model.
d. Complete the *ANOVA* summary table and interpret the results.
e. Graph the interaction effects (whether or not they are significant).
f. If *post hoc* tests are justified, explain which sets of means should be compared.

3. A student comes to you for help in applying the *ANOVA* techniques presented in this chapter to a problem in the behavioral sciences with the following factors and criterion data:

   Factor *R*: Intelligence (4 levels)
   Factor *C*: Age at first marriage (3 levels)
   Criterion variable: Scores on a marriage adjustment inventory

   On the basis of this information alone, what is the primary reason why these techniques are not applicable?
4. The average continuous-usage life in hours of three brands of batteries is compared in four models of hand calculators. The following data are obtained:

| | | Factor C: Hand Calculator Model | | | |
|---|---|---|---|---|---|
| | | 1 | 2 | 3 | 4 |
| Factor *R*: Battery Brand | 1 | *X*: 2, 2 | *X*: 1, 3 | *X*: 4, 4 | *X*: 3, 1 |
| | 2 | *X*: 9, 11 | *X*: 8, 7 | *X*: 13, 14 | *X*: 7, 5 |
| | 3 | *X*: 5, 2 | *X*: 10, 12 | *X*: 8, 6 | *X*: 3, 4 |

a. Formally outline this study, using an $\alpha$ level of .05 in the decision model. Notice that random samples of both batteries and calculators must be drawn.
b. Complete the *ANOVA* summary table and interpret the results.
c. Graph the interaction effects.

5. A psychologist is investigating the joint effects of letter size and letter color on readability. A random sample of 27 students is drawn from a large university and randomly assigned to one of the treatment groups. The following data are obtained:

| | | Factor $C$: Size | | |
|---|---|---|---|---|
| | | 1 | 2 | 3 |
| | 1 | $X$: 35, 40, 36 | $X$: 40, 46, 59 | $X$: 41, 48, 40 |
| Factor $R$: Color | 2 | $X$: 40, 41, 45 | $X$: 65, 56, 51 | $X$: 45, 44, 52 |
| | 3 | $X$: 38, 33, 35 | $X$: 44, 48, 57 | $X$: 45, 39, 37 |

a. Are the factors demographic, experimental, or a combination of types?
b. Formally outline this experiment, using an $\alpha$ level of .01 in the decision model.
c. Complete the *ANOVA* table and interpret the results.
d. Graph the interaction effects.

# III

# CORRELATION AND REGRESSION

# 12

# THE PEARSON COEFFICIENT, *r*

In this chapter we shall be concerned with a measure of relationship, the Pearson correlation coefficient, $r$. In dealing with correlation we have two variables, usually labeled $X$ and $Y$. Then we have a measure of each variable on a group of individuals. For example, we could take the age of a group of children ($X$) and correlate these ages with the height of the children ($Y$). Another common example is the correlation of scores on a test of mental ability ($X$) with grades obtained in a course or semester grades ($Y$). The amount of wheat produced ($X$) could be related to factors such as temperature or amount of rainfall ($Y$). There is no limit to the number of variables that can be correlated in the social, biological, and physical sciences. Note that a correlation coefficient is merely a measure of relationship and may or may not show that one variable is the cause of the other.

Distributions used in correlation work are referred to as being *bivariate* distributions, an example of which is shown in Table 12.1. In this table are given the scores of 10 students on a measure of intelligence ($X$) and on a test of arithmetic ($Y$).

In Figure 12.1 these scores have been plotted into a diagram that is referred to as a *scatterplot*. Note that the $Y$ scores are on the $y$ axis and the $X$ scores on the $x$ axis. On the scatterplot each dot represents the two scores of one individual. The scores of the first student are plotted by going out to 120 on the $x$ axis and up to 70 on the $y$ axis. In this fashion all 10 students are located on the scatterplot.

The value of the Pearson $r$ varies from plus one through zero to minus

**TABLE 12.1 SCORES OF 10 STUDENTS ON 2 TESTS**

| STUDENT | X MENTAL ABILITY | Y ARITHMETIC TEST |
|---|---|---|
| 1 | 120 | 70 |
| 2 | 115 | 60 |
| 3 | 110 | 65 |
| 4 | 107 | 55 |
| 5 | 100 | 50 |
| 6 | 98 | 48 |
| 7 | 95 | 45 |
| 8 | 90 | 55 |
| 9 | 88 | 40 |
| 10 | 85 | 45 |

one. The sign has nothing to do with the size of the coefficient, for a correlation of +.88 is equal in the amount of relationship to a coefficient of −.88. The positive relationship indicates that as the scores increase on the $X$ variable, they tend to increase on the $Y$ variable. The minus sign indicates an inverse relationship; individuals high on $X$ tend to be low on $Y$ and vice versa. Figures 12.2, 12.3, 12.4, 12.5, and 12.6 show the scatterplots of correlation coefficients of various sizes. These should be examined carefully. In Figure 12.2 we have an example of a perfect positive relationship, $r$ being +1.00. Note that, when the plotted points are connected, they fall on a straight line. This line is called a *regression line* and will be discussed in detail in the next chapter. Figure 12.3 shows

**FIG. 12.1 SCATTERPLOT OF THE SCORES IN TABLE 12.1.**

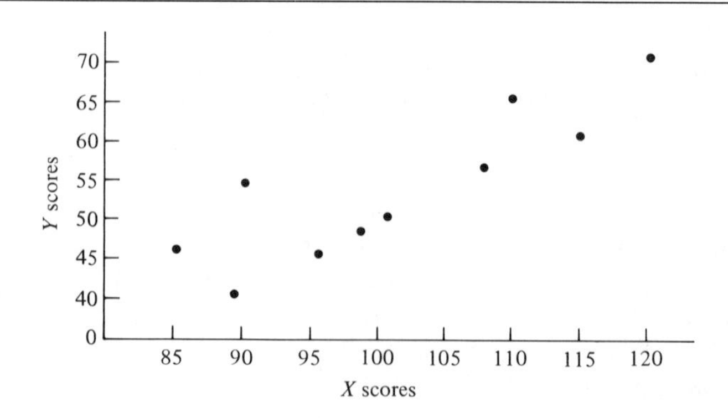

FIG. 12.2 A PERFECT POSITIVE CORRELATION, $r = 1.00$.

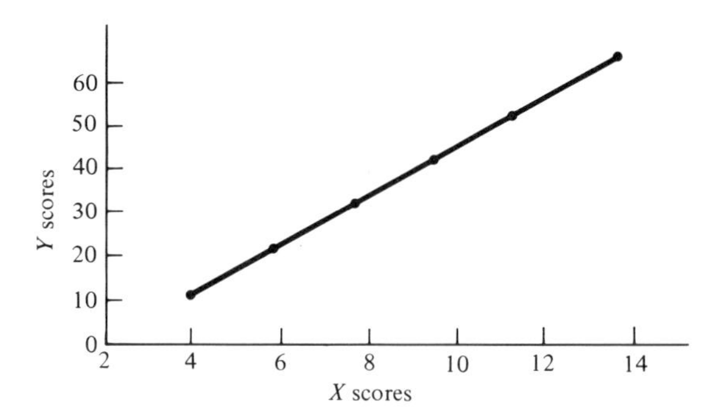


FIG. 12.3 A PERFECT NEGATIVE CORRELATION, $r = -1.00$.

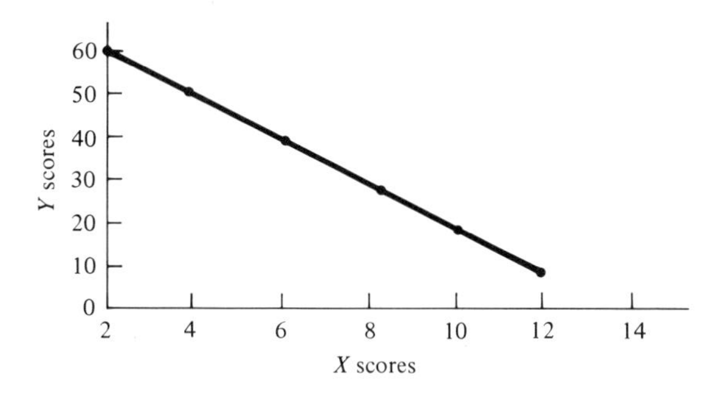


FIG. 12.4 A MODERATELY HIGH POSITIVE CORRELATION.

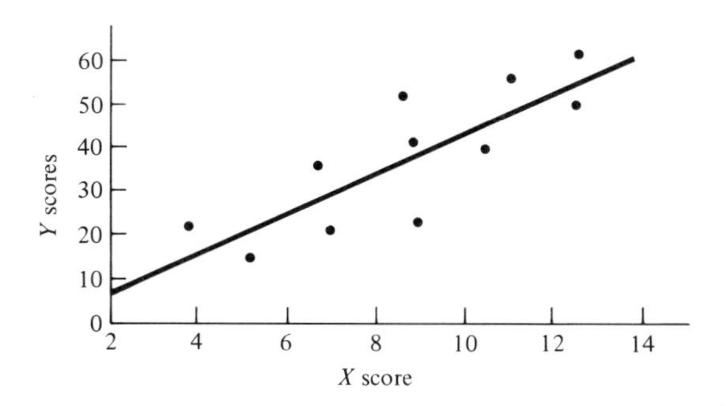

FIG. 12.5 A CORRELATION OF ABOUT .00.

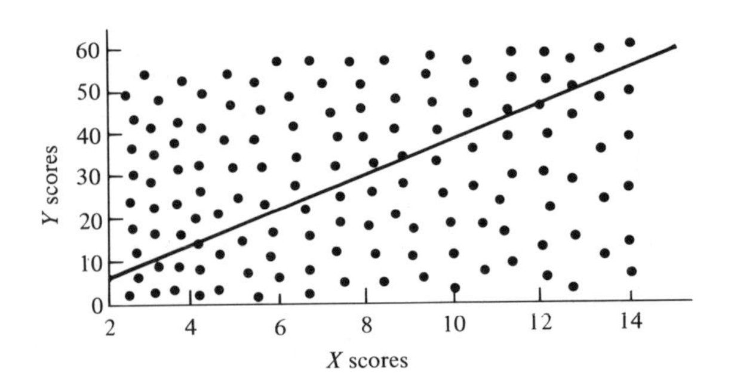


FIG. 12.6 A CURVILINEAR RELATIONSHIP.

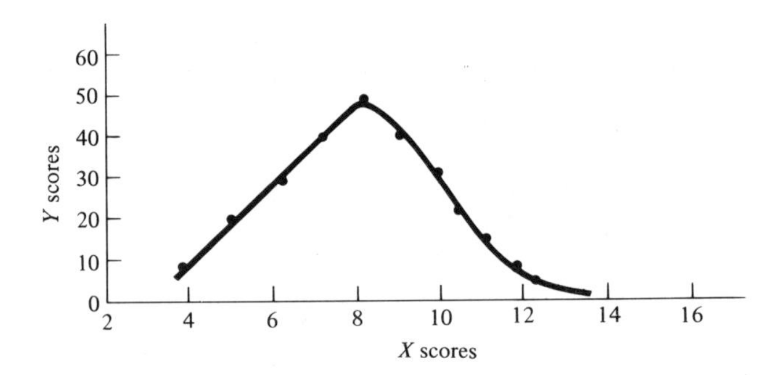


an example of a perfect negative relationship, $r$ being $-1.00$. The difference between these two lines is in the slope of the line. In Figure 12.2 we have a positive slope, the regression line going from the upper right to the lower left and in Figure 12.3 the reverse of this prevails and we have a negative slope. Figure 12.4 is an illustration of a high positive correlation with the points not on the line, but scattered closely about it. Figure 12.5 shows the situation when there is no relationship between $X$ and $Y$. Since the tallies are scattered all over the surface of the scatterplot, we see that for any value of $X$, $Y$ can take on any value, indicating no relationship between the two variables. In Figures 12.1 through 12.4 we have what is known as a linear relationship between the two variables; that is, the tallies fall or tend to fall along a straight line. Figure 12.6 illustrates a curvilinear relationship. In this case as scores on one variable increase, scores on the second variable also increase up to a point. Then the pattern reverses and as scores on the first variable continue to increase, scores on the second begin and continue to decrease. This type of relationship is not uncommon.

## CONDITIONS NECESSARY FOR THE USE OF THE PEARSON *r*

First, it is imperative that there be a *linear* relationship between $X$ and $Y$ before $r$ is used as a measure of relationship. If a set of data depart from linearity, the $r$ is lowered in size when $r$ is used in such cases. Hence, if we assume linearity and go ahead and calculate $r$, we may be making a gross error. We can always make a scatterplot to get the overall pattern of relationship between the two variables before going ahead and calculating $r$. A second condition is that the data possess *homoscedasticity*; that is, there is equal dispersion of the points about the regression line for all values of $X$. The presence and lack of homoscedasticity are illustrated in Figures 12.7 and 12.8. In Figure 12.7 notice that, for any value

FIG. 12.7 DATA THAT POSSESS HOMOSCEDASTICITY.

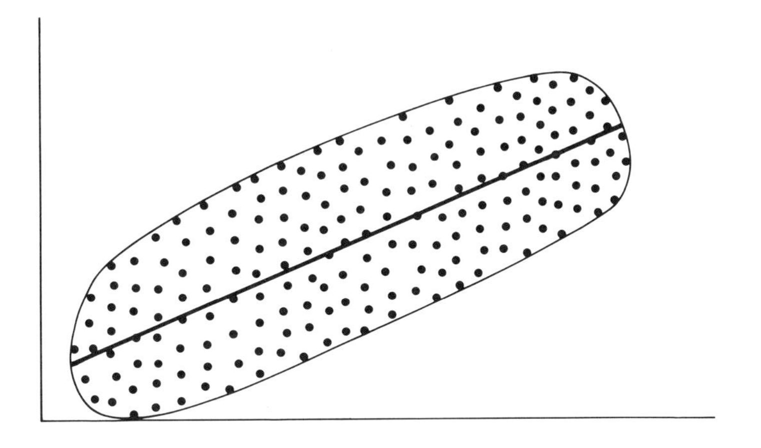

FIG. 12.8 DATA LACKING IN HOMOSCEDASTICITY.

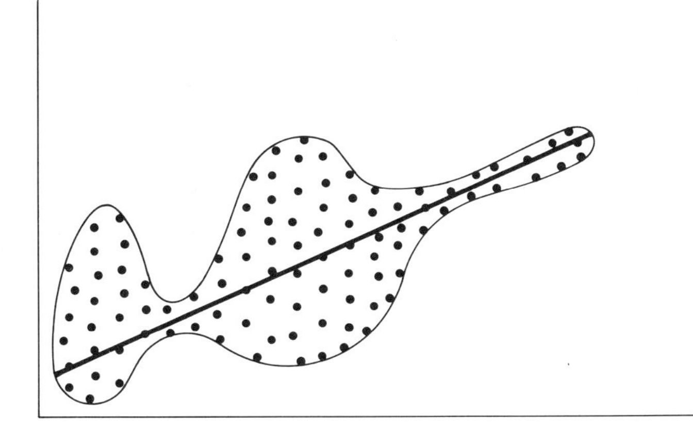

of $X$, there is a similar dispersion of the tallies about the regression line. This means that for the various values of $X$ the variances of the tallies about the regression line are approximately equal. In Figure 12.8 the variability of the tallies about the regression line changes for different values of $X$. Such data do not possess homoscedasticity.

## CALCULATION OF THE PEARSON r

The Pearson $r$ was originally defined as the mean $z$-score product:

$$r = \frac{\Sigma z_x z_y}{N} \tag{12.1}$$

This is not a practical formula to use, as is also the case with other formulas and methods that have been developed since Pearson initiated correlation work late in the nineteenth century. By a series of algebraic manipulations, formula (12.1) can be changed to the following, which is the one best adapted for use with the pocket calculator.

$$r = \frac{N\Sigma XY - (\Sigma X)(\Sigma Y)}{\sqrt{[N\Sigma X^2 - (\Sigma X)^2][N\Sigma Y^2 - (\Sigma Y)^2]}} \tag{12.2}$$

The use of this formula will be illustrated using the following model.

| (1) | (2) | (3) | (4) | (5) |
|---|---|---|---|---|
| X | Y | $X^2$ | $Y^2$ | XY |
| 20 | 12 | 400 | 144 | 240 |
| 15 | 6 | 225 | 36 | 90 |
| 10 | 8 | 100 | 64 | 80 |
| 8 | 4 | 64 | 16 | 32 |
| 5 | 0 | 25 | 0 | 0 |
| Σ = 58 | 30 | 814 | 260 | 442 |

Each $X$ value in column (1) is squared, the square placed in column (3), and both columns summed. This may be all done on your calculator and it is not necessary to show all the work. Then each $Y$ value in column (2) is squared, the squares placed in column (4), and both of these columns summed. Finally the cross-products, column (5), are obtained by multiplying each $X$ value in column (1) by its corresponding $Y$ value in column (2). These products are then summed. The five sums so obtained are then entered into formula (12.2). In correlational work $N$ is always the number of pairs. Then

$$r = \frac{(5)(442) - (58)(30)}{\sqrt{[(5)(814) - (58)^2][(5)(260) - (30)^2]}}$$

$$= \frac{2210 - 1740}{\sqrt{(4070 - 3364)(1300 - 900)}}$$

$$= \frac{470}{\sqrt{(706)(400)}}$$

$$= \frac{470}{\sqrt{282400}}$$

$$= \frac{470}{531.4}$$

$$= .88$$

The above example based on an $N$ of 5 is merely used as an illustration. Actually, the Pearson $r$ is a large sample statistic. Also, in many cases the values of the variables being correlated are much larger than those shown in the example and manipulating them may often tax the capacity of the calculator. In cases like this the solution is to code the data as shown in Table 12.2. In this table coding is done by subtracting a constant from each score in each distribution. Notice that the same constant does not have to be used for both variables. In this case the $X$ scores are reduced by 80 and the $Y$ scores by 40. Since the relative position of the scores are not changed by this type of coding, the value of $r$ is not affected. Even coding will sometimes not solve the problem of overtaxing the calculator. When $N$ is large, the student may have to break his prob-

**TABLE 12.2 FINDING THE PEARSON *r* BY CODING DATA**

| $X$ | $Y$ | $X - 80$ | $Y - 40$ | $X^2$ | $Y^2$ | $XY$ |
|---|---|---|---|---|---|---|
| 120 | 70 | 40 | 30 | 1600 | 900 | 1200 |
| 115 | 60 | 35 | 20 | 1225 | 400 | 700 |
| 110 | 65 | 30 | 25 | 900 | 625 | 750 |
| 107 | 55 | 27 | 15 | 729 | 225 | 405 |
| 100 | 50 | 20 | 10 | 400 | 100 | 200 |
| 98 | 48 | 18 | 8 | 324 | 64 | 144 |
| 95 | 45 | 15 | 5 | 225 | 25 | 75 |
| 90 | 55 | 10 | 15 | 100 | 225 | 150 |
| 88 | 40 | 8 | 0 | 64 | 0 | 0 |
| 85 | 45 | 5 | 5 | 25 | 25 | 25 |
| | | $\Sigma = 208$ | $\Sigma = 133$ | $\Sigma = 5592$ | $\Sigma = 2589$ | $\Sigma = 3649$ |

lem into parts, get the five required sums for each part, then add these five sums from the different parts before going into the formula.

Using the sums from Table 12.2 and substituting into the formula we have

$$
\begin{aligned}
r &= \frac{(10)(3649) - (208)(133)}{\sqrt{[(10)(5592) - (208)^2][(10)(2589) - (133)^2]}} \\
&= \frac{(36490) - (27664)}{\sqrt{[(55920) - (43264)][(25890) - 17689)]}} \\
&= \frac{8826}{\sqrt{(12656)(8201)}} \\
&= \frac{8826}{\sqrt{103791856}} \\
&= \frac{8826}{10187} \\
&= .866
\end{aligned}
$$

Values can be made even smaller by taking an approximation of each mean and using this as the constant to be subtracted. However, this would lead to some negative values that might complicate the procedure.

When we compute $r$, we usually obtain both means and both standard deviations from the data already at hand. This is done as follows with the data that have been coded by subtracting out the constants $C_x$ and $C_y$.

$$
\begin{aligned}
\bar{X} &= \frac{\Sigma X}{N} + C_x & \bar{Y} &= \frac{\Sigma Y}{N} + C_y \\
&= \frac{208}{10} + 80 & &= \frac{133}{10} + 40 \\
&= 20.8 + 80 & &= 13.3 + 40 \\
&= 100.8 & &= 53.3
\end{aligned}
$$

The easiest way to obtain the sample standard deviations is by using formula (4.3):

$$S_x = \frac{1}{N}\sqrt{N\Sigma X^2 - (\Sigma X)^2}$$

Note that we have already computed the values under the square root sign, this being the first term in the denominator of the Pearson $r$ formula. So we take the value of this term after it has been reduced to one number, obtain its square root, and multiply it by the reciprocal of $N$. Then

$$
\begin{aligned}
S_x &= \frac{1}{10}\sqrt{12656} \\
&= .1(112) \\
&= 11.2
\end{aligned}
$$

Similarly we obtain $S_y$:

$$S_y = \frac{1}{10}\sqrt{8201}$$
$$= .1(90.56)$$
$$= 9.1$$

Note that the correlation between $X$ and $Y$ is the same as the correlation between $Y$ and $X$. That is, $r_{xy} = r_{yx}$.

## THE PEARSON r AND THE RANGE

The size of a Pearson $r$ is directly related to the range of the two variables being correlated, the greater the variability in each variable, the higher the $r$. Or, stated in other words, the more homogeneous the subjects with respect to the traits being correlated, the lower the correlation coefficient. This is illustrated in Figure 12.9, where it is apparent that there is a high correlation between chronological age and height for these children. Consider the part of the scatterplot cut off by drawing lines to separate the area included between ages 6 through 8 from the rest of the scatterplot. By doing this we restrict the range to two years. The tallies in this part of the scatterplot, actually a small scatterplot within the larger one, are randomly scattered all over it. As pointed out earlier, such a condition leads to a correlation coefficient of around .00. Hence, by drastically

FIG. 12.9 SCATTERPLOT OF CHRONOLOGICAL AGE AND HEIGHT FOR A GROUP OF CHILDREN.

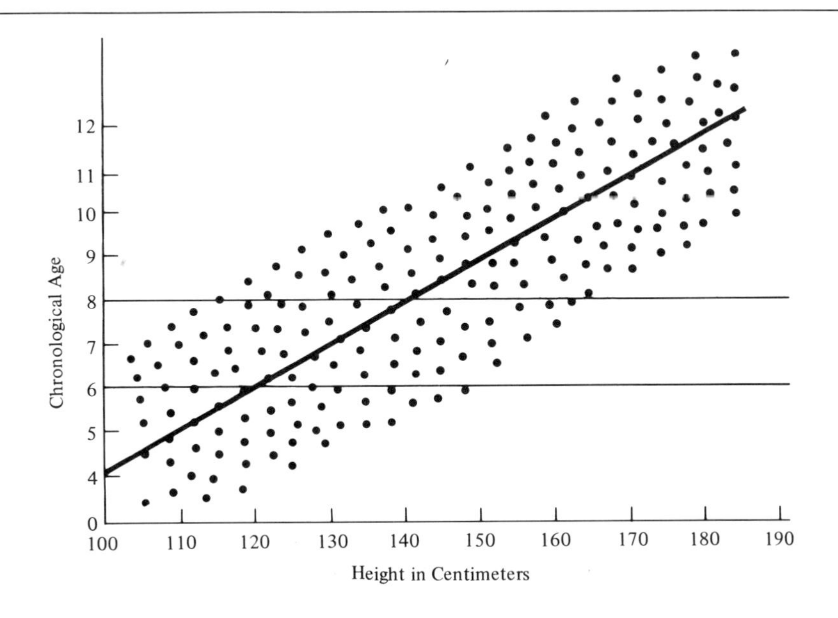

curtailing the range, we no longer have any relationship between chronological age and height. It follows, then, that when examining or interpreting a Pearson $r$, the variability of the two variables must be considered. The importance of computing the standard deviation of each variable should be apparent. Such restrictions are quite common, being brought about by age, mental ability, socioeconomic status, and the like.

## INTERPRETATION AND USE OF THE PEARSON *r*

It should be stressed, although our examples do not show it, that the Pearson $r$ is a large sample statistic. It will be shown later in this chapter that $r$'s computed on small samples are apt to be unreliable. It is almost impossible with small samples to demonstrate the presence of the conditions necessary for the use of the Pearson $r$. It is also true that one or two extreme cases in a small sample may greatly inflate the size of a correlation coefficient.

The Pearson $r$ is interpreted in a relative sort of a way. We talk about a high positive correlation, .80 and above, a moderately high positive correlation, .50 to .70, and so on. Pearson $r$'s are frequently employed in test construction where a group of individuals is given a test and later the same individuals take the same test or a similar one. The two sets of scores obtained are correlated and the resulting $r$ is used as a measure of the reliability (stability) of the test. With well-constructed intelligence and achievement tests, an $r$ of .90 or higher is usually the case. Reliability in the sense used here means consistency, the individuals scoring in about the same position on both administrations of the test.

Scores on a test such as the Scholastic Aptitude Test (SAT) of the College Examination Board (CEEB) are used to predict the grades of college freshmen. Scores on the test, the predictor, are correlated with freshmen grade-point averages, the criterion, and the resulting coefficient is referred to as a validity coefficient. Typically these tend to be about .50 or less, sometimes very much less. In this case .50 would be considered high. Prediction and problems related to it will be discussed in the next chapter.

## TESTING THE SIGNIFICANCE OF THE PEARSON *r*

The statistical significance of the Pearson $r$ may be tested by using the following formula:

$$t = \frac{r\sqrt{N-2}}{\sqrt{1-r^2}} \tag{12.3}$$

We shall take a case where $r = .48$ and $N = 20$. Substituting in formula (12.3) we have

$$t = \frac{.48\sqrt{20 - 2}}{\sqrt{1 - .48^2}}$$

$$= \frac{.48\sqrt{18}}{\sqrt{1 - .2304}}$$

$$= \frac{.48(4.243)}{\sqrt{.7696}}$$

$$= \frac{2.03664}{.877}$$

$$= 2.32$$

In this case we are testing the null hypothesis that this $r$ does not differ significantly from zero. Entering Appendix F with $N - 2$ or 18 degrees of freedom, we find that this $t$ of 2.32 is significant at the 5 percent level.

The above computational work can be avoided by using Appendix G, entering it directly with the number of degrees of freedom. From the table we find that for $r$ to be significant at the 5 percent level for 18 degrees of freedom, it must be equal to or exceed the table value of .4438. To be significant at the 1 percent level, it must be equal to or greater than .5614. Since our $r$ of .48 is greater than the .05 value and less than the .01 value, we can reject the null hypothesis at the 5 percent level. This appendix is very useful when we wish to test the significance of a large number of Pearson $r$'s.

## TESTING THE SIGNIFICANCE OF THE DIFFERENCE BETWEEN TWO PEARSON $r$'s

Suppose that we have two Pearson $r$'s between the same variables (1 and 2) based on sets of *independent* samples, as follows:

| Sample I | Sample II |
|---|---|
| $r_{12} = .55$ | $r_{12} = .45$ |
| $N_1 = 77$ | $N_2 = 86$ |

We are going to test the null hypothesis that there is no statistically significant difference between these two $r$'s or, in other words, these two $r$'s are based on samples from the same population. Before we do this, we change each $r$ to Fisher's $Z$ statistic and make the test of significance between the two $Z$'s. Whatever conclusion is made about the two $Z$'s is then inferred to be true about the two $r$'s. The transformation of $r$'s to $Z$'s is necessary because the sampling distribution of $r$ is not normal except when the parameter values are near zero. The further $r$ gets from zero, the greater the skew in the sampling distribution. This is shown in Figure 12.10. Fisher's $Z$ has the advantage that its sampling distribution is

FIG. 12.10 THE SAMPLING DISTRIBUTIONS OF THE CORRELATION COEFFICIENT WHEN THE PARAMETER VALUES ARE −.95, .00, AND +.94 (LARGE SAMPLES).

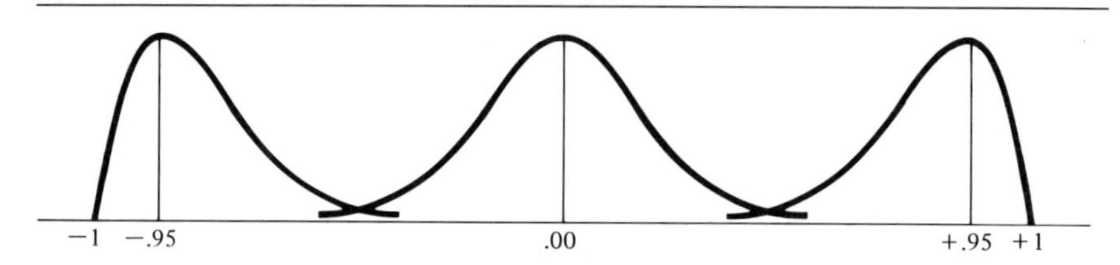


normal for all values of $Z$. The transformation from $r$ to $Z$ is accomplished by use of Appendix H. For these data:

| I | II |
|---|---|
| $r_{12} = .55$ | $r_{12} = .45$ |
| $Z_1 = .618$ | $Z_2 = .485$ |

For large samples, the formula used in testing the difference between two Fisher's $Z$'s is

$$z = \frac{Z_1 - Z_2}{\sqrt{\dfrac{1}{N_1 - 3} + \dfrac{1}{N_2 - 3}}} \qquad (12.4)$$

This formula is based on the fact that the standard error of $Z$ is

$$S_Z = \frac{1}{\sqrt{N - 3}} \qquad (12.5)$$

Substituting this in the formula for the standard error of the difference between two $Z$'s

$$S_{D_Z} = \sqrt{S^2_{Z_1} + S^2_{Z_2}}$$

results in the denominator of equation (12.4).

Using the above data we have

$$z = \frac{.618 - .485}{\sqrt{\dfrac{1}{77 - 3} + \dfrac{1}{86 - 3}}}$$

$$= \frac{.133}{\sqrt{\dfrac{1}{74} + \dfrac{1}{83}}}$$

$$= \frac{.133}{\sqrt{.013514 + .012048}}$$

$$= \frac{.133}{\sqrt{.025562}}$$

$$= \frac{.133}{.160}$$
$$= .83$$

In evaluating this null hypothesis we use a $z$ of 1.96 as the critical 5 percent value and a $z$ of 2.58 as the 1 percent value as with a two-tailed $z$ test between means. Since this $z$ of .83 is less than 1.96 the null hypothesis that there is no significant difference between the two $Z$'s stands. The same conclusion is then drawn concerning the two $r$'s.

When data are *not* independent, the above technique must not be used. An example of such data is when we have three variables: 1, 2, and 3. For the same individuals, variable 1 is first correlated with variable 2 ($r_{12}$) and then variable 1 is correlated with variable 3 ($r_{13}$). These data are thus correlated because variable 1 occurs in both $r$'s. In such cases we test the significance of the difference between the two $r$'s by use of the following formula:

$$z = \frac{(r_{12} - r_{13})\sqrt{(N-3)(1 + r_{23})}}{\sqrt{2(1 - r_{12}^2 - r_{13}^2 - r_{23}^2 + 2r_{12}\,r_{13}\,r_{23})}} \tag{12.6}$$

The above $z$ is interpreted in the usual manner using 1.96 and 2.58 as the critical values. If these $r$'s are based on small samples, $z$ is replaced by $t$ and the results interpreted in the $t$ table (Appendix F) with $N - 3$ degrees of freedom.

## ESTABLISHING THE CONFIDENCE INTERVAL FOR THE PEARSON *r*

Setting up the confidence for the Pearson $r$ is done by setting up the confidence interval for the corresponding $Z$ and transforming these values back to $r$'s. Suppose $r = .82$ with $N = 103$:

$$S_Z = \frac{1}{\sqrt{N-3}} = \frac{1}{\sqrt{103-3}} = \frac{1}{\sqrt{100}} = .1$$

$$Z_{99} = Z \pm 2.58(S_Z) = 1.157 \pm 2.58(.1) = 1.157 \pm .258 = .899\text{–}1.415$$

Changing back to $r$'s,

$$r_{99} = .715\text{–}.890$$

## AVERAGING PEARSON $r$'s

Unless $r$'s are small, less than .30, they should not be averaged arithmetically. The proper procedure is to change each $r$ to $Z$, average the $Z$'s, and then transform the mean $Z$ back to an $r$. This will then be the mean $r$. For example we have the following 4 $r$'s based on the same number of cases and wish to find the average $r$.

| $r$ | $Z$ |
|---|---|
| .90 | 1.472 |
| .87 | 1.333 |
| .77 | 1.020 |
| .55 | .618 |

$$\bar{Z} = \frac{\Sigma Z}{N} = \frac{4.443}{4} = 1.110$$

which converts to an average $r$ of .80.

If the $r$'s are based on different size samples, each $Z$ is multiplied by the number in the sample minus three. Then the products are summed and the sum divided by $(N_1 - 3) + (N_2 - 3) + (N_3 - 3)$, and so on. This average $Z$ is translated back to $r$ using Appendix H.

## EXERCISES

1. Below are the scores of 10 students on two tests of strength. Find $r$.

| TEST 1 | TEST 2 |
|---|---|
| 19 | 11 |
| 18 | 7 |
| 16 | 6 |
| 14 | 9 |
| 12 | 7 |
| 10 | 5 |
| 9 | 6 |
| 8 | 5 |
| 6 | 4 |
| 4 | 2 |

2. Below are the scores of 12 students on two statistics tests. By coding your data find $r$.

| TEST 1 | TEST 2 |
|---|---|
| 104 | 48 |
| 98 | 62 |
| 96 | 40 |
| 90 | 54 |
| 85 | 42 |
| 80 | 52 |
| 77 | 46 |
| 76 | 40 |
| 70 | 38 |
| 69 | 35 |
| 65 | 32 |
| 63 | 36 |

3. Given the SAT-Verbal scores and freshmen grades,

| SAT-V | GPA | SAT-V | GPA |
|---|---|---|---|
| 770 | 5.90 | 580 | 4.96 |
| 750 | 5.72 | 580 | 5.00 |
| 710 | 5.80 | 570 | 4.52 |
| 690 | 5.84 | 560 | 4.81 |
| 680 | 5.92 | 550 | 4.42 |
| 650 | 5.42 | 550 | 5.01 |
| 650 | 5.51 | 540 | 4.84 |
| 640 | 5.20 | 530 | 4.26 |
| 610 | 5.84 | 500 | 4.46 |
| 600 | 5.44 | 490 | 4.13 |

a. Compute $r$ for these data.
b. Find $\overline{X}$ and $S$ for each distribution.

4. Give an example of each of the following: Two variables that have
 a. A high positive correlation
 b. A moderately high correlation
 c. A low positive correlation
 d. A curvilinear relationship
 e. A negative relationship
 f. A correlation because both are correlated with some other variable
5. The following represent the scores of a group of individuals on two scales of a personality inventory:

| SCALE *P* | SCALE *S* |
|---|---|
| 14 | 18 |
| 13 | 16 |
| 12 | 15 |
| 12 | 12 |
| 11 | 10 |
| 10 | 6 |
| 10 | 4 |
| 9 | 3 |
| 8 | 6 |
| 7 | 8 |
| 6 | 13 |
| 5 | 14 |
| 4 | 16 |
| 3 | 17 |
| 2 | 18 |

a. Compute $r$.
b. Make a scatterplot for these data.
c. What can you infer from your scatterplot?

6. Test the significance of the following $r$'s:

| | $r$ | $N$ |
|---|---|---|
| a. | .55 | 18 |
| b. | .55 | 12 |
| c. | .59 | 47 |
| d. | .58 | 10 |
| e. | .22 | 92 |

7. With one group of 72 individuals an $r$ of .72 was found between two traits. On a second group of 85 an $r$ of .68 was found between the same two traits. With $\alpha$ set at .05, is there a significant difference between the two $r$'s?
8. The following shows the correlation among three variables for a sample of 50 students. Does the correlation between variables 1 and 2 differ significantly from that between 1 and 3?

| | 1 | 2 | 3 |
|---|---|---|---|
| 1 | — | .90 | .70 |
| 2 | | — | .80 |
| 3 | | | — |

9. Set up the 99 and 95 percent confidence intervals for an $r$ of .77, $N = 60$.
10. The following $r$'s were based on samples of equal size. What is the average $r$? .98, .88, .76, .72, .64, .50

# 13

# LINEAR REGRESSION

As pointed out in the last chapter a linear relationship is required for the correct use of the Pearson $r$. In this chapter we shall look into this linear relationship, establish the regression equations, and discuss the use of the regression equation in prediction.

We have seen that correlational procedures are used to express the degree of relationship between two variables. Once the existence of a linear relationship has been established, we can use regression analysis to enable us to predict measures on one variable from measures obtained the other. For example, if it is known that the $r$ between high school and college grade point averages is .46 for present students at a particular institution, the admissions officer can use linear regression techniques to predict the future academic performance of each new prospective student to help him make decisions regarding applicants.

## THE EQUATION FOR THE STRAIGHT LINE

A common mathematical equation is that for the straight line:

$$Y = a + bX \tag{13.1}$$

To illustrate the meaning of $a$ and $b$ in the above equation we shall use an example where $a = 2$ and $b = .75$. Since, if we have two sets of points, we can plot a straight line, we can now solve the equation $Y = 2 + .75X$ for any two sets of points as follows:

| X | Y |
|---|---|
| 0 | 2 |
| 4 | 5 |

These two sets of points are then plotted as shown in Figure 13.1. Their connection results in the straight line $Y = 2 + .75X$.

## THE *b* AND *a* COEFFICIENTS

The $b$ coefficient represents the slope of the line and is the ratio of the increase in $Y$ over the increase in $X$. In this case as $Y$ increases from 2 to 5, $X$ increases from 0 to 4. Hence the slope of this line is $(5 - 2)/(4 - 0) = 3/4$, or .75. As pointed out previously, the slope may be either positive or negative depending on the direction that it goes on the graph; a positive slope extending from the upper right to the lower left and a negative one from the upper left to the lower right. Obviously there is an infinite number of lines that may be drawn with a slope of .75. It follows, then, that we must be able to separate this particular line from all other lines with a slope of .75. This is done with the $a$ coefficient, the point at which the line *intercepts* the $y$ axis. The fact that this line crosses the $y$ axis at the $Y$ value of 2 separates this line from all other lines of slope .75. In summary, then, we need two values to determine the equation for any straight line; $a$, the *intercept,* and $b$, the *slope.*

## REGRESSION ANALYSIS

The equation for the straight line in regression analysis is written

$$\tilde{Y} = a + bX \tag{13.2}$$

here $\tilde{Y}$ is read as the predicted value of $Y$. Any time we are dealing with prediction we must introduce the concept of error, as many of the obtained results differ from what was predicted. This difference between the obtained score ($Y$) and the predicted score ($\tilde{Y}$), $(Y - \tilde{Y})$, is known as the

**FIG. 13.1 GRAPH OF THE LINE $Y = 2 + .75X$.**

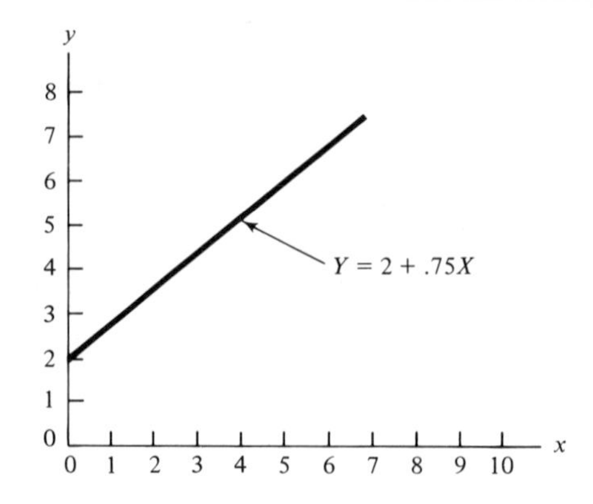

*error of prediction.* The regression line is that line about which the sum of the squares of the errors of prediction is at a minimum. The regression coefficients are developed from the equation for a straight line as follows:

$$\tilde{Y} = a + bX$$
$$Y - \tilde{Y} = Y - (a + bX)$$

where $Y - \tilde{Y}$ is the error of prediction.

These errors of prediction are squared and summed:

$$\Sigma(Y - \tilde{Y})^2 = \Sigma Y - (a + bX)^2$$

By the use of calculus the above equation is differentiated with respect to $a$ and $b$ and each derivative is set equal to zero. This results in the two following equations:

$$b_{YX} = \frac{N\Sigma XY - (\Sigma X)(\Sigma Y)}{N\Sigma X^2 - (\Sigma X)^2} \tag{13.3}$$

$$a_{YX} = \bar{Y} - b_{YX}(\bar{X}) \tag{13.4}$$

Since there are two regression equations except when $r$ is $\pm$ 1.00, the first used in predicting $Y$ from $X$ and the second in predicting $X$ from $Y$, it follows that there are two $b$ and two $a$ coefficients. To distinguish between them we write the $b$ coefficient used in predicting $Y$ from $X$ as $b_{YX}$, and as $b_{XY}$ when predicting $X$ from $Y$. The $a$ coefficients are written with similar subscripts.

To illustrate regression analysis, we shall use the data from Table 12.2. In that correlation problem, the following values were obtained:

| | |
|---|---|
| $\Sigma X = 208$ | $\bar{X} = 100.8$ |
| $\Sigma X^2 = 5592$ | $\bar{Y} = 53.3$ |
| $\Sigma Y = 133$ | $S_X = 11.25$ |
| $\Sigma Y^2 = 2589$ | $s_x = 11.85$ |
| $\Sigma XY = 3649$ | $S_Y = 9.06$ |
| $N = 10$ | $s_y = 9.55$ |
| | $r_{XY} = .866$ |

Substituting these values in equations (13.3) and (13.4) we obtain

$$\begin{aligned} b_{YX} &= \frac{(10)(3649) - (208)(133)}{(10)(5592) - (208)^2} \\ &= \frac{8826}{12656} \\ &= .70 \end{aligned}$$

$$\begin{aligned} a_{YX} &= 53.3 - (.70)100.8 \\ &= 53.3 - 70.56 \\ &= -17.26 \end{aligned}$$

The regression equation for predicting $Y$ from $X$ is then written

$$\tilde{Y} = -17.26 + .70X$$

As noted previously there are two regression equations except when $r = \pm 1.00$. In the second equation, where we are predicting $X$ from $Y$, the $a$ and $b$ coefficients are obtained by the following formulas:

$$b_{XY} = \frac{N\Sigma XY - (\Sigma X)(\Sigma Y)}{N\Sigma Y^2 - (\Sigma Y)^2} \quad (13.5)$$

$$a_{XY} = \bar{X} - (b_{XY})\bar{Y} \quad (13.6)$$

Substituting in these formulas we obtain

$$b_{XY} = \frac{(10)(3649) - (208)(133)}{(10)(2589) - (133)^2}$$
$$= \frac{8826}{8201}$$
$$= 1.08$$

$$a_{XY} = 100.8 - (1.08)(53.3)$$
$$= 100.8 - 57.6$$
$$= 43.2$$

The regression equation for predicting $X$ from $Y$ is then written

$$\tilde{X} = 43.2 + 1.08Y$$

Next we shall plot each of these two lines by obtaining several sets of points and plotting them, as in Figure 13.2. For the equation $\tilde{Y} = -17.26 + .70X$ we obtain

| $X$ | $\tilde{Y}$ |
|---|---|
| **30** | **3.7** |
| **50** | **17.7** |

FIG. 13.2 GRAPH SHOWING THE TWO REGRESSION LINES.

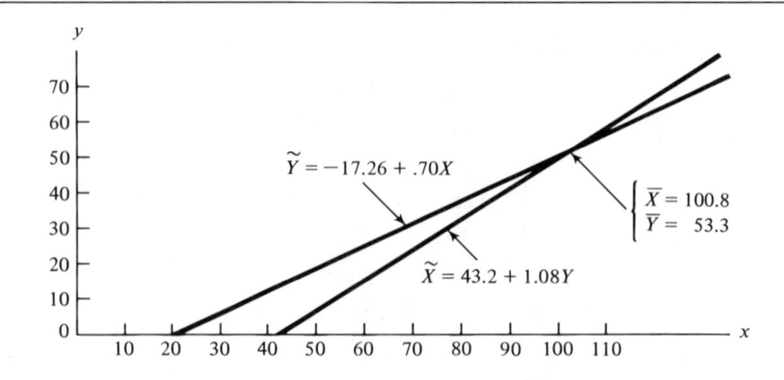

For the second equation, $\tilde{X} = 43.2 + 1.08Y$, the values are

| $Y$ | $\tilde{X}$ |
|---|---|
| 0 | 43.2 |
| 10 | 54.0 |

Notice that when plotted, the two regression lines should cross at a point that is the equivalent of the two means.

Another check on the accuracy of the work is that the product of the two $b$ coefficients should equal $r^2$:

$$(b_{YX})(b_{XY}) = r^2$$
$$(.70)(1.08) = (.866)^2$$
$$.756 = .750$$

The slight discrepancy is probably caused by rounding errors.

## AN ALTERNATE FORMULA FOR THE $b$ COEFFICIENT

The student may find that it is easier to find $b$ by use of the following, the derivation of which we will not go into here.

$$b_{YX} = \frac{S_Y}{S_X}(r) = \frac{s_Y}{s_X}(r) \tag{13.7}$$

Note that either the sample statistic $S$ or population estimate $s$ may be used.

Using the above data we have

$$b_{YX} = \frac{9.1}{11.2}(.866)$$
$$= (.81)(.866)$$
$$= .70$$

which is the same value that was obtained previously.

The $a$ coefficient is then obtained by use of the usual formula:

$$a_{YX} = \bar{Y} - (b_{YX})(\bar{X})$$

## AN ALTERNATE SOLUTION

The following equation includes the combined equations for the $b$ and $a$ coefficients:

$$\tilde{Y} = \bar{Y} + \frac{S_Y}{S_X}(r)(X - \bar{X}) = \bar{Y} + \frac{s_y}{s_x}(r)(X - \bar{X}) \tag{13.8}$$

This equation is obtained by starting with the usual regression equation

$$\tilde{Y} = a_{YX} + b_{YX}$$

and substituting for $a_{YX}$ equation (13.4) and for $b_{YX}$ equation (13.7).

THE PEARSON $r$ AND THE $b$ COEFFICIENT

Beginning with equation (13.7)

$$b_{YX} = \frac{S_Y}{S_X}(r)$$

Solving this equation for $r$, we obtain

$$b_{YX}(S_X) = (S_Y)r$$

$$r = \frac{S_X}{S_Y}(b_{YX}) \qquad (13.9)$$

This shows that the Pearson $r$ is a function of the $b$ coefficient, the slope of the line, and the standard deviations of the two variables.

## THE STANDARD ERROR OF ESTIMATE

As previously noted, in any prediction scheme there are always errors of prediction except when $r$ is $\pm 1.00$, a situation which almost never exists. These errors of prediction are measured by a statistic known as the *standard error of estimate*. When $r = \pm 1.00$, all of the predicted scores would fall on the unique regression line and there would be no errors of measurement; that is, our predicted $\tilde{Y}$'s would be the same as the obtained $Y$'s. At the other extreme, when $r = .00$, the standard error of estimate would be at its maximum. When $r = .00$ the regression line has slope equal to zero and may be written

$$\tilde{Y} = a_{YX}$$

If $b_{YX} = 0$, then

$$a_{YX} = \bar{Y}$$

In this case, $\tilde{Y} = \bar{Y}$. Or, in other words, when $r = .00$, the predicted value of $Y$ is equal to $\bar{Y}$. When this regression line is plotted, it will be a straight line parallel to the $x$ axis and crossing the $y$ axis at $\bar{Y}$. As will be seen below, the standard error of estimate for predicting $Y$ from $X$, $s_{Y \cdot X}$, is equal to the standard deviation of the $Y$ variable when $r = .00$. Hence, the size of the standard error of estimate ranges from .00 to the value of $s_Y$.

As was pointed out earlier, each obtained score differs from its predicted score, this difference being the error of prediction. The errors of prediction are squared, summed, divided by $N - 2$, and the square root taken, giving the standard error of estimate:

$$s_{Y \cdot X} = \sqrt{\frac{\Sigma(Y - \tilde{Y})^2}{N - 2}} \qquad (13.10)$$

The standard error of estimate, then, is the standard deviation of the errors of prediction, and provides an indication of their variability about the regression line in the population in which predictions are being made.

An easier way to get the standard error of estimate using data already at hand for the solution of the regression equations is

$$s_{Y \cdot X} = \sqrt{\left[ N\Sigma Y^2 - (\Sigma Y)^2 - \frac{[N\Sigma XY - (\Sigma X)(\Sigma Y)]^2}{N\Sigma X^2 - (\Sigma X)^2} \right] \left[ \frac{1}{N(N-2)} \right]} \quad (13.11)$$

Using the values already obtained, we have

$$\begin{aligned} s_{Y \cdot X} &= \sqrt{\left[ 8201 - \frac{8826^2}{12656} \right] \left[ \frac{1}{(10)(8)} \right]} \\ &= \sqrt{(8201 - 6155)(.0125)} \\ &= \sqrt{(2046)(.0125)} \\ &= \sqrt{25.5750} \\ &= 5.06, \text{ or } 5.1 \end{aligned}$$

AN ALTERNATE FORMULA

There are times when it is easier to obtain the standard error of estimate by the following formula:

$$s_{Y \cdot X} = s_Y \sqrt{1 - r^2} \quad (13.12)$$

Using the data previously computed, we have

$$\begin{aligned} s_{Y \cdot X} &= 9.55\sqrt{1 - .866^2} \\ &= 9.55\sqrt{1 - .749956} \\ &= 9.55\sqrt{.25004} \\ &= (9.55)(.5000) \\ &= 4.77 \end{aligned}$$

When this formula is used and $N < 50$, a correction should be applied by multiplying the obtained standard error of estimate by

$$\sqrt{\frac{N-1}{N-2}} \quad (13.13)$$

In this case the correction is

$$\sqrt{(9/8)} = \sqrt{1.125} = 1.06$$

Then

$$s_{Y \cdot X} = (4.77)(1.06)$$
$$= 5.06, \text{ or } 5.1$$

the same value obtained by the other method.

An examination of formula (13.12) shows that when $r = 1.00$, $s_{Y \cdot X} = 0$, and that when $r = .00$, $s_{Y \cdot X} = s_Y$ as previously mentioned.

## USE OF THE STANDARD ERROR OF ESTIMATE

For the regression equation $\tilde{Y} = -17.26 + .70X$ we can say that, for any value of $X$, the chances are two out of three (or more exactly 68 in 100) that the true value of $Y$ will fall in a band that is the predicted score plus or minus one standard error of estimate. For example, when $X = 50$, we can say that the chances are two out of three, or 68 in 100, that the actual value of $Y$ will fall in a band $17.7 \pm 5.1$, or between 12.6 and 22.8. Similarly, we can say that the chances are 95 in 100 the $Y$ score will fall in a band of $17.7 \pm 2(5.1)$. Finally, we can be even more certain (99 chances in 100) that $Y$ will fall in a band $17.7 \pm 3(5.1)$. All of this is possible because the standard error of estimate is a standard deviation and may be interpreted as such.

We assume that the standard error of estimate is constant throughout the range. Hence we can measure off 5.1 units on each side of the regression line for all values of $X$. See Figure 13.3. When these points are connected we have two lines parallel to the regression line and 5.1 units from it. In doing this we assume that the distributions about the regression line are normal and possess the property of homoscedasticity, that is, have equal variances. The use of this technique facilitates making predictions

**FIG. 13.3 REGRESSION LINE FOR PREDICTING *Y* FROM *X*. THE PARALLEL LINES ARE EACH A DISTANCE OF ONE STANDARD ERROR OF ESTIMATE FROM THE REGRESSION LINE.**

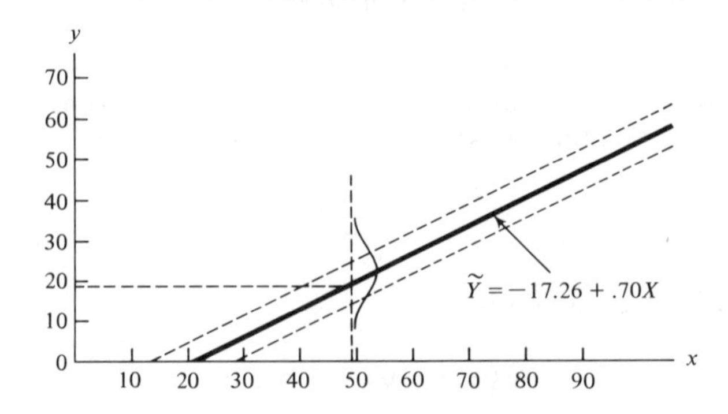

by putting the predictions in a confidence interval band rather than predicting a single point.

## THE COEFFICIENT OF DETERMINATION

Suppose that the correlation between freshmen grades ($Y$) and a test of mental ability ($X$) is .40. One way of interpreting this $r$ of .40 is to square it and to change the result to a percent. Doing this results in .16, or 16 percent. This is interpreted by saying that in the prediction of freshmen grades from this test, 16 percent of the variance in the criterion ($Y$) is predicted by the test ($X$). Other factors beyond this test account for the 84 percent of the variance which is not predicted. This is an important concept to keep in mind, especially when dealing with low correlations. Suppose that in using a large sample, one obtains an $r$ of .08 between intelligence test scores and the uric acid concentration of the blood. Squaring this $r$ and changing it to a percent results in .64 percent of the variance being predicted. Thus over 99 percent of the variance is not accounted for, being due to other factors. Let us further suppose that this $r$ of .08 is statistically significant at the 1 percent level. Here, then, is an example of a statistic that, although it is statistically significant, is utterly worthless in actual usage and has no practical significance or validity.

This $r^2$ is called a *coefficient of determination*. It is particularly useful in evaluating the utility of a validity coefficient expressed as $r$ in variance terms. Sometimes $k^2$ is used as the symbol representing the percent of the variance *not* accounted for and is called a *coefficient of nondetermination*. This can be written

$$r^2 + k^2 = 100 \tag{13.14}$$

If this equation is solved for varying values of $r$, we see that as $r$ increases, $k$ decreases, but not at the same rate. This is shown in the following:

| $r$ | $k$ |
|---|---|
| .00 | 1.00 |
| .20 | .98 |
| .40 | .95 |
| .60 | .80 |
| .80 | .60 |
| .90 | .44 |
| .95 | .31 |
| 1.00 | .00 |

## MULTIPLE PREDICTION

The use of the standard error of estimate with a single predictor shows that, with the typical validity coefficient (between .40 and .50), the predicted scores fall within a wide range for individual predictions. Also the use of the coefficient of determination shows that in the typical situation the major portion of the variance is not predicted. Individuals who use prediction techniques regularly, such as admission officers of a college or university, industrial personnel workers, guidance counselors, and the like, improve their predictions by using not one but several predictors. For example, the predictions of freshmen grades in a university may be improved by using scores on the verbal part of the Scholastic Aptitude Test, scores on the mathematics part of this test, high school rank, and perhaps a measure of motivation. In many cases the $r$ goes from .40 or .50 to the low or mid-.70's using these additional predictors. This coefficient based on several predictors is called a *multiple correlation coefficient* ($R$) and will be discussed in the next chapter. The relationship between criterion and predictors continues to increase as long as each predictor added contributes something unique to the prediction scheme. In most cases four to five predictors cover almost all the predictable variance possible with our present state of knowledge.

The setting up of multiple regression equations is a topic that is beyond the scope of this book.

## EXERCISES

1. Given $s_X = 12$, $s_Y = 8$, $\bar{X} = 60$, $\bar{Y} = 40$, $r_{XY} = .50$, set up the regression equations for predicting $Y$ from $X$ and $X$ from $Y$. Check your work.
2. Given the following data:

| X | Y |
|---|---|
| 12 | 20 |
| 10 | 12 |
| 10 | 18 |
| 8 | 10 |
| 7 | 12 |
| 6 | 14 |
| 6 | 6 |
| 5 | 7 |
| 4 | 3 |
| 2 | 1 |

   a. Set up the regression equation for predicting $Y$ from $X$.
   b. From your equation obtain $Y$ for values of $X = 11$, 13, and 3.
   c. Compute $s_{Y \cdot X}$ for these data.
   d. Interpret this $s_{Y \cdot X}$ by using $Y$ for an $X$ of 11 in part b.

3. Given the following data:

| PREDICTOR X | CRITERION Y | |
|---|---|---|
| $\bar{X} = 500$ | $\bar{Y} = 4.5$ | $r_{XY} = .40$ |
| $s_X = 100$ | $s_Y = .5$ | $N = 100$ |

   a. When $X = 750$, what is $Y$?
   b. When $X$ is 250, what is $Y$?
   c. Compute $s_{Y \cdot X}$.
4. Given $s_Y = 10$, $s_X = 8$, $r_{XY} = .45$, $N = 30$, compute $s_{Y \cdot X}$.
5. a. Given $r = .00$, $\bar{X} = 40$, $\bar{Y} = 60$, what is the best prediction of $Y$ from $X$?
   b. When $r = 1.00$, why is there only one regression line?
   c. Why are there usually two regression lines?
   d. Why in actual practice do we usually only establish one regression line?
   e. What is the relationship of the size of $r$ to the size of the angle formed by the intersection of the regression lines?
   f. Explain why $s_{X \cdot Y}$ is a standard deviation.
   g. Can you explain that a regression line is a line made by connecting a series of means?
6. Given $r_{XY} = .55$:
   a. What is the coefficient of determination?
   b. Explain the meaning of this in your own words.
   c. For this $r$, what is $k^2$?
   d. Explain the meaning of $k^2$.

# 14

# SPECIAL CORRELATION TECHNIQUES

In Chapter 12 we discussed the use of the Pearson $r$ and noted its limitations. There are often situations for which we desire a measure of relationship, but, because the data do not fit the Pearson $r$ model, $r$ cannot be used. The purpose of this chapter is to introduce measures of relationship that can be used when the conditions necessary for the calculation of the Pearson $r$ are not present. First rank correlation methods will be considered, then special cases of the Pearson $r$, the partial and multiple coefficients, and finally a measure of nonlinear relationship.

## RANK CORRELATION METHODS

### THE SPEARMAN RANK-ORDER CORRELATION COEFFICIENT, rho, $r_S$, OR $\rho$

Of those included in this group, the *Spearman rank-order correlation coefficient* is by far the one most frequently encountered. It is especially useful when the number of pairs is small, less than 30. As $N$ increases, its use becomes cumbersome as we will see. To illustrate this statistic we shall use the data in Table 14.1, where the scores of 10 students on two statistics tests are shown.

To produce $r_S$ we go through the following steps:

1. The scores on test 1, $X$, are first ranked. The highest score, 98, is given a rank of 1 and this ranking is placed in column (3). Then the second score of 90 receives a rank of 2. Individuals 3 and 4 are tied with a score of 84. Ties are handled by averaging the posi-

tions that the tied scores hold and giving each the average rank. In this case the average of 3 and 4 is 3.5. So individuals 3 and 4 each are given a rank of 3.5. Continuing, individual 5 with a score of 83 obtains a rank of 5. And so we go on. Individuals 7, 8, and 9 also have tied scores. These positions are averaged and each receives a rank of 8. The scores in column (1) are in their natural order, from high to low. However, it is not necessary to have them arranged in this order to calculate the Spearman coefficient.

2. Next the scores on test 2, $Y$, are ranked in the same manner. These ranks appear in column (4).
3. Then the difference between ranks, $D$, is obtained and these differences placed in column (5). Because we are not summing this column, no attention is paid to the sign of the difference.
4. The differences in column (5) are each squared and the squares placed in column (6).
5. Column (6) is summed.
6. Then the sum of these squares is used in the following formula:

$$r_S = 1 - \frac{6\Sigma D^2}{N(N^2 - 1)} \tag{14.1}$$

$$= 1 - \frac{6(42.5)}{10(100 - 1)}$$

$$= 1 - \frac{255}{10(99)}$$

$$= 1 - \frac{255}{990}$$

$$= 1 - .26$$

$$= .74$$

## INTERPRETATION OF $r_S$

The statistic $r_S$ is a Pearson correlation coefficient for ranked data and for all practical purposes may be interpreted as $r$. However, if there are many ties there will be discrepancies between $r$ and $r_S$. Corrections for tied ranks are available and may be found in Siegel (1956). However, in most cases the use of these correction formulas is not worth the effort.

## TESTING THE SIGNIFICANCE OF $r_S$

When $N$ is less than 10, special tables for testing the significance of $r_S$ may be found in Siegel (1956, p. 294) or Edwards (1967, p. 343). The student should be reminded that when $N$ is small, $r_S$ has to be large to be significant. Siegel's table shows the 5 percent and 1 percent values for an $N$ of 5 to be .90 and 1.00. When $r_S$ is greater than 10, the following formula is used:

TABLE 14.1 THE SPEARMAN RANK-ORDER CORRELATION COEFFICIENTS, $r_S$

| | (1) | (2) | (3) | (4) | (5) | (6) |
|---|---|---|---|---|---|---|
| IND. | X, SCORES ON TEST 1 | Y, SCORES ON TEST 2 | RANK X | RANK Y | D | $D^2$ |
| 1 | 98 | 48 | 1 | 3 | 2 | 4 |
| 2 | 90 | 46 | 2 | 4 | 2 | 4 |
| 3 | 84 | 52 | 3.5 | 1 | 2.5 | 6.25 |
| 4 | 84 | 50 | 3.5 | 2 | 1.5 | 2.25 |
| 5 | 83 | 38 | 5 | 7 | 2 | 4 |
| 6 | 80 | 36 | 6 | 9 | 3 | 9 |
| 7 | 75 | 37 | 8 | 8 | 0 | 0 |
| 8 | 75 | 39 | 8 | 6 | 2 | 4 |
| 9 | 75 | 40 | 8 | 5 | 3 | 9 |
| 10 | 70 | 32 | 10 | 10 | 0 | 0 |
| | | | | | | $\Sigma D^2 = 42.50$ |

$$t = \frac{r_S\sqrt{N-2}}{\sqrt{1-r_S{}^2}} \tag{14.2}$$

this being the same as formula (12.3).

Recall that a table exists which makes the use of this formula unnecessary. Hence when $N > 10$, Appendix G may be used in evaluating the significance of $r_S$ just as it was for evaluating the Pearson $r$.

## THE COEFFICIENT OF CONCORDANCE, *W*

The *coefficient of concordance* is actually an average of a group of Spearman rank-order correlation coefficients. In Table 14.2 appear the rankings at a 4H Fair of 5 pies ($n$) by 4 judges ($m$). The task is to find out how the judges agree in their ranking of these objects. This would be an indicator of the reliability of the rankings. We could correlate the rankings of judge 1 with judge 2, judge 1 with 3, and so on, until six different Spearman rank-order coefficients have been obtained. The average of these six coefficients would then be a measure of the reliability of the rankings.

The use of the statistic $W$ makes it unnecessary to obtain all these separate coefficients. $W$ is obtained as follows:

1. The rankings of the 4 judges of each pie are summed and the results placed in column (1).

**TABLE 14.2 USING THE COEFFICIENT OF CONCORDANCE IN DETERMINING THE RELIABILITY OF THE RATINGS OF 5 PIES BY 4 JUDGES**

| | Judges' Rank | | | | (1) | (2) | (3) |
|---|---|---|---|---|---|---|---|
| PIE | 1 | 2 | 3 | 4 | SUM OF RANKS (R) | $D$ | $D^2$ |
| 1 | 5 | 4 | 5 | 5 | 19 | 7 | 49 |
| 2 | 3 | 3 | 2 | 3 | 11 | 1 | 1 |
| 3 | 1 | 2 | 1 | 2 | 6 | 6 | 36 |
| 4 | 2 | 1 | 3 | 1 | 7 | 5 | 25 |
| 5 | 4 | 5 | 4 | 4 | 17 | 5 | 25 |
| | | | | | $\Sigma = 60$ | | $\Sigma = 136$ |

$$\Sigma R = \frac{m(n)(n+1)}{2} = \frac{(4)(5)(6)}{2} = \frac{120}{2} = 60$$

2. Column (1) is then summed. If there is no relationship among the rankings, the sum of the ranks for each object would be the same. In this case the sum of each would equal $60/5 = 12$.
3. Obtain the difference between the sum of each group of rankings and this average of 12. Place the results ($D$) in column (2).
4. Square each of these differences and place the results in column (3).
5. Compute $W$ as follows:

$$W = \frac{12\Sigma D^2}{m^2n(n^2-1)} \tag{14.3}$$
$$= \frac{12(136)}{(4^2)(5)(5^2-1)}$$
$$= \frac{1632}{(16)(5)(24)}$$
$$= \frac{1632}{1920}$$
$$= .85$$

This $W$ of .85 shows that there is high agreement among the 4 judges in evaluating the goodness of these 5 pies.

At the bottom of Table 14.2 there is a check on some of the work. Note that the sum of the ranks is equal to the product of the number of judges, the number of objects rated, and the number of objects rated plus 1, all divided by 2; in this case 60 = 60.

### TESTING THE SIGNIFICANCE OF *W*

The statistical significance of $W$ is evaluated by the use of Appendix K. Going into this table with $m = 4$ (judges) and $n = 5$ (objects rated), we find that the .01 value is .67. Since our $W$ of .85 exceeds this, we can say that this $W$ is significant beyond the 1 percent level and that our judges did a reliable job in evaluating the merit of these 5 pies.

## TAU (*T*) RANK CORRELATION

The *tau coefficient,* developed by Kendall, is preferred by some over the widely used $r_S$ as a measure of rank correlation. In general tau is slightly more difficult to compute, especially when there are ties. First we shall illustrate how to obtain $T$ when there are no ties. In Table 14.3 we have the scores of 9 students on tests $X$ and $Y$.

1. First each distribution is ranked separately as in computing $r_S$. The ranks are placed in columns (1) and (2). In computing $T$ one set of ranks must be in a natural order, from high to low or vice versa.
2. Then for each individual count the number of individuals who are higher and lower than he on the second variable, $Y$. The first individual in distribution $X$ has 3 individuals higher than he on the $Y$ variable and 5 lower. The second individual in the $X$ distribution has 1 higher and 6 lower on the $Y$ distribution. These values of higher than or lower than are placed in columns (3) and (4) and this process is continued until all individuals have been taken into account.
3. Sum column (3) and call the sum $P$. Do the same for column (4) and call this sum $Q$.
4. Find $T$ by using the following formula:

$$T = \frac{P - Q}{N(N - 1)/2} \tag{14.4}$$

$$= \frac{31 - 5}{(9)(8)/2}$$

$$= \frac{26}{36}$$

$$= .72$$

**TABLE 14.3 OBTAINING THE TAU COEFFICIENT FOR DATA WITH NO TIES**

| | | | (1) | (2) | (3) | (4) |
|---|---|---|---|---|---|---|
| IND. | TEST X | TEST Y | $R_x$ | $R_y$ | NO. RANKS HIGHER | NO. RANKS LOWER |
| 1 | 84 | 60 | 1 | 4 | 5 | 3 |
| 2 | 80 | 64 | 2 | 2 | 6 | 1 |
| 3 | 78 | 71 | 3 | 1 | 6 | 0 |
| 4 | 76 | 61 | 4 | 3 | 5 | 0 |
| 5 | 70 | 58 | 5 | 5 | 4 | 0 |
| 6 | 64 | 57 | 6 | 6 | 3 | 0 |
| 7 | 62 | 54 | 7 | 8 | 1 | 1 |
| 8 | 50 | 55 | 8 | 7 | 1 | 0 |
| 9 | 47 | 52 | 9 | 9 | 0 | 0 |
| | | | | | $\Sigma = P = 31$ | $\Sigma = Q = 5$ |

When ties appear, certain adjustments as shown below have to be made. The data of Table 14.1 have been copied into Table 14.4 and will be used to illustrate the computation of $T$ when there are ties. The method is as follows:

1. The number of cases ranking higher and lower on the $Y$ variable is determined as in the previous example. We find that $P = 33$ and $Q = 12$.
2. We first take the $X$ distribution and for each set of ties, determine $(x)(x - 1)$ where $x$ is the number tied for a particular rank. These are summed and divided by 2. For the data in Table 14.4 we have

$$\frac{(2)(2-1) + (3)(3-1)}{2} = \frac{2+6}{2} = \frac{8}{2} = 4$$

   The same process is repreated for the $Y$ distribution. Since there are no ties in the $Y$ distribution in this example, this factor is zero.
3. Calculate

$$\frac{N(N-1)}{2} = \frac{(10)(9)}{2} = \frac{90}{2} = 45$$

4. Subtract the correction found in step 2 from this value:

   $X$ distribution: $45 - 4 = 41$
   $Y$ distribution: $45 - 0 = 45$

**TABLE 14.4 COMPUTATION OF THE TAU COEFFICIENT WHEN TIES EXIST**

| IND. | X SCORES ON TEST 1 | Y SCORES ON TEST 2 | (1) $R_x$ | (2) $R_y$ | (3) NO. RANKS HIGHER | (4) NO. RANKS LOWER |
|---|---|---|---|---|---|---|
| 1 | 98 | 48 | 1 | 3 | 7 | 2 |
| 2 | 90 | 46 | 2 | 4 | 6 | 2 |
| 3 | 84 | 52 | 3.5 | 1 | 7 | 0 |
| 4 | 84 | 50 | 3.5 | 2 | 6 | 0 |
| 5 | 83 | 38 | 5 | 7 | 3 | 2 |
| 6 | 80 | 36 | 6 | 9 | 1 | 3 |
| 7 | 75 | 37 | 8 | 8 | 1 | 2 |
| 8 | 75 | 39 | 8 | 6 | 1 | 1 |
| 9 | 75 | 40 | 8 | 5 | 1 | 0 |
| 10 | 70 | 32 | 10 | 10 | 0 | 0 |
| | | | | | $\Sigma = P = 33$ | $\Sigma = Q = 12$ |

5. Multiply the above two terms:

$$(41)(45) = 1845$$

6. Extract the square root of this product:

$$\sqrt{1845} = 43$$

7. Then find $T$:

$$T = \frac{P - Q}{43} = \frac{33 - 12}{43} = \frac{21}{43} = .49$$

INTERPRETATION OF TAU

*Tau* is interpreted in the same manner as $r_S$. When both coefficients are computed for the same data, $T$, as illustrated, is the smaller.

TESTING THE SIGNIFICANCE OF TAU

When $N \geqslant 10$, the statistical significance of $T$ may be evaluated by the following formula:

$$z = \frac{T}{S_T} = \frac{T}{\sqrt{2(2N + 5)/9N(N - 1)}} \tag{14.5}$$

For the above data we have

$$\begin{aligned} z &= \frac{.49}{\sqrt{2(20 + 5)/(90)(9)}} \\ &= \frac{.49}{\sqrt{2(25)/810}} \\ &= \frac{.49}{\sqrt{50/810}} \\ &= \frac{.49}{\sqrt{.06173}} \\ &= \frac{.49}{.248} \\ &= 1.98 \end{aligned}$$

Since this $z$ of 1.98 is greater than 1.96, this $T$ is significant at the 5 percent level. Although $r_S$ and $T$, when computed for the same data, differ in magnitude, a test of significance when applied to both will lead to similar conclusions. That is, if $r_S$ is significant at the 5 percent level, $T$ computed for the same data will also be significant at the 5 percent level.

## LINEAR RELATIONSHIPS

Included here are two coefficients that are essentially Pearson correlation coefficients. However, instead of having two continuous variables, one or both of the variables have been reduced to or exists as two categories. In statistical terms we say that the variable has been *dichotomized* or exists as a dichotomy. Common dichotomies are Pass–Fail, Right–Wrong, and the like. Currently two coefficients are widely used with one or both of the variables dichotomized. The first of these, the *point-biserial,* is used when there is one continuous variable and one dichotomy. With two dichotomous variables, the *phi coefficient* is used as a measure of correlation.

## THE POINT-BISERIAL CORRELATION COEFFICIENT, $r_{pb}$

In the building and application of educational and psychological tests, we often have a situation where one variable is continuous and the other is a dichotomy. As an example, we might be correlating scores on a test with a criterion that has been dichotomized to Pass–Fail. In this case the criterion might be the completion of a course of study. In test construction, test scores of individuals are correlated with their response to each item on a certain test. In this case the items are scored dichotomously, Right–Wrong. Today most universities and producers of tests have computer programs that make the calculation of a point-biserial coefficient a simple matter. Below we shall illustrate how $r_{pb}$ is obtained by using the scores of 100 individuals on test $X$ and correlating these scores with their responses to a single item on the test. This is shown in Table 14.5.

In column (1) of Table 14.5 are the various scores obtained by individuals on the given test ($X$). In column (2) is the number of individuals ($f$) who received the various scores listed in column (1). In column (3) is the frequency of those with a score of 40 who responded to this item correctly ($f_r$). So it is for each of the scores in column (1). In column (4) is the frequency of those with different scores who responded incorrectly to the item ($f_w$). In column (5), ($fX$), each product of the values in column (1) multiplied by the frequencies in column (2) is recorded. In column (6), ($fX^2$), are the products of each of the values in column (5) multiplied by its respective value in column (1). In column (7) are the values in column (1), ($X$), multiplied by their corresponding values in column (3) ($f_r$). Columns (2) through (7) are then summed and the following formula is used to obtain the point-biserial coefficient:

$$r_{pb} = \frac{N(\Sigma f_r X) - N_r(\Sigma fX)}{\sqrt{(N_r N_w)[N\Sigma fX^2 - (\Sigma fX)^2]}} \tag{14.6}$$

The above equation is a modification of equation (12.2) for dichotomous data. (See Edwards, 1967.)

$$\begin{aligned} r_{pb} &= \frac{100(1987) - 63(2932)}{\sqrt{(63)(37)[100(89898) - 2932^2]}} \\ &= \frac{198700 - 184716}{\sqrt{(2331)(8989800 - 7596624)}} \\ &= \frac{13984}{\sqrt{(2331)(393176)}} \\ &= \frac{13984}{\sqrt{916493256}} \\ &= \frac{13894}{30274} \\ &= .46 \end{aligned}$$

LIMITATION OF THE SIZE OF THE POINT-BISERIAL

When the continuous variable is normally distributed and the dichotomous variable is split 50–50, that is, $p = .50$, it is generally accepted that $r_{pb}$ has a maximum value of .80. Research has shown (Karabinus, 1975) that when the shape of the distribution of the continuous variable departs from normal, the computed $r_{pb}$ may exceed this value of .80. If the con-

**TABLE 14.5 POINT-BISERIAL CORRELATION COEFFICIENT BETWEEN TEST SCORES AND RESPONSES TO A SINGLE ITEM ON A TEST**

| (1) | (2) | (3) | (4) | (5) | (6) | (7) |
|---|---|---|---|---|---|---|
| $X$ | $(f)$ | $f_r$ | $f_w$ | $fX$ | $fX^2$ | $f_rX$ |
| 40 | 2 | 2 | 0 | 80 | 3200 | 80 |
| 38 | 4 | 4 | 0 | 152 | 5776 | 152 |
| 37 | 6 | 5 | 1 | 222 | 8214 | 185 |
| 36 | 12 | 10 | 2 | 432 | 15552 | 360 |
| 32 | 12 | 9 | 3 | 384 | 12288 | 288 |
| 31 | 10 | 8 | 2 | 310 | 9610 | 248 |
| 30 | 12 | 7 | 5 | 360 | 10800 | 210 |
| 28 | 10 | 6 | 4 | 280 | 7840 | 168 |
| 27 | 10 | 7 | 3 | 270 | 7290 | 189 |
| 25 | 4 | 1 | 3 | 100 | 2500 | 25 |
| 24 | 4 | 1 | 3 | 96 | 2304 | 24 |
| 22 | 3 | 1 | 2 | 66 | 1452 | 22 |
| 20 | 3 | 1 | 2 | 60 | 1200 | 20 |
| 18 | 3 | 0 | 3 | 54 | 972 | 0 |
| 16 | 2 | 1 | 1 | 32 | 512 | 16 |
| 12 | 2 | 0 | 2 | 24 | 288 | 0 |
| 10 | 1 | 0 | 1 | 10 | 100 | 0 |
| | $\Sigma = 100$ | $\Sigma = 63$ | $\Sigma = 37$ | $\Sigma = 2932$ | $\Sigma = 89898$ | $\Sigma = 1987$ |

tinuous variable is platykurtic or rectangular, coefficients in the .80's may be expected. When the continuous variable is bimodal with little or no overlap, we may obtain coefficients above .90.

## TESTING THE SIGNIFICANCE OF THE POINT-BISERIAL COEFFICIENT

Since the point-biserial is a Pearson $r$, the statistical significance of such a coefficient may be obtained by using Appendix G. In our illustration, there are 98 degrees of freedom. The closest number of degrees of freedom in the table is 100. Using this value, we see that the obtained point-biserial is significant at the .001 level. In item analysis work, of which this is an example, coefficients of this magnitude are not uncommon. We conclude then that there is a significant correlation between total test scores and responses to this item. This means that those who scored highest on the test tended to get the item right and those who scored low tended to miss the item. Then the item is doing what the test is supposed to do, to separate the good students from the poor ones.

## THE BISERIAL $r$, $r_b$

The student may encounter in the literature a correlation coefficient called the *biserial correlation coefficient.* For many years this statistic was widely used in test construction work and many charts and other aids were devised to assist in obtaining it or estimations of it rapidly. The biserial $r$ is an estimate of the Pearson $r$ and is a less reliable statistic. In this period of the modern computers there is little need for this statistic.

## THE PHI OR FOURFOLD CORRELATION COEFFICIENT

When both variables, $X$ and $Y$, are dichotomies, it is possible to determine the correlation between the two variables by using the *phi coefficient,* $\phi$.

Table 14.6 shows the responses of 200 males and 200 females to an item on a scale built to measure attitudes. The question is whether or not there is a correlation between the sex of the individual and his response. For the data in Table 14.6, $\phi$ is obtained as follows:

$$\phi = \frac{ad - bc}{\sqrt{(a+b)(a+c)(b+d)(c+d)}} \tag{14.7}$$

$$= \frac{(160)(160) - (40)(40)}{\sqrt{(200)(200)(200)(200)}}$$

$$= \frac{25600 - 1600}{\sqrt{1600000000}}$$

$$= \frac{24000}{40000}$$

$$= .60$$

TABLE 14.6 RESPONSES OF 400 INDIVIDUALS TO AN ITEM ON AN ATTITUDINAL SCALE

| | AGREE | DISAGREE | |
|---|---|---|---|
| MALE | 160 (a) | 40 (b) | 200 |
| FEMALE | 40 (c) | 160 (d) | 200 |
| | 200 | 200 | 400 = $N$ |

The phi coefficient is also a form of a Pearson $r$ (Edwards, 1967) and because of this it has become very widely used in the construction and analysis of tests and test data. One factor that affects the size of the phi coefficient is the manner in which the two variables have been split. When there is an equal number of cases in both variables as in the example, the maximum size of $\phi$ is $\pm 1.00$. When the marginal totals vary, the maximum value of phi will be less than 1.00 and will be different from case to case. As an illustration, suppose we have the following data:

| | | |
|---|---|---|
| **30** (a) | **20** (b) | **50** |
| **10** (c) | **40** (d) | **50** |
| **40** | **60** | **100** |

Using the formula for phi, we obtain a value of .41. The maximum value for such marginal values would be obtained when we have a distribution such as the following:

| | | |
|---|---|---|
| **40** (a) | **10** (b) | **50** |
| **0** (c) | **50** (d) | **50** |
| **50** | **60** | **100** |

Solving these data produces a phi of .82. A comparison of the phi of .41 with this phi of .82 gives an indication of the size phi may take when certain cells have no frequency.

### TESTING THE SIGNIFICANCE OF THE PHI COEFFICIENT

One of the advantages of using the phi coefficient is that a ready test of significance can be made because of the relationship between phi and chi-square. This relationship can be deduced from the two following formulas:

$$\phi = \frac{(ad - bc)}{\sqrt{(a + b)(a + c)(c + d)(b + d)}}$$

$$\chi^2 = \frac{N(ad - bc)^2}{(a + b)(a + c)(a + d)(b + d)}$$ (Formula for chi-square when data are in a $2 \times 2$ table)

From the above we see that

$$\chi^2 = N\phi^2 \tag{14.8}$$

In our illustrative problem for the phi coefficient we obtained phi equal to .60 with $N$ equal to 400. Substituting in formula (14.8) we have

$$\begin{aligned}\chi^2 &= (400)(.60)^2\\ &= (400)(.36)\\ &= 144\end{aligned}$$

With *df* of 1, this chi-square is significant beyond the .001 level and hence the obtained phi coefficient is significant at the same level.

Very often a student may have a large number of phi coefficients based on the same sample. Rather than compute a chi-square for each separate phi coefficient, the values of phi that are significant at both the 1 and 5 percent levels can be established and then by comparing each of the phi coefficients with these values determine at once whether or not each phi coefficient is significant. As an example suppose there are a series of phi coefficients based on an $N$ of 150. Using values from the chi-square table (Appendix D) for 1 degree of freedom, we proceed as follows:

| 1% LEVEL | 5% LEVEL |
|---|---|
| $\chi^2 = N\phi^2$ | $\chi^2 = N\phi^2$ |
| $6.635 = 150\phi^2$ | $3.841 = 150\phi^2$ |
| $\phi^2 = .044233$ | $\phi^2 = .02561$ |
| $\phi = .21$ | $\phi = .16$ |

Then the obtained phi coefficients are taken one at a time and compared with these two values. For example, a phi coefficient of .18 is significant at the 5 percent level because its value falls between the 1 and 5 percent value of phi. By similar reasoning, a phi coefficient of .14 is not significant, while one of .28 is significant at the 1 percent level.

## THE TETRACHORIC *r*, $r_t$

In the past this statistic was used as a measure of relationship when data were reduced to two dichotomies. The *tetrachoric r* is, like the biserial *r*, an estimate of a Pearson *r*. It is a rather unreliable statistic, having a large standard error. At the present time, there is little justification for the use of this statistic since the phi coefficient is a far superior one.

## THE CONTINGENCY COEFFICIENT, *C*

*The contingency coefficient* is similar to the phi coefficient except that data are in more than two categories. In Table 14.7 we have an example of data, the average letter grade made by 350 freshmen who have been grouped on the basis of their verbal score on the SAT, in which this statistic may be used. First we compute chi-square:

| $O$ | $E$ | $O - E$ | $(O - E)^2$ | $(O - E)^2/E$ |
|---|---|---|---|---|
| 100 | 54.9 | 45.1 | 2034.01 | 37.0 |
| 50 | 68.6 | 18.6 | 345.96 | 5.0 |
| 10 | 36.6 | 26.6 | 707.56 | 19.3 |
| 15 | 41.1 | 26.1 | 681.21 | 16.6 |
| 85 | 51.4 | 33.6 | 1128.96 | 22.0 |
| 20 | 27.4 | 7.4 | 54.76 | 2.0 |
| 5 | 24.0 | 19.0 | 361.00 | 15.0 |
| 15 | 30.0 | 15.0 | 25.00 | 7.5 |
| 50 | 16.0 | 34.0 | 1156.00 | 72.2 |
| | | | | $\chi^2 = 196.6$ |

In Table 14.7 the expected frequencies appear in parentheses below the observed frequency of each cell.

The contingency coefficient is then obtained by using the following formula:

$$C = \sqrt{\frac{\chi^2}{N + \chi^2}} \tag{14.9}$$

$$= \sqrt{\frac{196.6}{350 + 196.6}}$$

$$= \sqrt{\frac{196.6}{546.6}}$$

$$= \sqrt{.359678}$$

$$= .60$$

One of the problems associated with the contingency coefficient is that it does not have 1 as an upper limit, the upper limit in each case being determined by the number of columns and rows. When the number of columns and rows is the same, the upper limit is determined by using the formula $\sqrt{(k - 1)/k}$, where $k$ is the number of columns or rows. In this case we have $\sqrt{(3 - 1)/3} = \sqrt{2/3} = \sqrt{.6667} = .82$. When the number of columns and rows differ, the upper limit is obtained by inserting for $k$ the number of columns or rows that is the smaller.

$C$ has no sign; however, the direction of the relationship may be obtained by inspection of the data. Examination of Table 14.7 reveals that the relationship is a positive one. Also the contingency coefficient requires that no assumptions be made about the shape of the distribution. Hence it can be applied to any data that can be reduced to frequencies, normal or not. Another aspect of $C$ is that it is not directly comparable

TABLE 14.7 SAT-VERBAL SCORES AND LETTER GRADES FOR 350 FRESHMEN

| | Grades | | | |
|---|---|---|---|---|
| Sat-V | A, B | C | D, F | |
| ABOVE 600 | 100 (54.9) | 15 (41.1) | 5 (24) | 120 |
| 500–599 | 50 (68.6) | 85 (51.4) | 15 (30) | 150 |
| BELOW 500 | 10 (36.6) | 20 (27.4) | 50 (16) | 80 |
| | 160 | 120 | 70 | 350 |

to any other correlation coefficient nor with any other contingency coefficient unless they are based on similar contingency tables.

TESTING THE SIGNIFICANCE OF C

If the chi-square used in the computation of $C$ is significant, then it follows that the $C$ is also statistically significant. In this case chi-square of 196.6 with $df$ of 2 is significant beyond the .001 level.

## CRAMÉR'S STATISTIC, $\phi'$

Cramér's statistic is similar to $C$, differing in that it has 1 as an upper limit. For the data in Table 14.7, we have

$$\phi' = \sqrt{\frac{\chi^2}{N(L-1)}} \qquad (14.10)$$

where $L$ is the smaller of either the number of columns or rows.

$$\phi' = \sqrt{\frac{196.6}{(350)^2}}$$

$$= \sqrt{\frac{196.6}{700}}$$

$$= \sqrt{.280857}$$

$$= .53$$

If the chi-square used in computing Cramér's statistic is statistically significant, then $\phi'$ is also statistically significant.

## *POST HOC* PROCEDURES FOR CHI-SQUARE TESTS OF INDEPENDENCE

Phi, the contingency coefficient, or Cramér's statistic may be used as *post hoc indices of association* for chi-square tests of independence. A significant $\chi^2_{\text{obs}}$ merely permits a researcher to reject the hypothesis of independence between variables. It provides no numerical estimate of the *strength* of this relationship. Referring back to Table 7.5, we find data with which a researcher has concluded that two attitudes are associated in the population of interest. *Post hoc* analyses using the above indices can now be demonstrated with this example of a test of independence:

$$\phi = \frac{(47)(116) - (85)(102)}{\sqrt{(132)(149)(201)(218)}}$$
$$= -.11$$

$$C = \sqrt{\frac{4.21}{350 + 4.21}}$$
$$= .11$$

$$\phi' = \sqrt{\frac{4.21}{350(2 - 1)}}$$
$$= .11$$

All three indices happen to provide the same value with these particular data. The conclusion reached with any one of these *post hoc* analysis would be that the *strength* of association is very low between attitudes A and B in this population. There is a significant, but very weak, tendency for people who favor one project to oppose the other.

## TWO CORRELATION COEFFICIENTS BASED ON PEARSON *r*'s

### THE MULTIPLE CORRELATION COEFFICIENT, *R*

A *multiple correlation coefficient* is the correlation between one variable and the combined effects of two or more other variables. As an example, suppose that we have the correlation between freshmen grades, the criterion, and the two parts of the Scholastic Aptitude Test (SAT), verbal and mathematics, the predictors. Multiple correlation and its counterpart, multiple regression, are frequently encountered in prediction schemes.

To begin with we shall give each variable a number:

1 = Freshman grades
2 = SAT-verbal
3 = SAT-math

On a certain sample, Pearson $r$'s as shown below were obtained between these variables:

$$r_{12} = .38$$
$$r_{13} = .40$$
$$r_{23} = .30$$

The use of the multiple correlation coefficient makes it possible to obtain the correlation between these freshmen grades and the combined effects of the verbal and mathematics scores. This is done as follows:

$$R_{1\cdot 23} = \sqrt{\frac{r_{12}^2 + r_{13}^2 - (2r_{12}\, r_{13}\, r_{23})}{1 - r_{23}^2}} \quad (14.11)$$

$$= \sqrt{\frac{.38^2 + .40^2 - 2(.38)(.40)(.30)}{1 - (.30)^2}}$$

$$= \sqrt{\frac{.1444 + .1600 - .0912}{1 - .09}}$$

$$= \sqrt{\frac{.3044 - .0912}{.91}}$$

$$= \sqrt{\frac{.2132}{.91}}$$

$$= \sqrt{.23429}$$
$$= .48$$

Notice that the multiple correlation of .48 is larger than each of the correlations of the predictors with the criterion. This is usually the case when both predictors are measuring different factors or traits. In actual practice, we can usually use four or five predictor variables before the increment in $R$ is so small that it is not worth the effort to add more predictor variables.

Formula (14.11) is limited to use with two predictors. When there are more than two, working a multiple $R$ by hand is both very tedious and cumbersome. The easiest way to obtain such a multiple $R$ is to use one of the available programs for use with a digital computer.

The multiple $R$ is interpreted as to its magnitude and is used in the same fashion as the Pearson $r$. Like the Pearson $r$, this statistic also has a standard error of estimate that is interpreted in the same fashion as that of the Pearson $r$:

$$S_{R_{1\cdot 23}} = S_1\sqrt{1 - R^2} \quad (14.12)$$

## THE PARTIAL $r$

Suppose we have the following three variables:

1 = scores on a spelling test
2 = scores on a test of English usage
3 = scores on an intelligence test

Pearson $r$'s have been run between each pair of variables with the following results:

$$r_{12} = .60$$
$$r_{13} = .60$$
$$r_{23} = .70$$

From the above it is apparent that intelligence test scores are significantly related to both the spelling test scores and to the English usage scores. Also note that there is a high correlation between scores on the spelling test and those on the language usage test. At this point we might want to find out if this last relationship is actually as high as it seems to be or whether it is an artifact brought about because both the spelling test scores and the language usage scores are related to the third variable, intelligence. Now the question is what would be the correlation between these spelling test scores and the language usage scores if the effect of intelligence is removed, or partialed out as it is spoken of statistically.

To remove the effects of a third variable from the correlation between two other variables, we use the following formula for the *partial r*:

$$r_{12 \cdot 3} = \frac{r_{12} - (r_{13}\, r_{23})}{\sqrt{(1 - r_{13}{}^2)(1 - r_{23}{}^2)}} \tag{14.13}$$

This is read as the correlation between variables 1 and 2 with the effects of variable 3 partialed out. For this problem this becomes the correlation between spelling test scores and language usage scores with the effects of intelligence partialed out. Substituting in this formula, we have

$$r_{12 \cdot 3} = \frac{.60 - (.60)(.70)}{\sqrt{(1 - .60^2)(1 - .70^2)}}$$
$$= \frac{.60 - .42}{\sqrt{(1 - .36)(1 - .49)}}$$
$$= \frac{.18}{\sqrt{(.64)(.51)}}$$
$$\frac{.18}{\sqrt{.3264}}$$
$$= \frac{.18}{.57}$$
$$= .32$$

By removing the effects of intelligence test scores we have considerably reduced the size of the correlation coefficient between the spelling and language usage test scores.

The *partial* $r$ is a Pearson $r$ and it may be used and tested for significance as is the Pearson $r$.

## CURVILINEAR DATA

With most of the variables used in the social sciences, a linear relationship is found between the variables studied. This may be confirmed generally by noting whether each of the variables is normally distributed. However, there are cases, as when age is studied against various aspects of physical development, where the relationship is curvilinear. Taking any physical trait of performance or strength, it is found that as one grows older, one tends to perform better on that trait. Then a plateau on which one may stay for a long period of time is reached. And then with advancing age there is a decline in performance.

When it is apparent that data depart from linearity as in the above example, the Pearson $r$ should not be used as a measure of relationship. In such cases a statistic called *eta,* or the correlation ratio, is appropriate. The computation of eta is beyond the scope of this book and formulas for and discussions related to it will be found in advanced statistics texts. The interested student will find this statistic covered in Guilford and Fruchter (1973, pp. 284ff.).

There are several aspects of this statistic that should be noted. First eta is always positive. Secondly there are always two eta's, one for each variable distribution. One gives the relationship between $X$ and $Y$, and the other between $Y$ and $X$. Recall that the correlation between $X$ and $Y$ is the same as that between $Y$ and $X$ when $r$ is used as a measure of relationship. Thirdly the greater the departure from linearity, that is, the larger the bend in the regression line, the greater the difference between $r$ and eta when $r$ is used as a measure of relationship. At times when $r$ is so used, it may become so small that it may be assumed that there is no relationship between $X$ and $Y$ when in actuality there may be a high relationship. A general statement may be made that when $r$ is used as a measure of relationship with data that are curvilinear, the $r$ so obtained is an underestimate of the real relationship between $X$ and $Y$. It follows then that before a Pearson $r$ is computed, the data should be plotted to determine by inspection whether or not the data are linearly related.

## EXERCISES

1. On tests $X$ and $Y$, 8 students score as shown below. Compute the Spearman rank-order correlation coefficient and test it for significance.

| X | Y |
|---|---|
| 18 | 18 |
| 15 | 16 |
| 17 | 11 |
| 9 | 20 |
| 12 | 22 |
| 8 | 26 |
| 7 | 28 |
| 3 | 30 |

2. The following represent the scores of a group of students on two tests. Calculate the Spearman rank-order correlation coefficient and test it for significance.

| TEST 1 | TEST 2 |
|---|---|
| 36 | 18 |
| 34 | 26 |
| 33 | 24 |
| 29 | 18 |
| 29 | 15 |
| 28 | 12 |
| 28 | 14 |
| 28 | 13 |
| 24 | 12 |
| 22 | 10 |
| 19 | 8 |
| 12 | 9 |

3. Calculate Kendall's $T$ for the data in problem 1.
4. The following represent the scores of 15 students on two tests. Calculate $T$ and test its significance.

| TEST 1 | TEST 2 |
|---|---|
| 54 | 23 |
| 50 | 24 |
| 50 | 22 |
| 48 | 20 |
| 47 | 19 |
| 44 | 17 |
| 43 | 16 |
| 42 | 15 |
| 42 | 13 |
| 42 | 14 |
| 41 | 18 |
| 39 | 12 |
| 38 | 10 |
| 37 | 8 |
| 37 | 9 |

5. Eight essays were ranked by 3 judges with the following results. Calculate the coefficient of concordance and test it for significance.

| Essay | Judge 1 | Judge 2 | Judge 3 |
|---|---|---|---|
| 1 | 8 | 7 | 8 |
| 2 | 6 | 5 | 6 |
| 3 | 4 | 6 | 5 |
| 4 | 1 | 2 | 1 |
| 5 | 3 | 3 | 2 |
| 6 | 2 | 1 | 3 |
| 7 | 5 | 4 | 4 |
| 8 | 7 | 8 | 7 |

6. Below are summary stanine scores on an aptitude test and the number of individuals with each score who completed successfully or did not complete a military training program. For the data compute the point-bisearial *r* and test it for significance.

| STANINE | COMPLETED | NOT COMPLETED |
|---|---|---|
| 9 | 10 | 0 |
| 8 | 8 | 1 |
| 7 | 7 | 2 |
| 6 | 12 | 4 |
| 5 | 10 | 10 |
| 4 | 4 | 8 |
| 3 | 2 | 10 |
| 2 | 1 | 10 |
| 1 | 0 | 1 |

7. Below are the responses of a sample of Democrats and Republicans to a question on an attitude scale. Is there a correlation between party affiliation and the given response? Use the phi coefficient and test it for significance.

| | AGREE | DISAGREE |
|---|---|---|
| DEMOCRAT | 100 | 80 |
| REPUBLICAN | 60 | 120 |

8. The following phi coefficients were obtained from a sample of 100 individuals. Test each to see if it is statistically significant.

a. .14 b. .22 c. .31 d. .20
e. .27 f. .12 g. .25 h. .40

9. a. In the table below the responses of 3 groups to an item on an attitude scale item are found. Compute a contingency coefficient for these data.

| | AGREE | NO OPINION | DISAGREE |
|---|---|---|---|
| DEMOCRAT | 60 | 10 | 30 |
| INDEPENDENT | 20 | 5 | 75 |
| REPUBLICAN | 80 | 5 | 15 |

   b. Compute Cramér's statistic for the same data.

10. The correlation between grades in a typewriting course and a test of manual dexterity was found to be .50 and with a test of spatial relations, .25. The correlation between the latter two tests was .15. Compute the multiple correlation coefficient between these typing grades and the combined effect of the two tests.

11. The following correlation coefficients were obtained in a research study:

| | INTELLIGENCE TEST SCORES | CHRONOLOGICAL AGE | WEIGHT |
|---|---|---|---|
| INTELLIGENCE TEST SCORES | — | .40 | .45 |
| CHRONOLOGICAL AGE | | — | .90 |
| WEIGHT | | | — |

   a. Partial out the effects of weight and determine the correlation between intelligence test scores and chronological age.
   b. Partial out the effects of chronological age and determine the correlation between intelligence test scores and weight.

# IV
# FURTHER TOPICS

# 15

# NONPARAMETRIC STATISTICAL TESTS

In the past forty years there has been extensive development of *non-parametric* or *distribution-free tests*. With the use of tests of this type, no assumptions are made or are even necessary about the shape in the population of the distribution of the trait being measured. While there are often conditions that have to be met with certain tests, such as having a continuous criterion variable, none requires that sampling be from a normal distribution.

Distribution-free statistics have certain advantages over parametric tests, among which are the relative ease of computation (many are based on ranks), speed of application, and utility with small samples, especially samples of less than 10. In such cases it is often difficult to justify the use of parametric tests. Parametric tests also require that data be collected with interval or ratio scales, whereas with distribution-free tests ordinal data are typically sufficient. In some cases even nominal data may be analyzed, as previously demonstrated with the nonparametric chi-square tests. On the negative side, distribution-free tests are less powerful than parametric tests, sometimes noticeably so. The researcher should make it a general practice, therefore, to use a parametric test whenever the assumptions are justified and time is available for the more complex calculations that are involved with the use of parametric tests such as the $t$ or $F$ tests.

In the pages that follow we shall consider a few of the widely used distribution-free tests. While most of these have been developed for use with uncorrelated data, several have also been specifically designed for use with correlated data. One of these will be found at the end of the chapter.

## TESTS FOR UNCORRELATED DATA

### THE MEDIAN TEST

To illustrate the use of the *median test* we shall use data obtained from a test given to two groups of students who were taught a unit of statistics by different methods. The obtained scores are given in Table 15.1.

The first step is to combine all the scores into one distribution and to find the median of the combined distribution of scores. Doing this for the 37 scores produces a median of 32. Then, for each distribution, the number of cases at and above the median and the number below the median are counted. These results are shown in Table 15.2. Chi-square is then computed for these data using the formula for a 2 × 2 table corrected by the Pirie-Hamden method. The procedure is as follows:

$$\chi^2 = \frac{N[|ad - bd| - .5]^2}{(a+b)(a+c)(b+d)(c+d)}$$

$$= \frac{37[|(8)(6) - (11)(12)| - .5]^2}{(19)(20)(17)(18)}$$

$$= \frac{37[|48 - 132| - .5]^2}{116280}$$

$$= \frac{37[(84) - .5]^2}{116280}$$

$$= \frac{37(83.5)^2}{116280}$$

$$= \frac{37(6872.25)}{116280}$$

$$= \frac{257979.25}{116280}$$

$$= 2.22$$

This chi-square of 2.22 with $df = 1$ is not significant at the predetermined $\alpha$ level of .05. In this case we are testing the null hypothesis that these two medians came from the same population. Since the chi-square is not significant, we have no reason for rejecting this hypothesis.

The median test may also be applied when there are more than two groups. The procedure is as outlined above. The median for all cases in all groups combined is first found. Then the number of cases at and above the median and the number below the median is counted for each group. If we had five groups this would result in a 2 × 5 table. The expected frequency in each cell is half of the number of the cases in each group.

**TABLE 15.1 SCORES OF TWO GROUPS OF STUDENTS ON A STATISTICS TEST, TAUGHT BY DIFFERENT METHODS**

| METHOD I | | METHOD 2 | |
|---|---|---|---|
| 48 | 27 | 54 | 36 |
| 47 | 27 | 50 | 30 |
| 46 | 26 | 48 | 25 |
| 46 | 25 | 47 | 25 |
| 44 | 24 | 46 | 22 |
| 40 | 18 | 45 | 20 |
| 38 | 18 | 45 | 14 |
| 32 | 16 | 45 | |
| 30 | 12 | 43 | |
| 28 | 10 | 42 | |
| | $N_1 = 20$ | | $N_2 = 17$ |

When the number of expected frequencies is small, some of them being 2 or less, the use of *chi*-square is inappropriate and Fisher's exact method should be used. (See Ferguson, 1971, pp. 340–342.)

As a statistical test the median test is deficient in power because the size of the difference of each score from the median is disregarded. The next two nonparametric tests to be described are more powerful, because the distance of each score from the center is used by taking its rank into consideration.

**TABLE 15.2 USE OF THE MEDIAN TEST WITH TWO SAMPLES**

| | METHOD 1 | METHOD 2 |
|---|---|---|
| AT AND ABOVE MEDIAN | 8 (a) | 11 (b) |
| BELOW MEDIAN | 12 (c) | 6 (d) |
| | 20 | 17 |

## THE MANN-WHITNEY *U* TEST

When the Mann-Whitney $U$ test is used we have two samples independently drawn from one or several populations with or without an equal number of cases. To illustrate an application of this test we shall use the data in Table 15.3 where there are the scores of two groups of students on a spelling test. As usual we are testing a null hypothesis that these two samples are drawn from the same population. For this test we shall set alpha at .01 and make a nondirectional test.

The first step is to rank the two distributions into one distribution, with the lowest score being given a rank of 1. These rankings appear in the last two columns of Table 15.3. Then the two columns of ranks are summed. At this point a check may be made on the work by using the following formula:

$$\frac{N(N+1)}{2} = \Sigma R_1 + \Sigma R_2 \qquad (15.1)$$

where $N$ equals the total number of cases in both samples:

$$\frac{27(28)}{2} = 251 + 127$$
$$378 = 378$$

Then we find the two statistics $U_1$ and $U_2$:

$$\begin{aligned} U_1 &= N_1N_2 + \frac{N_1(N_1+1)}{2} - \Sigma R_1 \qquad (15.2a) \\ &= (14)(13) + \frac{(14)(15)}{2} - 251 \\ &= 182 + 105 - 251 \\ &= 36 \end{aligned}$$

$$\begin{aligned} U_2 &= N_1N_2 + \frac{N_2(N_2+1)}{2} - \Sigma R_2 \qquad (15.2b) \\ &= (14)(13) + \frac{(13)(14)}{2} - 127 \\ &= 182 + 91 - 127 \\ &= 146 \end{aligned}$$

In testing for significance only the smaller of the two $U$'s is considered. In this case we use the $U$ of 36 and enter Appendix J, which is made up of four parts: $\alpha$ values of .001, .01, .025, and .05 for a nondirectional test, and $\alpha$ values of .002, .02, .05, and .10 for a directional test. Since we have set $\alpha$ at .01 and since we are making a two-tailed or nondirectional test, we enter part (b) of the table with $N_1 = 14$ and $N_2 = 13$; that is, for this difference to be significant $U$ must be 43 or less. Since the obtained $U$ is 36, we can reject the null hypothesis at the 1 percent level.

**TABLE 15.3 ILLUSTRATING THE USE OF THE MANN-WHITNEY *U* TEST WITH SCORES OF TWO GROUPS OF STUDENTS ON A SPELLING TEST**

| GROUP I | GROUP II | $R_1$ | $R_2$ |
|---|---|---|---|
| 54 | | 27.0 | |
| 50 | 45 | 26.0 | 21.5 |
| 48 | 40 | 25.0 | 18.0 |
| 47 | 32 | 24.0 | 15.0 |
| 46 | 30 | 23.0 | 12.0 |
| 45 | 30 | 21.5 | 12.0 |
| 43 | 30 | 20.0 | 12.0 |
| 42 | 29 | 19.0 | 10.0 |
| 38 | 28 | 16.5 | 8.0 |
| 38 | 28 | 16.5 | 8.0 |
| 31 | 26 | 14.0 | 4.5 |
| 28 | 25 | 8.0 | 3.0 |
| 27 | 22 | 6.0 | 2.0 |
| 26 | 21 | 4.5 | 1.0 |
| $N_1 = 14$ | $N_2 = 13$ | $\Sigma R_1 = 251.0$ | $\Sigma R_2 = 127.0$ |

When the sample size is small, that is, when either or both $N_1$ and $N_2$ are 8 or less, tables will be found in Siegel (1956) that will enable such data to be tested by this method. The $U$ test is regarded as a powerful one and a good substitute for the $t$ test in cases where the assumptions for the use of the $t$ test cannot be met.

THE *U* TEST FOR LARGE SAMPLES

None of the tables in Appendix J is usable when $N_2$ is 20 or more. In cases where $N_2$ is greater than 20 the sampling distribution of $U$ approximates that of a normal distribution with the following estimates of parameters:

$$\text{Mean} = \mu_u = N_1N_2/2$$

$$\sigma_u = \sqrt{N_1N_2(N_1 + N_2 + 1)/12}$$

Combining these we have

$$z = \frac{U_1 - \mu_u}{\sigma_u} = \frac{U_1 - N_1N_2/2}{\sqrt{N_1N_2(N_1 + N_2 + 1)/12}} \qquad (15.3)$$

Table 15.4 gives the scores of two groups of students on the same statistics test. These shall be used in calculating the *Mann-Whitney U* test for large samples in a nondirectional test with an $\alpha$ level of .01. As in the solution for small samples, the data are ranked into one distribution. Then the two columns of ranks are summed. First we find $U_1$:

$$\begin{aligned} U_1 &= N_1N_2 + \frac{N_1(N_1 + 1)}{2} - \Sigma R_1 \\ &= (23)(24) + \frac{23(24)}{2} - 400.5 \\ &= 552 + 276 - 400.5 \\ &= 427.5 \end{aligned}$$

Then $z$ is computed:

$$\begin{aligned} z &= \frac{2U_1 - N_1N_2}{\sqrt{N_1N_2(N_1 + N_2 + 1)/12}} \\ &= \frac{2(427.5) - (23)(24)}{\sqrt{(23)(24)(23 + 24 + 1)/12}} \\ &= \frac{855 - 552}{\sqrt{(552)(48)/12}} \\ &= \frac{303}{\sqrt{26496/12}} \\ &= \frac{303}{\sqrt{2208}} \\ &= \frac{303}{47} \\ &= 6.45 \end{aligned}$$

Since this $z$ is larger than the critical value of 2.58 we reject the null hypothesis of no difference well beyond the 1 percent level and accept the alternate hypothesis that the two groups differ. If $U_2$ had been used instead of $U_1$ the same $z$ would have been obtained but with the opposite sign. If a directional hypothesis had been set up with an $\alpha$ level of .01, the critical $z$ would be 2.33.

## THE KRUSKAL-WALLIS *H* TEST

*The Kruskal-Wallis H test* is an analysis of variance using ranks. To illustrate this technique we shall use the data in Table 15.5., which give the scores of four groups of students on a short aptitude test. The first step is to rank all of the data into one distribution giving the lowest score a rank of 1 with the ties being treated as for all ranked data. There is a formula for the correction of ties, but it is the consensus of statisticians that little is gained by using such a correction formula (Siegel, 1956). The

**TABLE 15.4 USE OF THE MANN-WHITNEY $U$ TEST WITH LARGE SAMPLES**

| TEST 1 | TEST 2 | $R_1$ | $R_2$ |
|---|---|---|---|
| 37 | 42 | 40.0 | 47.0 |
| 35 | 41 | 37.5 | 45.5 |
| 35 | 41 | 35.0 | 45.5 |
| 34 | 40 | 33.0 | 44.0 |
| 31 | 39 | 28.5 | 43.0 |
| 30 | 38 | 26.5 | 42.0 |
| 29 | 37 | 24.5 | 40.0 |
| 28 | 37 | 22.5 | 40.0 |
| 27 | 36 | 20.0 | 37.5 |
| 27 | 35 | 20.0 | 35.0 |
| 26 | 35 | 17.5 | 35.0 |
| 25 | 33 | 15.0 | 32.0 |
| 25 | 32 | 15.0 | 30.5 |
| 24 | 32 | 12.5 | 30.5 |
| 23 | 31 | 11.0 | 28.5 |
| 22 | 30 | 9.5 | 26.5 |
| 21 | 29 | 8.0 | 24.5 |
| 20 | 28 | 7.0 | 22.5 |
| 19 | 27 | 5.5 | 20.0 |
| 19 | 26 | 5.5 | 17.5 |
| 18 | 25 | 3.5 | 15.0 |
| 17 | 24 | 2.0 | 12.5 |
| 14 | 22 | 1.0 | 9.5 |
| | 20 | | 3.5 |
| $N_1 = 23$ | $N_2 = 24$ | $\Sigma R_1 = 400.5$ | $\Sigma R_2 = 727.5$ |

rationale behind this test is that if the sum of the ranks differs significantly, then the samples being studied are from different populations. In making the $H$ test, we first sum each column of ranks and then apply the following formula:

$$H = \frac{12}{N(N+1)}\left[\sum \frac{R_i^2}{N_k}\right] - 3(N+1) \qquad (15.4)$$

where:

$R_i$ = the sum of the ranks
$N_k$ = the number of measures in each of the $k$ samples

TABLE 15.5 SCORES OF INDIVIDUALS IN 4 GROUPS ON A SHORT APTITUDE TEST

| Group | | | | Ranks | | | |
|---|---|---|---|---|---|---|---|
| *A* | *B* | *C* | *D* | *A* | *B* | *C* | *D* |
| 12 | 14 | 18 | 14 | 21.5 | 26.5 | 29.0 | 26.5 |
| 8 | 12 | 15 | 13 | 15.0 | 21.5 | 28.0 | 25.0 |
| 6 | 12 | 12 | 12 | 8.5 | 21.5 | 21.5 | 21.5 |
| 4 | 9 | 10 | 12 | 5.0 | 17.0 | 18.0 | 21.5 |
| 2 | 7 | 8 | 7 | 2.5 | 12.0 | 15.0 | 12.0 |
| | 6 | 8 | 6 | | 8.5 | 15.0 | 8.5 |
| | 2 | 7 | 6 | | 2.5 | 12.0 | 8.5 |
| | | | 5 | | | | 6.0 |
| | | | 3 | | | | 4.0 |
| | | | 1 | | | | 1.0 |
| $N_k = 5$ | 7 | 7 | 10 | $\Sigma R = 52.5$ | 109.5 | 138.5 | 134.5 |

Then using the data in Table 15.5 we have

$$
\begin{aligned}
H &= \frac{12}{29(29+1)}\left(\frac{52.5^2}{5} + \frac{109.5^2}{7} + \frac{138.5^2}{7} + \frac{134.5^2}{10}\right) - 3(30) \\
&= \frac{12}{870}\left(\frac{2756.5}{5} + \frac{11990.25}{7} + \frac{19182.25}{7} + \frac{18090.5}{10}\right) - 90 \\
&= .014(551.25 + 1712.89 + 2740.32 + 1809.05) - 90 \\
&= .014(6813.51) - 90 \\
&= 95.39 - 90 \\
&= 5.39
\end{aligned}
$$

$H$ is interpreted as a chi-square with degrees of freedom equal to the number of samples minus one ($k - 1$). If $\alpha$ has been set at .05 for this test, the value of chi-square with 3 degrees of freedom is 7.815 and hence, the null hypothesis that the 4 groups come from the same population stands. If the number of cases in any sample is less than 5, $H$ may be evaluated for its statistical significance by using Appendix L.

## A TEST FOR DEPENDENT OR CORRELATED DATA

### THE WILCOXON MATCHED-PAIRS SIGNED RANKS TEST

The three tests discussed above are to be used only with unrelated data. Several nonparametric tests have been developed for use with data that are correlated, in pairs, or repeated measures on the same individuals. Of these several tests one of the most useful and also one of the most powerful is the *Wilcoxon matched-pairs signed ranks test.* The test shall be illustrated using the data in Table 15.6. In this table are the scores of 15 individuals on two tests. In applying this test the following procedure is used:

1. The difference is found between each pair of scores and the sign recorded. It is important to continue taking the differences in the same direction for each pair. An individual with the same score on both tests is automatically *dropped* from the analysis.
2. The differences are then ranked, the sign being disregarded in making these ranks, that is, the absolute difference is used in making these rankings.
3. The ranks are then sorted according to sign and these two columns are summed.

**TABLE 15.6 SCORES OF 15 INDIVIDUALS ON TWO TESTS ILLUSTRATING THE WILCOXON MATCHED-PAIRS SIGNED RANKS TEST**

| TEST X | TEST Y | DIFFERENCE | ABSOLUTE RANK | PLUS RANKS | MINUS RANKS |
|---|---|---|---|---|---|
| 12 | 20 | −8 | 9.0 | | −9.0 |
| 24 | 18 | 6 | 5.0 | 5.0 | |
| 28 | 17 | 11 | 13.0 | 13.0 | |
| 26 | 16 | 10 | 11.5 | 11.5 | |
| 24 | 16 | 8 | 9.0 | 9.0 | |
| 27 | 15 | 12 | 14.0 | 14.0 | |
| 22 | 14 | 8 | 9.0 | 9.0 | |
| 19 | 13 | 6 | 5.0 | 5.0 | |
| 18 | 12 | 6 | 5.0 | 5.0 | |
| 10 | 10 | 0* | | | |
| 12 | 9 | 3 | 2.0 | 2.0 | |
| 16 | 9 | 7 | 7.0 | 7.0 | |
| 18 | 8 | 10 | 11.5 | 11.5 | |
| 8 | 4 | 4 | 3.0 | 3.0 | |
| 4 | 2 | 2 | 1.0 | 1.0 | |
| | | | | $\Sigma = 96.0$ | $\Sigma = -9$ $T = 9$ |

* Dropped from the analysis at this point because of having the same score on both tests. $N$ is now reduced to 14.

4. The smaller of these sums is taken as $T$, which, in this example, is 9.
5. We enter Appendix I with $N = 14$ to evaluate the results. Suppose that we have set alpha at .01 and that we are making a nondirectional test. An inspection of the table shows that for $N = 14$ a $T$ less than 13 is significant at the 1 percent level for a nondirectional hypothesis. We thus reject $H_0$ at the 1 percent level, accepting the alternate hypothesis, $H_1$, of a difference in the scores of these groups on this test.

If there were no difference between the two sets of scores, $T$ would equal the mean of $T$, which is obtained by the following formula:

$$\bar{T} = \frac{N(N + 1)}{4} = \frac{14(15)}{4} = 52.2$$

It is obvious that a $T$ of 9 is a long way from this mean of 52.5 and, as we have shown, this $T$ is indicative of a significant difference at the .01 level.

With larger samples, $N > 25$, the sum of the ranks takes the shape of the normal distribution and we compute the $z$ statistic using formula (15.5) to obtain first the standard deviation:

$$\sigma_T = \frac{N(N + 1)(2N + 1)}{24} \qquad (15.5)$$

Then

$$z = \frac{T - \mu_T}{\sigma_T} = \frac{T - [N(N + 1)/4]}{\sqrt{\frac{N(N + 1)(2N + 1)}{24}}} \qquad (15.6)$$

The obtained $z$ is then interpreted in the usual manner.

## EXERCISES

1. Two groups, one of farmers and the other comprised of members of a trade union, were given a short attitude test on the use of food stamps. Below are their scores. By use of the median test, see if there is a significant difference.

| FARMERS | UNION MEMBERS |
|---|---|
| 12 | 28 |
| 10 | 26 |
| 10 | 25 |
| 9 | 24 |
| 8 | 24 |
| 7 | 19 |
| 7 | 18 |
| 6 | 10 |
| 4 | 9 |
| 2 | 6 |

2. The following are the scores of 4 groups of students on a test. By use of the median test, test to see if the differences are significant.

| *A* | *B* | *C* | *D* |
|---|---|---|---|
| 16 | 14 | 14 | 13 |
| 13 | 10 | 13 | 16 |
| 10 | 9 | 10 | 17 |
| 10 | 8 | 11 | 13 |
| 10 | 6 | 10 | 12 |
| 19 | 8 | 9 | 12 |
| 9 | 8 | 11 | 11 |
| 8 | 7 | 7 | 7 |
| 7 | 5 | 8 | 8 |
| 6 | 7 | 8 | 9 |

3. Apply the Mann-Whitney $U$ test to the data in exercise 1. What conclusion can you make?
4. The following are the scores of 2 groups of students on a short test.

Test for a significant difference in performance by use of the Mann-Whitney $U$ test.

| TEST 1 | TEST 2 |
|---|---|
| 38 | 32 |
| 32 | 30 |
| 30 | 28 |
| 28 | 28 |
| 23 | 27 |
| 23 | 26 |
| 22 | 25 |
| 20 | 25 |
| 12 | 24 |
| 6 | 23 |
| | 20 |
| | 18 |
| | 17 |
| | 16 |
| | 16 |
| | 12 |
| | 8 |
| | 7 |
| | 6 |
| | 4 |
| | 2 |

5. Below are the scores of 3 groups on an attitude scale. Test to see if their responses are different using the Kruskal-Wallis $H$ test.

| GROUP I | GROUP II | GROUP III |
|---|---|---|
| 38 | 42 | 18 |
| 26 | 40 | 12 |
| 18 | 38 | 10 |
| 16 | 36 | 10 |
| 16 | 32 | 8 |
| 14 | 20 | 6 |

6. Apply the Kruskal-Wallis $H$ test to the data in exercise 1.
7. Below are the scores of 3 groups of individuals on a short perceptual test. Test for differences using the Kruskal-Wallis $H$ test.

| *A* | *B* | *C* |
|---|---|---|
| 20 | 12 | 10 |
| 18 | 8 | 8 |
| 8 | 6 | 7 |
| 2 | | 4 |

8. Eleven students were given a short test and then two weeks later were given the same test. Using the Wilcoxon matched-pairs signed rank test, determine if there was a significant increase in scores.

| PRETEST | POSTTEST |
|---|---|
| 40 | 45 |
| 38 | 48 |
| 36 | 47 |
| 36 | 42 |
| 35 | 40 |
| 32 | 38 |
| 30 | 36 |
| 29 | 35 |
| 28 | 28 |
| 27 | 24 |
| 24 | 26 |

9. A group of individuals was given a test to determine their ability to add 8-digit numbers. Each subject was given three martinis and the test was then readministered. Using Wilcoxon's matched-pairs signed rank test, see if the drinks affected performance (fictitious data).

| PRETEST | POSTTEST |
|---|---|
| 122 | 106 |
| 118 | 94 |
| 100 | 102 |
| 95 | 87 |
| 84 | 72 |
| 83 | 75 |
| 80 | 64 |
| 78 | 60 |
| 72 | 40 |
| 70 | 56 |
| 65 | 60 |
| 62 | 64 |

# 16

# TEST STATISTICS AND ANALYSES

In this chapter certain applications of statistics to educational and psychological measurement will be considered. Four major topics will be included: the nature of a test score, reliability, validity, and item analysis techniques.

## THE NATURE OF A TEST SCORE

It has long been the practice to assume that any given test score, $X$, is made up of two parts: a true component and an error component. This may also be stated as a *true score*, $X_t$, and an *error score*, $X_e$. As an equation this may be expressed as

$$X = X_t + X_e \tag{16.1}$$

The true score is that score an individual would make if he were being tested with a perfectly reliable measuring instrument. Of course in actual practice such tests do not exist; so we never know what an individual's true score actually is. The error component is that part of the score associated with day-to-day fluctuations of an individual's physical and psychological condition, and aspects of the situation under which the test was administered, such as heat, light, noise effects of the examiner, and of the actual room itself. Chance or luck in happening to have studied a particular fact or concept measured by an item on one test but not on another test also affect the error score. These errors of measurement are considered to be random and have associated with them three assumptions: (1) the mean score is 0; (2) there is no correlation between true scores and error scores (this means that there is no reason for assuming that a high true score will be accompanied with a high error score or a low true score with a low error score, and so on); and (3) error scores of parallel tests (discussed below) are not correlated. Opposed to random errors we have constant errors. If, because of a mistake made by either man or machine, a constant of 5 is added to every score when a set of tests is being graded, the obtained scores would be similarly affected, and the mean of the test would be raised by 5 points. If these were random errors, the mean would be little affected, especially if the number of cases is large. In this chapter we are not interested in these constant errors.

Since the mean error score is zero, we can write the following equation:

$$\bar{X} = \bar{X}_t \tag{16.2}$$

This shows that the mean of the obtained scores is equal to the mean of the true scores. Just as any score can be broken down into two components, one associated with true scores and one associated with error scores, the variance of any test can also be divided into two parts:

$$S_x^2 = S_t^2 + S_e^2 \tag{16.3}$$

From this we see that the variance of any test is equal to the sum of the variance associated with true scores and the variance associated with error scores.

## RELIABILITY

Equation (16.3) is important because it leads to the definition of reliability:

$$S_x^2 = S_t^2 + S_e^2$$

Dividing by $S_x^2$,

$$\frac{S_x^2}{S_x^2} = \frac{S_t^2}{S_x^2} + \frac{S_e^2}{S_x^2} = 1$$

Then

$$\frac{S_t^2}{S_x^2} = 1 - \frac{S_e^2}{S_x^2} \tag{16.4}$$

*Reliability* is defined as that part of the variance that is true variance:

$$r_{xx} = \frac{S_t^2}{S_x^2} \tag{16.5}$$

or by substituting $r_{xx}$ in equation (16.4), we have

$$r_{xx} = 1 - \frac{S_e^2}{S_x^2} \tag{16.6}$$

Reliability is thus also defined as one minus that part of the variance that is error variance.

## THE STANDARD ERROR OF MEASUREMENT

Algebraic manipulation of equation (16.6) easily reduces it to

$$S_e = S_x\sqrt{1 - r_{xx}} \tag{16.7}$$

The above formula is that of the *standard error of measurement* or the standard error of a score. This statistic is widely used both in theory and in practice. Suppose that for a particular test we compute $S_e$ and find it to be 3. Let us also suppose that a person attains a score of 92 on this test. An accepted way of interpreting this standard error of measurement is to say that the chances are two out of three that this obtained score is not more than 3 units from his true score. Notice that we do not make a probability statement about the true score because this is never actually known. Theoretically the standard error of measurement may be looked upon as the variability of a series of administrations of the same test to the same individual. The mean of all of these obtained scores would be a good estimate of his true score and $S_e$, the standard deviation of all of the obtained scores about this true score.

## MEASURING RELIABILITY

While there are numerous methods of obtaining the reliability of a group of scores, we shall consider only the four most widely used techniques.

### THE TEST-RETEST METHOD

This is the oldest of the methods of computing reliability coefficients and it is quite simple. A certain test is administered and then an hour, a week, a month, or even many years later, the same test is readministered to the same individuals. A Pearson correlation coefficient is then computed between the two sets of scores, the obtained $r$ indicating the reliability of the two scores. Today this method is not widely used for several reasons. Often the time between the two administrations of the test is short and the examinees remember their first responses to items and put the same responses down the second time, whether right or wrong. This leads to an inflation of the reliability coefficient. On the other hand, if the period between administrations is long, the individuals change in many ways. They learn, they forget, they may deteriorate physically or psychologically, or become somewhat different individuals for other reasons. Such changes would affect scores and lead to a decline in the reliability of the scores.

### THE METHOD OF PARALLEL FORMS

This is a widely used method that meets some of the objections raised above. When this method is used, test $X$ is administered first, and then later test $Y$, a parallel test, is taken by the same individuals. The optimal time between the two testings is 3–4 days. In this case, also, a Pearson $r$ is computed between the two sets of scores.

The chief problem associated with this method is related to the construction of tests that are actually parallel or equivalent. Two tests are said to be parallel when they have equal means, equal variances, correlate equally with the same criterion or other measure, and are made up of similar items measuring the same objectives, achievement, abilities, or traits.

Of all the methods used in obtaining reliability coefficients, this one has the most universal applicability. It can be used with speed tests or those in which speed is an important factor in the score. This will be seen not to be true with the next two methods discussed. When compared with the test-retest method, the effects of the memory factor and practice with the items are removed, but the effects of time and other factors are still present. However when, as noted, the period between testings is several days, these effects are minimized.

### THE SPLIT-HALF METHOD

The split-half method has the advantage in that a reliability coefficient may be obtained with only one administration of a test. Each test score is broken down into two parts: a score based on the number of odd-numbered items answered correctly and a score based on the number of even-numbered items also responded to correctly. A Pearson $r$ is then run between these odd and even scores. Since test reliability is related to the length of a test, by dividing this test into two parts, each half the length of the original test, we automatically lowered the size of the reliability coefficient. At this point we make a correction by using the *Spearman-Brown formula* to obtain the reliability of a test equal in length to the original test. For this particular case the *Spearman-Brown formula* is as follows:

$$r_{xx} = \frac{2r_{oe}}{1 - r_{oe}} \qquad (16.8)$$

where $r_{oe}$ is the correlation between the odd and even scores on the test and $r_{xx}$ is the reliability of the original test. As an example, suppose we find for a certain test $r_{oe} = .80$. Then by the Spearman-Brown formula the reliability of these scores becomes .89.

This type of reliability coefficient is referred to as an internal consistency coefficient. The odd-even split is just one of the many ways in which the items of a test may be split and analyzed. For example we could split the test by putting items 1 and 2 in form 1, 3 and 4 in form 2, 5 and 6 in form 1, and so on. It makes little difference how we split a test as long as it is not the first half versus the second half.

If a test is speeded, a reliability coefficient computed by this method is inflated. In a real speed test the items are usually so easy that, given enough time, every individual should get every item right. With the split-half method applied to such a set of test scores, we should expect a coefficient of about 1. Actually few tests are built as the one just described, but many tests are timed and speed becomes a factor in an individual's score. In such cases the split-half method is not applicable. Other factors such as day-to-day fluctuations of the individuals, testing conditions, and the like, also inflate this type of coefficient because variations brought about by these factors are not treated as sources of error.

### THE SPEARMAN-BROWN FORMULA

In the discussion of the split-half method we saw one application of the split-half formula, the case where a test is doubled in length. Actually the Spearman-Brown formula can be written in a general form:

$$\frac{Nr}{1 + (N - 1)r} \tag{16.9}$$

where $N$ is the number of times that a test is to be increased or decreased. As an illustration, suppose we have a test that has a reliability of .94 and that this test is made up of 100 items. It has been decided that the test is too long and that, for the purposes of a more economical administration, the test should be cut to half of its original length. In this case $N = .5$. Using formula (16.9) we have

$$\frac{(.5)(.94)}{1 + (-.5)(.94)} = \frac{.47}{.53} = .89$$

Another use would be found in the case where, given the reliability of a 20-item test to be .60, how many items would have to be added to increase the reliability to .90. In this case we solve equation (16.9) for $N$:

$$\begin{aligned} .90 &= \frac{N(.60)}{1 + (N - 1).60} \\ &= \frac{.6N}{1 + .6N - .6} \\ &= \frac{.6N}{.6N + .4} \\ .54N + .36 &= .6N \\ .36 &= .60N - .54N \\ &= .06N \\ N &= 6 \end{aligned}$$

The new test then must be 6 times the length of the original test or be made up of 6 × 20 or 120 items to have a reliability coefficient of .90.

Note that the Spearman-Brown formula applies to ratings as well as to test scores. As an example, the reliability of a group of 5 judges may be .82. By the use of the Spearman-Brown formula it is possible to determine the reliability coefficient that would be obtained if there were 10 equally trained and competent judges.

### KUDER-RICHARDSON FORMULA NO. 20

A second method of obtaining a reliability coefficient from the single administration of a test is through the use of the Kuder-Richardsom formula No. 20, which is as follows:

$$KR_{20} = \frac{k}{k-1}\left[1 - \frac{\Sigma PQ}{S_x{}^2}\right] \tag{16.10}$$

where

$k$ = the number of items
$P$ = proportion responding correctly to an item
$Q = 1 - P$
$S_x{}^2$ = the variance of test $X$

Actually, $PQ$ is the variance of a test item. Hence the numerator of the last term is the sum of the variances of all the items.

In terms of test theory, the variance of any test may be written as follows:

$$S^2 = \Sigma P_i Q_i + 2\Sigma r_{ij}\sqrt{P_i Q_i P_j Q_j} \tag{16.11}$$

where $j > i$. We read this equation as follows: the variance of any test is equal to the sum of the variances of each item plus two times the sum of the covariances, the last term in equation (16.11). Examination of the covariance term shows that its size is directly related to the size of the $r$; $r$ being the correlation between any two items. If $r$ is zero, then the test variance is equal to the sum of the item variances, and $KR_{20}$ would be .00. As the size of the item correlations increases, the size of the covariance term increases, and it follows that the test variance also increases. As a result of the increase of the item covariance, $KR_{20}$ also increases. $KR_{20}$ then, is, in part, a measure of the homogeneity of the test; the more homogeneous the items (the higher the item intercorrelations) the higher $KR_{20}$.

$KR_{20}$ is an internal consistency measure of reliability considered to be an average of all possible splits that can be arranged with any set of items. Like the split-half method, $KR_{20}$ should not be used with speeded tests. Such factors as daily fluctuations in human behavior, testing conditions, and the like are not treated as sources of error variance and thus the size of $KR_{20}$ is inflated. It is possible for a test to have a high test-retest reliability and low $KR_{20}$ reliability as in the case where the test is heterogeneous in its content. On the other hand, a test may be high on both test-retest reliability and internal consistency reliability. Then it follows that when a test is measuring a single factor, that is, all items measuring the same thing or high item intercorrelations, $KR_{20}$ is an acceptable measure of the reliability of the test.

## GENERAL COMMENTS ON RELIABILITY

A well-made standardized achievement or aptitude test, when administered under conditions for which it was intended, should result in scores that have a reliability coefficient of .90 or over. Very often reliability coefficients based on the scores of personality or interest inventories are much lower. This is especially true when one examines the reliability of the scores of the subtests that go into the making of an individual's test profile. To be useful diagnostically, tests and subtests should be highly reliable, for when reliability coefficients are low it means that the error variance is large and, as this increases, the test becomes less and less useful. The amount of error associated with any reliability coefficient may be obtained by subtracting the coefficient from one. For example, when $r = .88$, the amount of error variance is .12, or 12 percent.

## VALIDITY

Based on the monograph, *Standards for Educational and Psychological Tests,* 1971, published jointly by the American Psychological Association and the American Educational Research Association, we will discuss three types of validity. A survey of the literature will reveal that there are many types of validity mentioned, but all of these may be fitted into one of the three types discussed here.

### CONTENT VALIDITY

*Content validity* is a nonstatistical type of validity mostly associated with achievement and aptitude tests. When a test is examined to see if it possesses content validity, one must study the test items to see if they sample adequately the material to be tested and, in addition, to see whether the objectives of the learning that the test is supposed to measure are also tested. This can be exemplified in the construction of a chemistry test for college freshmen. To demonstrate content validity for this test, one would first see that all the topics included in the course were adequately covered and in proportion to the amount of time spent on each. Second, one would see if the test items were constructed to measure all the objectives of the course, such as application of principles, problem solving, and interpretation of data, as well as the usually measured knowledge of facts and principles. Content validity then is established by selecting items that cover the material, objectives, and skills that the test is supposed to measure.

### CONSTRUCT VALIDITY

*Construct validity* is determined by examining the psychological traits or constructs that a test is supposed to measure. For example, if one built an attitude scale to measure attitudes toward the church and it was later found that regular churchgoers had significantly higher scores on the scale than non-churchgoers, one would say that construct validity had been demonstrated.

Another type of construct validity is demonstrated through the use of a statistical procedure called *factor analysis*. By the use of this technique, one can take an intelligence test or a personality inventory and show that it is made up of a group of elements or factors. For example, a personality inventory may be shown to measure factors such as dominance, achievement, deference, and the like. If scores on another test correlate with one or more of these factors, construct validity has been demonstrated.

In summary, construct validity is shown when one evaluates a test in the light of a specific construct. This may be done by examining the procedures followed in building a test, by factor analysis, by studying the results of contrasting groups on the test, or by correlation with other tests. Usually the evidence of construct validity is based on the collection of data through research over a period of time. Only in very simple situations can this type of validity be demonstrated in a single administration of a test.

### CRITERION-RELATED VALIDITY

Using *criterion-related* validity, a test or some other predictor is used to infer from obtained scores an individual's probable position on another variable. The most frequently encountered example of this is found in the prediction of freshmen grade-point averages, the criterion, from a predictor test such as the two parts of the Scholastic Aptitude Test. In this case scores made on the two parts of the test are either separately or jointly correlated with a student's scholastic indices. The copious research of this type has shown that these validity coefficients tend to fall between .40 and .60. Although such coefficients are low when compared with reliability coefficients, the reasons for this difference should be readily apparent. This test of academic ability contains only two of the factors related to academic success, verbal and mathematical abilities. Such items as motivation, study habits, health, both physical and psychological, are all part of the picture and are not considered when predicting academic performance from these aptitude tests.

The student will recall that when regression was studied in Chapter 13 it was pointed out that prediction may usually be improved by adding predictors to a prediction scheme. Since criterion-related validity is usually determined by the use of correlation between a predictor and a criterion, the student should review Chapter 13, especially the part concerned with the construction of a regression line and its use in relation to the standard error of estimate.

There are several problems associated with this type of validity. First, a regression line based on a single piece of research may be unduly affected by errors of measurement and may lead to erroneous conclusions. Any test or other predictor that shows possibilities for use should be cross-validated using another sample. If this cross-validation produces results similar to the original administration, then one can have more confidence in using it. Second, we assume that the criterion we are using is both reliable and valid. There are many cases in which a criterion is valid but not reliable, or on the other hand, reliable but not valid. Our commonly used criterion of academic grades is often found to be quite unreliable and sometimes invalid also. The same situation prevails when the ratings of supervisors are used in establishing the validity of an industrial or personnel test. Ratings may be both unreliable and lacking in validity, or lacking only one of these properties.

Third, there is the problem of the representativeness of the sample being used in setting up the prediction scheme. The sample may be very typical of samples drawn for a short period of time after the original one. If an admission officer of a university sets up one or more regression equations to predict the performance of applicants, it may work quite well for several years. However, it is possible that the characteristics of the applicants may change, the curriculums may change, the grading practices of the faculty may produce higher or lower grades, or other aspects of the institution may change. Each or all of these factors may contribute to the invalidation of the original prediction schemes.

An finally, validation studies should be carried out using a large and representative sample. A student does not have to search far in the literature to find validity studies based on $N$'s of less than 20.

## ATTENUATION

Sometimes the question is posed as to what the correlation would be between $X$ and $Y$ if both the predictor and the criterion were perfectly reliable. Using the following formula a validity coefficient may be corrected for *attenuation* (as removing the effects of these errors of measurement is called):

$$r_{x_t y_t} = \frac{r_{xy}}{\sqrt{r_{xx} r_{yy}}} \tag{16.12}$$

Using formula (16.12) we have corrected for attenuation in both the predictor and the criterion ($y$). Since many of our tests used as predictors have high reliability, .90+, we can argue that it is of little value to correct predictor scores for attenuation. However, criterion measures, $y$, are often low in reliability and it is felt that there is more justification for correcting these measures for their lack of reliability. Formula (16.12), then, may be presented as follows with correction for attenuation in the criterion only:

$$r_{x_t y_t} = \frac{r_{xy}}{\sqrt{r_{yy}}} \tag{16.13}$$

Some test makers have pointed out that correction for attenuation is acceptable in theory, but of little use in actual practice. Rather than rely upon such a correction, it would seem that the use of a large $N$ based on good sampling techniques would produce validity coefficients that were of much more value than those corrected for attenuation.

## ITEM ANALYSIS

A person who builds a test usually wants to know how good his test is. Since any test is no better than the items that make up the test, it follows that the test items must be studied carefully to determine their merit. Three factors are considered in determining the goodness of an item: The first of these is the difficulty of the item, $p$, which has already been defined as the proportion of the individuals who responded to the item and answered it correctly. Secondly, we have the correlation between responses to the item with total test scores. With large samples this statistic is either a point-biserial $r$ or a phi coefficient (Chapter 14). Such a statistic is called an *index of discrimination* and it shows how well an item is doing what the test is supposed to be doing in spreading individuals out on the basis of their knowledge or ability. For the typical achievement test we have then for each item a measure of its difficulty and an indicator of its ability to discriminate among individuals. In addition to this it is valuable to know, finally, the number or proportion of individuals who responded to the various distractors (incorrect responses) of an item. With tests used in prediction, such as intelligence tests, we also have another item statistic, the correlation between the responses to the item with some outside criterion measure such as grades or ratings. This statistic is an indicator of the validity of the item.

By using a digital computer, item analyses are easily and rapidly made and the results include all the items mentioned above in addition to the reliability of the test and its standard error of measurement. Unfortunately many individuals do not have access to such computers and still wish to analyze their tests. The following procedure will show how this is done. To illustrate this we use a 50-item test that has been administered to 60 students.

1. The papers are divided into two groups of 30 each, using the median as the point of separation. In some cases there may be a tie at the median and papers are assigned to each group of a chance method. We shall refer to the group above the median as the upper group, and the one below the lower group.
2. Then, for each group, we count the number of correct responses to each item and change these responses to proportions, calling these $P_u$ and $P_l$.
3. Subtract $P_l$ from $P_u$ to obtain the upper-lower discrimination index, $D$.

We illustrate the above in the following:

| Item | (1) | (2) | (3) | (4) | (5) | (6) |
|---|---|---|---|---|---|---|
| | *N* CORRECT UPPER *Gp* | *N* CORRECT LOWER *Gp* | $P_u$ | $P_l$ | $P$ | $D$ |
| 1 | 20 | 10 | .67 | .33 | .50 | .34 |
| 2 | 30 | 10 | 1.00 | .33 | .67 | .70 |
| 3 | 24 | 18 | .80 | .60 | .70 | .20 |
| 4 | 12 | 4 | .40 | .13 | .27 | .27 |
| 5 | 24 | 24 | .80 | .80 | .80 | .00 |
| 6 | 20 | 24 | .67 | .80 | .73 | −.13 |
| 7 | . . . | | | | | |
| 8 | . . . | | | | | |
| . | . . . | | | | | |
| . | . . . | | | | | |
| . | . . . | | | | | |
| 50 | . . . | | | | | |

In column (5) we have the difficulty values of the items. These are readily obtained by averaging the $P$ values in columns (3) and (4). In selecting items for use on an achievement we concern ourselves with the values in columns (5) and (6). Usually we try to have the average $P$ value of a test close to .50 since it has been demonstrated that a test discriminates best among individuals when such items are used. Items above .50 are chosen (the easier items), and some below .50 (the more difficult items), hoping in the end that the average will come close to .50. After studying the $P$ values, the $D$ indices are then examined. Suppose that there are two items, each with $P$ values of .60, one with a $D$ value of .10 and the other with a $D$ of .50. The second item would be selected over the first one. Items 5 and 6 should be rejected because item 5 has no discrimination value and item 6 has negative discrimination. This means that those students with the better grades are answering it incorrectly and those with the poorer scores are getting it right. With an aptitude test, the third statistic, the item validity index, is present, and items are selected on the basis of a low correlation with the total test score and a high correlation with the criterion. By doing this, the criterion-related validity of the test is determined.

To improve items that have been used and thus to improve future tests in which the item may be reused, we need to know not only the difficulty of the item, the index of discrimination of the item, but in addition, information on how each of the distractors works. To illustrate this, below are some data from the 50-item test that was considered above.

| | | Responses | | | | |
|---|---|---|---|---|---|---|
| | | *A* | *B* | *C* | *D** | *E* |
| **Item 1** | UPPER GROUP | 0 | 0 | 2 | 24 | 4 |
| | LOWER GROUP | 2 | 0 | 8 | 12 | 8 |

| | | Responses | | | | |
|---|---|---|---|---|---|---|
| | | *A* | *B** | *C* | *D* | *E* |
| **Item 2** | UPPER GROUP | 16 | 14 | 0 | 0 | 0 |
| | LOWER GROUP | 14 | 6 | 4 | 2 | 2 |

* = Correct response

Examination of item 1 reveals that no one in either group responded to distractor B. It should be replaced or revised. Distractor A also needs revision to be a more attractive response for the lower group. In item 2 we have a situation where response A is more popular than the correct answer. Perhaps the item is keyed incorrectly or that it contains ambiguities. In this manner each of the 50 items is examined. After an item is revised, it has to be tried out again to obtain new $P$ and $D$ values.

It should be noted in conclusion that all that has been said about tests in this chapter refers to the traditional type of test and testing procedures where the purpose of the test is to spread students out on the basis of their achievement or abilities. In recent years a whole new type of testing has received increasing emphasis. This goes under the name of criterion-referenced testing, mastery testing, and other names. A large amount of the material presented here on reliability, validity, and item analysis becomes irrelevant when applied to criterion-referenced tests. Or if not irrelevant, substantial changes have to be made in using these concepts.

## EXERCISES

1. In the matrix below are the responses of 16 students to a 20-item test.
   a. Find the reliability of these scores by the split-half method.

b. Using the reliability coefficient found in a, compute the standard error of measurement for this test. Interpret this statistic for an individual with a score of 13 on this test.

c. Compute the Kuder-Richardson No. 20 reliability coefficient for the following test.

| | Item | | | | | | | | | | | | | | | | | | | | |
|---|---|---|---|---|---|---|---|---|---|---|---|---|---|---|---|---|---|---|---|---|---|
| STUDENT | 1 | 2 | 3 | 4 | 5 | 6 | 7 | 8 | 9 | 10 | 11 | 12 | 13 | 14 | 15 | 16 | 17 | 18 | 19 | 20 | SCORE |
| *A* | + | + | + | + | + | + | + | + | 0 | + | + | + | + | 0 | + | + | + | + | + | + | 18 |
| *B* | + | + | + | + | + | + | + | + | + | + | + | + | + | + | + | + | + | + | + | 0 | 19 |
| *C* | + | + | + | + | 0 | + | + | 0 | + | + | 0 | 0 | 0 | 0 | 0 | 0 | 0 | 0 | 0 | 0 | 8 |
| *D* | + | + | + | + | + | + | 0 | + | + | 0 | + | 0 | 0 | + | 0 | 0 | 0 | 0 | 0 | 0 | 10 |
| *E* | + | 0 | + | + | + | + | + | + | + | + | + | 0 | + | + | 0 | + | 0 | + | + | 0 | 15 |
| *F* | + | + | + | 0 | 0 | + | 0 | + | + | 0 | 0 | 0 | 0 | 0 | 0 | 0 | 0 | 0 | 0 | 0 | 6 |
| *G* | + | + | + | + | + | + | + | + | + | + | + | 0 | 0 | 0 | 0 | 0 | 0 | 0 | 0 | 0 | 11 |
| *H* | + | + | 0 | 0 | + | + | 0 | 0 | + | 0 | 0 | + | 0 | + | + | 0 | 0 | 0 | 0 | 0 | 8 |
| *I* | + | + | 0 | + | + | + | + | + | + | + | + | + | + | + | + | + | + | + | + | 0 | 18 |
| *J* | + | + | + | + | + | + | + | + | + | + | + | + | + | + | + | + | + | + | + | + | 20 |
| *K* | + | 0 | 0 | + | 0 | + | 0 | 0 | 0 | 0 | 0 | 0 | 0 | 0 | 0 | 0 | 0 | 0 | 0 | 0 | 3 |
| *L* | + | + | + | + | + | + | 0 | + | 0 | + | + | 0 | + | 0 | + | 0 | 0 | 0 | 0 | 0 | 11 |
| *M* | + | + | + | + | + | + | + | + | + | + | + | 0 | 0 | 0 | 0 | 0 | 0 | 0 | 0 | 0 | 11 |
| *N* | + | 0 | 0 | + | + | 0 | 0 | + | + | 0 | 0 | + | + | 0 | + | 0 | + | 0 | 0 | 0 | 9 |
| *O* | + | + | + | + | + | + | + | + | + | + | + | + | + | + | 0 | + | 0 | 0 | 0 | 0 | 15 |
| *P* | + | + | + | + | + | + | + | 0 | + | 0 | 0 | 0 | + | 0 | 0 | 0 | + | 0 | 0 | 0 | 10 |
| NUMBER CORRECT | 16 | 13 | 12 | 14 | 13 | 15 | 10 | 12 | 13 | 10 | 10 | 7 | 9 | 7 | 7 | 6 | 6 | 5 | 5 | 2 | |
| *p* | | | | | | | | | | | | | | | | | | | | | |
| *q* | | | | | | | | | | | | | | | | | | | | | |
| *pq* | | | | | | | | | | | | | | | | | | | | | |

$\Sigma pq =$ ______

2. a. Suppose that a test has a reliability coefficient of .83. What would the reliability be if the test were tripled in length?

   b. An industrial psychologist is using a test that takes 60 minutes to administer. This test consists of 100 items and has a reliability coefficient of .92 associated with it. He decides to cut the test into half of its original length so that he can administer it in 30 minutes. What is the reliability of the shortened test?

   c. Using the data in problem 1a, determine how many more items would have to be added to make the reliability of the test .95.

3. Given: $r_{xx} = .85$, $r_{yy} = .60$, $r_{xy} = .70$.

   a. Correct the validity coefficient for attenuation in both the predictor and in the criterion.

   b. Correct the validity coefficient in the criterion only.

4. If possible, obtain a set of test papers and make an item analysis following the method proposed in this chapter.

# APPENDIXES

## APPENDIX A

# REVIEW OF ARITHMETIC, ALGEBRA, AND OTHER THINGS

### BASIC OPERATIONS

#### ADDITION

When the numbers to be added are all of the same sign, add in the usual way:

$$\begin{array}{r} +8 \\ +16 \\ \hline \Sigma = +24 \end{array} \qquad \begin{array}{r} -9 \\ -7 \\ \hline \Sigma = -16 \end{array}$$

When the numbers are of different sign, the best procedure is to add all the positive ones, then the negative ones, and finally combine these two sums.

$$\begin{array}{r} -8 \\ +13 \\ -17 \\ -16 \\ +24 \\ -9 \end{array}$$

Combining the positive numbers gives $+37$

Combining the negative numbers gives $-50$

$$\Sigma = -13$$

#### SUBTRACTION

When subtracting a negative number, change the sign of the subtrahend and add.

$$\begin{array}{r} 209 \\ -108 \\ \hline +317 \end{array} \qquad \begin{array}{r} -171 \\ -71 \\ \hline -100 \end{array} \qquad \begin{array}{r} -86 \\ 12 \\ \hline -98 \end{array}$$

#### MULTIPLICATION

When a positive number is multiplied by a positive number the resulting product is positive. When a negative number is multiplied by a positive number, the result is negative. When two negative numbers are multiplied, the product is positive.

$$\begin{array}{r} 73 \\ \times\ 3 \\ \hline 219 \end{array} \qquad \begin{array}{r} -73 \\ \times -4 \\ \hline 292 \end{array} \qquad \begin{array}{r} -73 \\ \times\ \ 6 \\ \hline -438 \end{array} \qquad \begin{array}{r} 73 \\ \times -7 \\ \hline -511 \end{array}$$

DIVISION

The quotient of dividing two numbers of similar signs is positive. When the dividend and divisor are of opposite signs, the results are negative:

$$\frac{144}{12} = 12 \qquad \frac{-72}{-8} = 9 \qquad \frac{77}{-11} = -7 \qquad \frac{-81}{9} = -9$$

## DECIMALS AND FRACTIONS

DECIMALS

When two decimals are multiplied, the number of decimal places in the product is equal to the sum of the number of decimal places in the numbers being multiplied.

$$1.09 \times 2.1 = 2.289 \qquad .0897 \times 1.01 = .090597$$

When two decimals are divided, the number of decimal places in the quotient is equal to the number of decimal places in the dividend minus the number of decimal places in the denominator when there is no remainder.

$$\frac{.2289}{2.1} = .109 \qquad \frac{326.444}{.004} = 81611$$

FRACTIONS

Fractions are added by using the reciprocals instead of getting a least common denominator.

$$\frac{1}{39} + \frac{1}{37} = .025641 + .027027 = .052668$$

or

$$.0256 \quad + .0270 \quad = .0526$$

$$\begin{aligned} \frac{5}{12} + \frac{3}{19} &= 5(.08333) + 3(.052632) \\ &= .41665 + .157896 \\ &= .574546 \end{aligned}$$

Similarly, with subtraction

$$\frac{3}{77} - \frac{1}{91} = 3(.0129870) - .010989$$
$$= .038961 - .010989$$
$$= .027972$$

Multiplication may be done using reciprocals; or when the values are small, a cancellation procedure such as the following may be used:

$$\frac{\not{3}}{\not{4}} \times \frac{\not{4}}{7} \times \frac{5}{\underset{3}{\not{9}}} = \frac{5}{21} = 5\left(\frac{1}{21}\right)$$
$$= 5(.047619) = .238095$$

Division of fractions is carried out by inverting the fraction and proceeding as in multiplication:

$$\frac{3}{4} \div \frac{5}{12} = \frac{3}{\not{4}} \times \frac{\overset{3}{\not{12}}}{5}$$
$$= \frac{9}{5} = 1.8$$

## REMOVING PARENTHESES

The general rule for removing parentheses is to remove the innermost ones first and then work from there to the outer ones:

$$3 + [6 - (2 + 1)] + [7(12 - 14)]$$
$$3 + (6 - 3) + 7(-2)$$
$$3 + 3 - 14$$
$$-8$$

## USE OF ZERO

Any number multiplied by zero is equal to zero.

## ROUNDING NUMBERS

Rounding each of the following to the nearest tenth we have (see Chapter 1):

$$8.36 = 8.4$$
$$72.843 = 72.8$$
$$.087 = .1$$
$$45.45 = 45.4$$
$$66.75 = 66.8$$

## EXPONENTS

$X^4$ means that this $X$ is to be multiplied by itself 4 times:

$$X \cdot X \cdot X \cdot X$$

$$\left(\frac{1}{4}\right)^4 = \frac{1}{4} \times \frac{1}{4} \times \frac{1}{4} \times \frac{1}{4} = \frac{1}{256}$$

## ALGEBRAIC MANIPULATIONS

### CLEARING OF FRACTIONS AND SOLVING FOR ANOTHER TERM

$$\bar{X} = \frac{\Sigma X}{N}$$

$$N\bar{X} = \frac{\Sigma X}{\cancel{N}} \cdot \cancel{N}$$ Each side of the equation is multiplied by $N$

$$N\bar{X} = \Sigma X$$ Then it is simplified

$$\Sigma X = N\bar{X}$$ And finally transposed

When terms are transposed from one side of the equation to the other, change the sign of the term transposed.

$$7X - 2 = 2X + 4$$
$$7X - 2X = 4 + 2$$
$$5X = 6$$
$$X = \frac{6}{5} = 1.20$$

## PROPORTIONS

A proportion ($p$) is a part of a whole and hence can never be greater than 1. When a proportion is multiplied by 100 it is changed into a percentage ($P$). Percentages may be greater than 100 as is readily seen in the daily papers. In order to have reliability when using percentages, a good rule to follow is that percentages should be based on at least 100 cases.

## CALCULATOR EXERCISES

1. Add the following:
   a. $12 + 21$ b. $-9 + -11$
   c. $22 + -38$ d. $-18, 43, -23, -26, 24, 9$
2. Subtract the following:
   a. $177 - 106$ b. $177 - (-106)$
   c. $-177 - 106$ d. $-177 - (-106)$

3. Multiply the following:
   a. (89)(9)
   b. (75)(−4)
   c. (−87)(−9)
   d. (−33)(4)
4. Divide the following:
   a. 169 by 13
   b. −221 by 11
   c. −225 by −15
   d. 625 by −21
5. Multiply the following:
   a. (1.39)(.31)
   b. (.0013)(7.6)
   c. (55.32)(.16)
   d. (4.004)(.007)
   e. $(16)^2$
   f. $(353.1)^2$
   g. $(.049)^2$
6. Divide the following:
   a. 21.222 by .43
   b. .000037 by 1.24
   c. 826.4 by .007
   d. 11.36 by .22
7. Add the following:
   a. 1/42 + 1/35
   b. 3/26 + 8/35
      12/49 + 13/71 + 12/21
8. Subtract the following:
   a. 4/89 − 1/93
   b. 16/25 − 3/37
   c. 72/79 − 3/11 − 33/37
9. Multiply the following:
   a. (2/3)(3/4)
   b. (1/6)(4/7)(5/8)
   c. (2/3)(4/7)(12/15)
10. Divide the following:
    a. 3/8 by 5/16
    b. 7/8 by 2/3
    c. 21/6 by 4/11
11. Reduce the following:
    a. $\sqrt{5867}$
    b. $\sqrt{13.081}$
    c. $\sqrt{.0029}$
12. Reduce the following by removing parentheses
    a. $(16)^2 + (21)^2 + (19)^2$
    b. $9[(8)(7) - (6)(5)]^2$
    c. $10(72 - 53) + 8(23 + 11) - 12(72 - 20)$
    d. $\sqrt{[(10)480 - 32^2][(10)580 - 30^2]}$
    e. $\dfrac{13(12 + 16)^2 - (42)(39)}{\sqrt{\left(\dfrac{123 + 148}{16 + 23 - 2}\right)\left(\dfrac{1}{8} + \dfrac{1}{17}\right)}}$

13. Simplify
    a. $\frac{1}{2^5}$
    b. $8 \times 0 \times 7$
    c. $(1/3)^5 + (1/4)^6$
    d. $(1/2) \times (3/4)^2 - (3/8)^2$
    e. 12/87
    f. $1/85 + 3/89$
14. Round to the nearest tenth
    a. 7.48
    b. 67.45
    c. 67.55
    d. 1.0426
    e. .094
    f. 67/57
15. Solve each for the respective unknown:
    a. $15Y + 3Y = 2Y - 7$
    b. $12 - 2Z = 4Z - 28$
    c. $7X - 13 - 12X = 5X + 48$
16. Write in symbols:
    a. $X$ is not equal to $Y$
    b. $X$ is greater than $Y$
    c. $X$ is equal to or less than $Y$
    d. The sum of the scores is 42
    e. $X$ is equal to or greater than $Y$
    f. The sum of the $X$'s squared minus the sum of the $Y$'s squared equals 29

# APPENDIX B

# SQUARES, SQUARE ROOTS, AND RECIPROCALS OF INTEGERS FROM 1 TO 1000

| $n$ | $n^2$ | $\sqrt{n}$ | $\frac{1}{n}$ | $\frac{1}{\sqrt{n}}$ |
|---|---|---|---|---|
| **1** | 1 | 1.0000 | 1.000000 | 1.0000 |
| **2** | 4 | 1.4142 | .500000 | .7071 |
| **3** | 9 | 1.7321 | .333333 | .5774 |
| **4** | 16 | 2.0000 | .250000 | .5000 |
| **5** | 25 | 2.2361 | .200000 | .4472 |
| **6** | 36 | 2.4495 | .166667 | .4082 |
| **7** | 49 | 2.6458 | .142857 | .3780 |
| **8** | 64 | 2.8284 | .125000 | .3536 |
| **9** | 81 | 3.0000 | .111111 | .3333 |
| **10** | 100 | 3.1623 | .100000 | .3162 |
| **11** | 121 | 3.3166 | .090909 | .3015 |
| **12** | 144 | 3.4641 | .083333 | .2887 |
| **13** | 169 | 3.6056 | .076923 | .2774 |
| **14** | 196 | 3.7417 | .071429 | .2673 |
| **15** | 225 | 3.8730 | .066667 | .2582 |
| **16** | 256 | 4.0000 | .062500 | .2500 |
| **17** | 289 | 4.1231 | .058824 | .2425 |
| **18** | 324 | 4.2426 | .055556 | .2357 |
| **19** | 361 | 4.3589 | .052632 | .2294 |
| **20** | 400 | 4.4721 | .050000 | .2236 |
| **21** | 441 | 4.5826 | .047619 | .2182 |
| **22** | 484 | 4.6904 | .045455 | .2132 |
| **23** | 529 | 4.7958 | .043478 | .2085 |
| **24** | 576 | 4.8990 | .041667 | .2041 |
| **25** | 625 | 5.0000 | .040000 | .2000 |
| **26** | 676 | 5.0990 | .038462 | .1961 |
| **27** | 729 | 5.1962 | .037037 | .1925 |
| **28** | 784 | 5.2915 | .035714 | .1890 |
| **29** | 841 | 5.3852 | .034483 | .1857 |
| **30** | 900 | 5.4772 | .033333 | .1826 |

**APPENDIX B** *(Continued)*

| $n$ | $n^2$ | $\sqrt{n}$ | $\frac{1}{n}$ | $\frac{1}{\sqrt{n}}$ |
|---|---|---|---|---|
| **31** | 961 | 5.5678 | .032258 | .1796 |
| **32** | 1024 | 5.6569 | .031250 | .1768 |
| **33** | 1089 | 5.7446 | .030303 | .1741 |
| **34** | 1156 | 5.8310 | .029412 | .1715 |
| **35** | 1225 | 5.9161 | .028571 | .1690 |
| **36** | 1296 | 6.0000 | .027778 | .1667 |
| **37** | 1369 | 6.0828 | .027027 | .1644 |
| **38** | 1444 | 6.1644 | .026316 | .1622 |
| **39** | 1521 | 6.2450 | .025641 | .1601 |
| **40** | 1600 | 6.3246 | .025000 | .1581 |
| **41** | 1681 | 6.4031 | .024390 | .1562 |
| **42** | 1764 | 6.4807 | .023810 | .1543 |
| **43** | 1849 | 6.5574 | .023256 | .1525 |
| **44** | 1936 | 6.6332 | .022727 | .1508 |
| **45** | 2025 | 6.7082 | .022222 | .1491 |
| **46** | 2116 | 6.7823 | .021739 | .1474 |
| **47** | 2209 | 6.8557 | .021277 | .1459 |
| **48** | 2304 | 6.9282 | .020833 | .1443 |
| **49** | 2401 | 7.0000 | .020408 | .1429 |
| **50** | 2500 | 7.0711 | .020000 | .1414 |
| **51** | 2601 | 7.1414 | .019608 | .1400 |
| **52** | 2704 | 7.2111 | .019231 | .1387 |
| **53** | 2809 | 7.2801 | .018868 | .1374 |
| **54** | 2916 | 7.3485 | .018519 | .1361 |
| **55** | 3025 | 7.4162 | .018182 | .1348 |
| **56** | 3136 | 7.4833 | .017857 | .1336 |
| **57** | 3249 | 7.5498 | .017544 | .1325 |
| **58** | 3364 | 7.6158 | .017241 | .1313 |
| **59** | 3481 | 7.6811 | .016949 | .1302 |
| **60** | 3600 | 7.7460 | .016667 | .1291 |
| **61** | 3721 | 7.8102 | .016393 | .1280 |
| **62** | 3844 | 7.8740 | .016129 | .1270 |
| **63** | 3969 | 7.9373 | .015873 | .1260 |
| **64** | 4096 | 8.0000 | .015625 | .1250 |
| **65** | 4225 | 8.0623 | .015385 | .1240 |
| **66** | 4356 | 8.1240 | .015152 | .1231 |
| **67** | 4489 | 8.1854 | .014925 | .1222 |
| **68** | 4624 | 8.2462 | .014706 | .1213 |
| **69** | 4761 | 8.3066 | .014492 | .1204 |
| **70** | 4900 | 8.3666 | .014286 | .1195 |

**APPENDIX B** *(Continued)*

| $n$ | $n^2$ | $\sqrt{n}$ | $\frac{1}{n}$ | $\frac{1}{\sqrt{n}}$ |
|---|---|---|---|---|
| **71** | 5041 | 8.4261 | .014085 | .1187 |
| **72** | 5184 | 8.4853 | .013889 | .1179 |
| **73** | 5329 | 8.5440 | .013699 | .1170 |
| **74** | 5476 | 8.6023 | .013514 | .1162 |
| **75** | 5625 | 8.6603 | .013333 | .1155 |
| **76** | 5776 | 8.7178 | .013158 | .1147 |
| **77** | 5929 | 8.7750 | .012987 | .1140 |
| **78** | 6084 | 8.8318 | .012821 | .1132 |
| **79** | 6241 | 8.8882 | .012658 | .1125 |
| **80** | 6400 | 8.9443 | .012500 | .1118 |
| **81** | 6561 | 9.0000 | .012346 | .1111 |
| **82** | 6724 | 9.0554 | .012195 | .1104 |
| **83** | 6889 | 9.1104 | .012048 | .1098 |
| **84** | 7056 | 9.1652 | .011905 | .1091 |
| **85** | 7225 | 9.2195 | .011765 | .1085 |
| **86** | 7396 | 9.2736 | .011628 | .1078 |
| **87** | 7569 | 9.3274 | .011494 | .1072 |
| **88** | 7744 | 9.3808 | .011364 | .1066 |
| **89** | 7921 | 9.4340 | .011236 | .1060 |
| **90** | 8100 | 9.4868 | .011111 | .1054 |
| **91** | 8281 | 9.5394 | .010989 | .1048 |
| **92** | 8464 | 9.5917 | .010870 | .1043 |
| **93** | 8649 | 9.6437 | .010753 | .1037 |
| **94** | 8836 | 9.6954 | .010638 | .1031 |
| **95** | 9025 | 9.7468 | .010526 | .1026 |
| **96** | 9216 | 9.7980 | .010417 | .1021 |
| **97** | 9409 | 9.8489 | .010309 | .1015 |
| **98** | 9604 | 9.8995 | .010204 | .1010 |
| **99** | 9801 | 9.9499 | .010101 | .1005 |
| **100** | 10000 | 10.0000 | .010000 | .1000 |
| **101** | 10201 | 10.0499 | .009901 | .0995 |
| **102** | 10404 | 10.0995 | .009804 | .0990 |
| **103** | 10609 | 10.1489 | .009709 | .0985 |
| **104** | 10816 | 10.1980 | .009615 | .0981 |
| **105** | 11025 | 10.2470 | .009524 | .0976 |
| **106** | 11236 | 10.2956 | .009434 | .0971 |
| **107** | 11449 | 10.3441 | .009346 | .0967 |
| **108** | 11664 | 10.3923 | .009259 | .0962 |
| **109** | 11881 | 10.4403 | .009174 | .0958 |
| **110** | 12100 | 10.4881 | .009091 | .0953 |

**APPENDIX B** *(Continued)*

| $n$ | $n^2$ | $\sqrt{n}$ | $\frac{1}{n}$ | $\frac{1}{\sqrt{n}}$ |
|---|---|---|---|---|
| **111** | 12321 | 10.5357 | .009009 | .0949 |
| **112** | 12544 | 10.5830 | .008929 | .0945 |
| **113** | 12769 | 10.6301 | .008850 | .0941 |
| **114** | 12996 | 10.6771 | .008772 | .0937 |
| **115** | 13225 | 10.7238 | .008696 | .0933 |
| **116** | 13456 | 10.7703 | .008621 | .0928 |
| **117** | 13689 | 10.8167 | .008547 | .0925 |
| **118** | 13924 | 10.8628 | .008475 | .0921 |
| **119** | 14161 | 10.9087 | .008403 | .0917 |
| **120** | 14400 | 10.9545 | .008333 | .0913 |
| **121** | 14641 | 11.0000 | .008264 | .0909 |
| **122** | 14884 | 11.0454 | .008197 | .0905 |
| **123** | 15129 | 11.0905 | .008130 | .0902 |
| **124** | 15376 | 11.1355 | .008065 | .0898 |
| **125** | 15625 | 11.1803 | .008000 | .0894 |
| **126** | 15876 | 11.2250 | .007937 | .0891 |
| **127** | 16129 | 11.2694 | .007874 | .0887 |
| **128** | 16384 | 11.3137 | .007813 | .0884 |
| **129** | 16641 | 11.3578 | .007752 | .0880 |
| **130** | 16900 | 11.4018 | .007692 | .0877 |
| **131** | 17161 | 11.4455 | .007634 | .0874 |
| **132** | 17424 | 11.4891 | .007576 | .0870 |
| **133** | 17689 | 11.5326 | .007519 | .0867 |
| **134** | 17956 | 11.5758 | .007463 | .0864 |
| **135** | 18225 | 11.6190 | .007407 | .0861 |
| **136** | 18496 | 11.6619 | .007353 | .0857 |
| **137** | 18769 | 11.7047 | .007299 | .0854 |
| **138** | 19044 | 11.7473 | .007246 | .0851 |
| **139** | 19321 | 11.7898 | .007194 | .0848 |
| **140** | 19600 | 11.8322 | .007143 | .0845 |
| **141** | 19881 | 11.8743 | .007092 | .0842 |
| **142** | 20164 | 11.9164 | .007042 | .0839 |
| **143** | 20449 | 11.9583 | .006993 | .0836 |
| **144** | 20736 | 12.0000 | .006944 | .0833 |
| **145** | 21025 | 12.0416 | .006897 | .0830 |
| **146** | 21316 | 12.0830 | .006849 | .0828 |
| **147** | 21609 | 12.1244 | .006803 | .0825 |
| **148** | 21904 | 12.1655 | .006757 | .0822 |
| **149** | 22201 | 12.2066 | .006711 | .0819 |
| **150** | 22500 | 12.2474 | .006667 | .0816 |

**APPENDIX B** *(Continued)*

| $n$ | $n^2$ | $\sqrt{n}$ | $\frac{1}{n}$ | $\frac{1}{\sqrt{n}}$ |
|---|---|---|---|---|
| **151** | 22801 | 12.2882 | .006623 | .0814 |
| **152** | 23104 | 12.3288 | .006579 | .0811 |
| **153** | 23409 | 12.3693 | .006536 | .0808 |
| **154** | 23716 | 12.4097 | .006494 | .0806 |
| **155** | 24025 | 12.4499 | .006452 | .0803 |
| **156** | 24336 | 12.4900 | ..006410 | .0801 |
| **157** | 24649 | 12.5300 | ..006369 | .0798 |
| **158** | 24964 | 12.5698 | .006329 | .0796 |
| **159** | 25281 | 12.6095 | .006289 | .0793 |
| **160** | 25600 | 12.6491 | .006250 | .0791 |
| **161** | 25921 | 12.6886 | .006211 | .0788 |
| **162** | 26244 | 12.7279 | .006173 | .0786 |
| **163** | 26569 | 12.7671 | .006135 | .0783 |
| **164** | 26896 | 12.8062 | .006098 | .0781 |
| **165** | 27225 | 12.8452 | .006061 | .0778 |
| **166** | 27556 | 12.8841 | ..006024 | .0776 |
| **167** | 27889 | 12.9228 | .005988 | .0774 |
| **168** | 28224 | 12.9615 | .005952 | .0772 |
| **169** | 28561 | 13.0000 | .005917 | .0769 |
| **170** | 28900 | 13.0384 | .005882 | .0767 |
| **171** | 29241 | 13.0767 | .005848 | .0765 |
| **172** | 29584 | 13.1149 | .005814 | .0762 |
| **173** | 29929 | 13.1529 | .005780 | .0760 |
| **174** | 30276 | 13.1909 | .005747 | .0758 |
| **175** | 30625 | 13.2288 | .005714 | .0756 |
| **176** | 30976 | 13.2665 | .005682 | .0754 |
| **177** | 31329 | 13.3041 | .005650 | .0752 |
| **178** | 31684 | 13.3417 | .005618 | .0750 |
| **179** | 32041 | 13.3791 | .005587 | .0747 |
| **180** | 32400 | 13.4164 | .005556 | .0745 |
| **181** | 32761 | 13.4536 | .005525 | .0743 |
| **182** | 33124 | 13.4907 | .005495 | .0741 |
| **183** | 33489 | 13.5277 | .005464 | .0739 |
| **184** | 33856 | 13.5647 | .005435 | .0737 |
| **185** | 34225 | 13.6015 | .005405 | .0735 |
| **186** | 34596 | 13.6382 | .005376 | .0733 |
| **187** | 34969 | 13.6748 | .005348 | .0731 |
| **188** | 35344 | 13.7113 | .005319 | .0729 |
| **189** | 35721 | 13.7477 | .005291 | .0727 |
| **190** | 36100 | 13.7840 | .005263 | .0725 |

**APPENDIX B** *(Continued)*

| $n$ | $n^2$ | $\sqrt{n}$ | $\frac{1}{n}$ | $\frac{1}{\sqrt{n}}$ |
|---|---|---|---|---|
| **191** | 36481 | 13.8203 | .005236 | .0724 |
| **192** | 36864 | 13.8564 | .005208 | .0722 |
| **193** | 37249 | 13.8924 | .005181 | .0720 |
| **194** | 37636 | 13.9284 | .005155 | .0718 |
| **195** | 38025 | 13.9642 | .005128 | .0716 |
| **196** | 38416 | 14.0000 | .005102 | .0714 |
| **197** | 38809 | 14.0357 | .005076 | .0712 |
| **198** | 39204 | 14.0712 | .005051 | .0711 |
| **199** | 39601 | 14.1067 | .005025 | .0709 |
| **200** | 40000 | 14.1421 | .005000 | .0707 |
| **201** | 40401 | 14.1774 | .004975 | .0705 |
| **202** | 40804 | 14.2127 | .004950 | .0704 |
| **203** | 41209 | 14.2478 | .004926 | .0702 |
| **204** | 41616 | 14.2829 | .004902 | .0700 |
| **205** | 42025 | 14.3178 | .004878 | .0698 |
| **206** | 42436 | 14.3527 | .004854 | .0697 |
| **207** | 42849 | 14.3875 | .004831 | .0695 |
| **208** | 43264 | 14.4222 | .004808 | .0693 |
| **209** | 43681 | 14.4568 | .004785 | .0692 |
| **210** | 44100 | 14.4914 | .004762 | .0690 |
| **211** | 44521 | 14.5258 | .004739 | .0688 |
| **212** | 44944 | 14.5602 | .004717 | .0687 |
| **213** | 45369 | 14.5945 | .004695 | .0685 |
| **214** | 45796 | 14.6287 | .004673 | .0684 |
| **215** | 46225 | 14.6629 | .004651 | .0682 |
| **216** | 46656 | 14.6969 | .004630 | .0680 |
| **217** | 47089 | 14.7309 | .004608 | .0679 |
| **218** | 47524 | 14.7648 | .004587 | .0677 |
| **219** | 47961 | 14.7986 | .004566 | .0676 |
| **220** | 48400 | 14.8324 | .004545 | .0674 |
| **221** | 48841 | 14.8661 | .004525 | .0673 |
| **222** | 49284 | 14.8997 | .004505 | .0671 |
| **223** | 49729 | 14.9332 | .004484 | .0670 |
| **224** | 50176 | 14.9666 | .004464 | .0668 |
| **225** | 50625 | 15.0000 | .004444 | .0667 |
| **226** | 51076 | 15.0333 | .004425 | .0665 |
| **227** | 51529 | 15.0665 | .004405 | .0664 |
| **228** | 51984 | 14.0997 | .004386 | .0662 |
| **229** | 52441 | 14.1327 | .004367 | .0661 |
| **230** | 52900 | 15.1658 | .004348 | .0659 |

**APPENDIX B** *(Continued)*

| $n$ | $n^2$ | $\sqrt{n}$ | $\frac{1}{n}$ | $\frac{1}{\sqrt{n}}$ |
|---|---|---|---|---|
| **231** | 53361 | 15.1987 | .004329 | .0658 |
| **232** | 53824 | 15.2315 | .004310 | .0657 |
| **233** | 54289 | 15.2643 | .004292 | .0655 |
| **234** | 54756 | 15.2971 | .004274 | .0654 |
| **235** | 55225 | 15.3297 | .004255 | .0652 |
| **236** | 55696 | 15.3623 | .004237 | .0651 |
| **237** | 56169 | 15.3948 | .004219 | .0650 |
| **238** | 56644 | 15.4272 | .004202 | .0648 |
| **239** | 57121 | 15.4596 | .004184 | .0647 |
| **240** | 57600 | 15.4919 | .004167 | .0645 |
| **241** | 58081 | 15.5242 | .004149 | .0644 |
| **242** | 58564 | 15.5563 | .004132 | .0643 |
| **243** | 59049 | 15.5885 | .004115 | .0642 |
| **244** | 59536 | 15.6205 | .004098 | .0640 |
| **245** | 60025 | 15.6525 | .004082 | .0639 |
| **246** | 60516 | 15.6844 | .004065 | .0638 |
| **247** | 61009 | 15.7162 | .004049 | .0636 |
| **248** | 61504 | 15.7480 | .004032 | .0635 |
| **249** | 62001 | 15.7797 | .004016 | .0634 |
| **250** | 62500 | 15.8114 | .004000 | .0632 |
| **251** | 63001 | 15.8430 | .003984 | .0631 |
| **252** | 63504 | 15.8745 | .003968 | .0630 |
| **253** | 64009 | 15.9060 | .003953 | .0629 |
| **254** | 64516 | 15.9374 | .003937 | .0627 |
| **255** | 65025 | 15.9687 | .003922 | .0626 |
| **256** | 65536 | 16.0000 | .003906 | .0625 |
| **257** | 66049 | 16.0312 | .003891 | .0624 |
| **258** | 66564 | 16.0624 | .003876 | .0623 |
| **259** | 67081 | 16.0935 | .003861 | .0621 |
| **260** | 67600 | 16.1245 | .003846 | .0620 |
| **261** | 68121 | 16.1555 | .003831 | .0619 |
| **262** | 68644 | 16.1864 | .003817 | .0618 |
| **263** | 69169 | 16.2173 | .003802 | .0617 |
| **264** | 69696 | 16.2481 | .003788 | .0615 |
| **265** | 70225 | 16.2788 | .003774 | .0614 |
| **266** | 70756 | 16.3095 | .003759 | .0613 |
| **267** | 71289 | 16.3401 | .003745 | .0612 |
| **268** | 71824 | 16.3707 | .003731 | .0611 |
| **269** | 72361 | 16.4012 | .003717 | .0610 |
| **270** | 72900 | 16.4317 | .003704 | .0609 |

**APPENDIX B** *(Continued)*

| $n$ | $n^2$ | $\sqrt{n}$ | $\frac{1}{n}$ | $\frac{1}{\sqrt{n}}$ |
|---|---|---|---|---|
| **271** | 73441 | 16.4621 | .003690 | .0607 |
| **272** | 73984 | 16.4924 | .003676 | .0606 |
| **273** | 74529 | 16.5227 | .003663 | .0605 |
| **274** | 75076 | 16.5529 | .003650 | .0604 |
| **275** | 75625 | 16.5831 | .003636 | .0603 |
| **276** | 76176 | 16.6132 | .003623 | .0602 |
| **277** | 76729 | 16.6433 | .003610 | .0601 |
| **278** | 77284 | 16.6733 | .003597 | .0600 |
| **279** | 77841 | 16.7033 | .003584 | .0599 |
| **280** | 78400 | 16.7332 | .003571 | .0598 |
| **281** | 78961 | 16.7631 | .003559 | .0597 |
| **282** | 79524 | 16.7929 | .003546 | .0595 |
| **283** | 80089 | 16.8226 | .003534 | .0594 |
| **284** | 80656 | 16.8523 | .003521 | .0593 |
| **285** | 81225 | 16.8819 | .003509 | .0592 |
| **286** | 81796 | 16.9115 | .003497 | .0591 |
| **287** | 82369 | 16.9411 | .003484 | .0590 |
| **288** | 82944 | 16.9706 | .003472 | .0589 |
| **289** | 83521 | 17.0000 | .003460 | .0588 |
| **290** | 84100 | 17.0294 | .003448 | .0587 |
| **291** | 84681 | 17.0587 | .003436 | .0586 |
| **292** | 85264 | 17.0880 | .003425 | .0585 |
| **293** | 85849 | 17.1172 | .003413 | .0584 |
| **294** | 86436 | 17.1464 | .003401 | .0583 |
| **295** | 87025 | 17.1756 | .003390 | .0582 |
| **296** | 87616 | 17.2047 | .003378 | .0581 |
| **297** | 88209 | 17.2337 | .003367 | .0580 |
| **298** | 88804 | 17.2627 | .003356 | .0579 |
| **299** | 89401 | 17.2916 | .003344 | .0578 |
| **300** | 90000 | 17.3205 | .003333 | .0577 |
| **301** | 90601 | 17.3494 | .003322 | .0576 |
| **302** | 91204 | 17.3781 | .003311 | .0575 |
| **303** | 91809 | 17.4069 | .003300 | .0574 |
| **304** | 92416 | 17.4356 | .003289 | .0574 |
| **305** | 93025 | 17.4642 | .003279 | .0573 |
| **306** | 93636 | 17.4929 | .003268 | .0572 |
| **307** | 94249 | 17.5214 | .003257 | .0571 |
| **308** | 94864 | 17.5499 | .003247 | .0570 |
| **309** | 95481 | 17.5784 | .003236 | .0569 |
| **310** | 96100 | 17.6068 | .003226 | .0568 |

**APPENDIX B** *(Continued)*

| $n$ | $n^2$ | $\sqrt{n}$ | $\frac{1}{n}$ | $\frac{1}{\sqrt{n}}$ |
|---|---|---|---|---|
| **311** | 96721 | 17.6352 | .003215 | .0567 |
| **312** | 97344 | 17.6635 | .003205 | .0566 |
| **313** | 97969 | 17.6918 | .003195 | .0565 |
| **314** | 98596 | 17.7200 | .003185 | .0564 |
| **315** | 99225 | 17.7482 | .003175 | .0563 |
| **316** | 99856 | 17.7764 | .003165 | .0563 |
| **317** | 100489 | 17.8045 | .003155 | .0562 |
| **318** | 101124 | 17.8326 | .003145 | .0561 |
| **319** | 101761 | 17.8606 | .003135 | .0560 |
| **320** | 102400 | 17.8885 | .003125 | .0559 |
| **321** | 103041 | 17.9165 | .003115 | .0558 |
| **322** | 103684 | 17.9444 | .003106 | .0557 |
| **323** | 104329 | 17.9722 | .003096 | .0556 |
| **324** | 104976 | 18.0000 | .003086 | .0556 |
| **325** | 105625 | 18.0278 | .003077 | .0555 |
| **326** | 106276 | 18.0555 | .003067 | .0554 |
| **327** | 106929 | 18.0831 | .003058 | .0553 |
| **328** | 107584 | 18.1108 | .003049 | .0052 |
| **329** | 108241 | 18.1384 | .003040 | .0551 |
| **330** | 108900 | 18.1659 | .003030 | .0550 |
| **331** | 109561 | 18.1934 | .003021 | .0550 |
| **332** | 110224 | 18.2209 | .003012 | .0549 |
| **333** | 110889 | 18.2483 | .003003 | .0548 |
| **334** | 111556 | 18.2757 | .002994 | .0547 |
| **335** | 112225 | 18.3030 | .002985 | .0546 |
| **336** | 112896 | 18.3303 | .002976 | .0546 |
| **337** | 113569 | 18.3576 | .002967 | .0545 |
| **338** | 114244 | 18.3848 | .002959 | .0544 |
| **339** | 114921 | 18.4120 | .002950 | .0543 |
| **340** | 115600 | 18.4391 | .002941 | .0542 |
| **341** | 116281 | 18.4662 | .002933 | .0542 |
| **342** | 116964 | 18.4932 | .002924 | .0541 |
| **343** | 117649 | 18.5203 | .002915 | .0540 |
| **344** | 118336 | 18.5472 | .002907 | .0539 |
| **345** | 119025 | 18.5742 | .002899 | .0538 |
| **346** | 119716 | 18.6011 | .002890 | .0538 |
| **347** | 120409 | 18.6279 | .002882 | .0537 |
| **348** | 121104 | 18.6548 | .002874 | .0536 |
| **349** | 121801 | 18.6815 | .002865 | .0535 |
| **350** | 122500 | 18.7083 | .002857 | .0535 |

**APPENDIX B** *(Continued)*

| $n$ | $n^2$ | $\sqrt{n}$ | $\frac{1}{n}$ | $\frac{1}{\sqrt{n}}$ |
|---|---|---|---|---|
| **351** | 123201 | 18.7350 | .002849 | .0534 |
| **352** | 123904 | 18.7617 | .002841 | .0533 |
| **353** | 124609 | 18.7883 | .002833 | .0532 |
| **354** | 125316 | 18.8149 | .002825 | .0531 |
| **355** | 126025 | 18.8414 | .002817 | .0531 |
| **356** | 126736 | 18.8680 | .002809 | .0530 |
| **357** | 127449 | 18.8944 | .002801 | .0529 |
| **358** | 128164 | 18.9209 | .002793 | .0529 |
| **359** | 128881 | 18.9473 | .002786 | .0528 |
| **360** | 129600 | 18.9737 | .002778 | .0527 |
| **361** | 130321 | 19.0000 | .002770 | .0526 |
| **362** | 131044 | 19.0263 | .002762 | .0526 |
| **363** | 131769 | 19.0526 | .002755 | .0525 |
| **364** | 132496 | 19.0788 | .002747 | .0524 |
| **365** | 133225 | 19.1050 | .002740 | .0523 |
| **366** | 133956 | 19.1311 | .002732 | .0523 |
| **367** | 134689 | 19.1572 | .002725 | .0522 |
| **368** | 135424 | 19.1833 | .002717 | .0521 |
| **369** | 136161 | 19.2094 | .002710 | .0521 |
| **370** | 136900 | 19.2354 | .002703 | .0520 |
| **371** | 137641 | 19.2614 | .002695 | .0519 |
| **372** | 138384 | 10.2873 | .002688 | .0518 |
| **373** | 139129 | 19.3132 | .002681 | .0518 |
| **374** | 139876 | 19.3391 | .002674 | .0517 |
| **375** | 140625 | 19.3649 | .002667 | .0516 |
| **376** | 141376 | 19.3907 | .002660 | .0516 |
| **377** | 142129 | 19.4165 | .002653 | .0515 |
| **378** | 142884 | 19.4422 | .002646 | .0514 |
| **379** | 143641 | 19.4679 | .002639 | .0514 |
| **380** | 144400 | 19.4936 | .002632 | .0513 |
| **381** | 145161 | 19.5192 | .002625 | .0512 |
| **382** | 145924 | 19.5448 | .002618 | .0512 |
| **383** | 146689 | 19.5704 | .002611 | .0511 |
| **384** | 147456 | 19.5959 | .002604 | .0510 |
| **385** | 148225 | 19.6214 | .002597 | .0510 |
| **386** | 148996 | 19.6469 | .002591 | .0509 |
| **387** | 149769 | 19.6723 | .002584 | .0508 |
| **388** | 150544 | 19.6977 | .002577 | .0508 |
| **389** | 151321 | 19.7231 | .002571 | .0507 |
| **390** | 152100 | 19.7484 | .002564 | .0506 |

**APPENDIX B** *(Continued)*

| $n$ | $n^2$ | $\sqrt{n}$ | $\frac{1}{n}$ | $\frac{1}{\sqrt{n}}$ |
|---|---|---|---|---|
| **391** | 152881 | 19.7737 | .002558 | .0506 |
| **392** | 153664 | 19.7990 | .002551 | .0505 |
| **393** | 154449 | 19.8242 | .002545 | .0504 |
| **394** | 155236 | 19.8494 | .002538 | .0504 |
| **395** | 156025 | 19.8746 | .002532 | .0503 |
| **396** | 156816 | 19.8997 | .002525 | .0503 |
| **397** | 157609 | 19.9249 | .002519 | .0502 |
| **398** | 158404 | 19.9499 | .002513 | .0501 |
| **399** | 159201 | 19.9750 | .002506 | .0501 |
| **400** | 160000 | 20.0000 | .002500 | .0500 |
| **401** | 160801 | 20.0250 | .002494 | .0499 |
| **402** | 161604 | 20.0499 | .002488 | .0499 |
| **403** | 162409 | 20.0749 | .002481 | .0498 |
| **404** | 163216 | 20.0998 | .002475 | .0498 |
| **405** | 164025 | 20.1246 | .002469 | .0497 |
| **406** | 164836 | 20.1494 | .002463 | .0496 |
| **407** | 165649 | 20.1742 | .002457 | .0496 |
| **408** | 166464 | 20.1990 | .002451 | .0495 |
| **409** | 167281 | 20.2237 | .002445 | .0494 |
| **410** | 168100 | 20.2485 | .002439 | .0494 |
| **411** | 168921 | 20.2731 | .002433 | .0493 |
| **412** | 169744 | 20.2978 | .002427 | .0493 |
| **413** | 170569 | 20.3224 | .002421 | .0492 |
| **414** | 171396 | 20.3470 | .002415 | .0491 |
| **415** | 172225 | 20.3715 | .002410 | .0491 |
| **416** | 173056 | 20.3961 | .002404 | .0490 |
| **417** | 173889 | 20.4206 | .002398 | .0490 |
| **418** | 174724 | 20.4450 | .002392 | .0489 |
| **419** | 175561 | 20.4695 | .002387 | .0489 |
| **420** | 176400 | 20.4939 | .002381 | .0488 |
| **421** | 177241 | 20.5183 | .002375 | .0487 |
| **422** | 178084 | 20.5426 | .002370 | .0487 |
| **423** | 178929 | 20.5670 | .002364 | .0486 |
| **424** | 179776 | 20.5913 | .002358 | .0486 |
| **425** | 180625 | 20.6155 | .002353 | .0485 |
| **426** | 181476 | 20.6398 | .002347 | .0485 |
| **427** | 182329 | 20.6640 | .002342 | .0484 |
| **428** | 183184 | 20.6882 | .002336 | .0483 |
| **429** | 184041 | 20.7123 | .002331 | .0483 |
| **430** | 184900 | 20.7364 | .002326 | .0482 |

**APPENDIX B** *(Continued)*

| $n$ | $n^2$ | $\sqrt{n}$ | $\frac{1}{n}$ | $\frac{1}{\sqrt{n}}$ |
|---|---|---|---|---|
| **431** | 185761 | 20.7605 | .002320 | .0482 |
| **432** | 186624 | 20.7846 | .002315 | .0481 |
| **433** | 187489 | 20.8087 | .002309 | .0481 |
| **434** | 188356 | 20.8327 | .002304 | .0480 |
| **435** | 189225 | 20.8567 | .002299 | .0479 |
| **436** | 190096 | 20.8806 | .002294 | .0479 |
| **437** | 190969 | 20.9045 | .002288 | .0478 |
| **438** | 191844 | 20.9284 | .002283 | .0478 |
| **439** | 192721 | 20.9523 | .002278 | .0477 |
| **440** | 193600 | 20.9762 | .002273 | .0477 |
| **441** | 194481 | 21.0000 | .002268 | .0476 |
| **442** | 195364 | 21.0238 | .002262 | .0476 |
| **443** | 196249 | 21.0476 | .002257 | .0475 |
| **444** | 197136 | 21.0713 | .002252 | .0475 |
| **445** | 198025 | 21.0950 | .002247 | .0474 |
| **446** | 198916 | 21.1187 | .002242 | .0474 |
| **447** | 199809 | 21.1424 | .002237 | .0473 |
| **448** | 200704 | 21.1660 | .002232 | .0472 |
| **449** | 201601 | 21.1896 | .002227 | .0472 |
| **450** | 202500 | 21.2132 | .002222 | .0471 |
| **451** | 203401 | 21.2368 | .002217 | .0471 |
| **452** | 204304 | 21.2603 | .002212 | .0470 |
| **453** | 205209 | 21.2838 | .002208 | .0470 |
| **454** | 206116 | 21.3073 | .002203 | .0469 |
| **455** | 207025 | 21.3307 | .002198 | .0469 |
| **456** | 207936 | 21.3542 | .002193 | .0468 |
| **457** | 208849 | 21.3776 | .002188 | .0468 |
| **458** | 209764 | 21.4009 | .002183 | .0467 |
| **459** | 210681 | 21.4243 | .002179 | .0467 |
| **460** | 211600 | 21.4476 | .002174 | .0466 |
| **461** | 212521 | 21.4709 | .002169 | .0466 |
| **462** | 213444 | 21.4942 | .002165 | .0465 |
| **463** | 214369 | 21.5174 | .002160 | .0465 |
| **464** | 215296 | 21.5407 | .002155 | .0464 |
| **465** | 216225 | 21.5639 | .002151 | .0464 |
| **466** | 217156 | 21.5870 | .002146 | .0463 |
| **467** | 218089 | 21.6102 | .002141 | .0463 |
| **468** | 219024 | 21.6333 | .002137 | .0462 |
| **469** | 219961 | 21.6564 | .002132 | .0462 |
| **470** | 220900 | 21.6795 | .002128 | .0461 |

**APPENDIX B** *(Continued)*

| $n$ | $n^2$ | $\sqrt{n}$ | $\frac{1}{n}$ | $\frac{1}{\sqrt{n}}$ |
|---|---|---|---|---|
| **471** | 221841 | 21.7025 | .002123 | .0461 |
| **472** | 222784 | 21.7256 | .002119 | .0460 |
| **473** | 223729 | 21.7486 | .002114 | .0460 |
| **474** | 224676 | 21.7715 | .002110 | .0459 |
| **475** | 225625 | 21.7945 | .002105 | .0459 |
| **476** | 226576 | 21.8174 | .002101 | .0458 |
| **477** | 227529 | 21.8403 | .002096 | .0458 |
| **478** | 228484 | 21.8632 | .002092 | .0457 |
| **479** | 229441 | 21.8861 | .002088 | .0457 |
| **480** | 230400 | 21.9089 | .002083 | .0456 |
| **481** | 231361 | 21.9317 | .002079 | .0456 |
| **482** | 232324 | 21.9545 | .002075 | .0455 |
| **483** | 233289 | 21.9773 | .002070 | .0455 |
| **484** | 234256 | 22.0000 | .002066 | .0455 |
| **485** | 235225 | 22.0227 | .002062 | .0454 |
| **486** | 236196 | 22.0454 | .002058 | .0454 |
| **487** | 237169 | 22.0681 | .002053 | .0453 |
| **488** | 238144 | 22.0907 | .002049 | .0453 |
| **489** | 239121 | 22.1133 | .002045 | .0452 |
| **490** | 240100 | 22.1359 | .002041 | .0452 |
| **491** | 241081 | 22.1585 | .002037 | .0451 |
| **492** | 242064 | 22.1811 | .002033 | .0451 |
| **493** | 243049 | 22.2036 | .002028 | .0450 |
| **494** | 244036 | 22.2261 | .002024 | .0450 |
| **495** | 245025 | 22.2486 | .002020 | .0449 |
| **496** | 246016 | 22.2711 | .002016 | .0448 |
| **497** | 247009 | 22.2935 | .002012 | .0449 |
| **498** | 248004 | 22.3159 | .002008 | .0449 |
| **499** | 249001 | 22.3383 | .002004 | .0448 |
| **500** | 250000 | 22.3607 | .002000 | .0447 |
| **501** | 251001 | 22.3830 | .001996 | .0447 |
| **502** | 252004 | 22.4054 | .001992 | .0446 |
| **503** | 253009 | 22.4277 | .001988 | .0446 |
| **504** | 254016 | 22.4499 | .001984 | .0445 |
| **505** | 255025 | 22.4722 | .001980 | .0445 |
| **506** | 256036 | 22.4944 | .001976 | .0445 |
| **507** | 257049 | 22.5167 | .001972 | .0444 |
| **508** | 258064 | 22.5389 | .001969 | .0444 |
| **509** | 259081 | 22.5610 | .001965 | .0443 |
| **510** | 260100 | 22.5832 | .001961 | .0443 |

**APPENDIX B** (*Continued*)

| $n$ | $n^2$ | $\sqrt{n}$ | $\frac{1}{n}$ | $\frac{1}{\sqrt{n}}$ |
|---|---|---|---|---|
| **511** | 261121 | 22.6053 | .001957 | .0442 |
| **512** | 262144 | 22.6274 | .001953 | .0442 |
| **513** | 263169 | 22.6495 | .001949 | .0442 |
| **514** | 264196 | 22.6716 | .001946 | .0441 |
| **515** | 265225 | 22.6936 | .001942 | .0441 |
| **516** | 266256 | 22.7156 | .001938 | .0440 |
| **517** | 267289 | 22.7376 | .001934 | .0440 |
| **518** | 268324 | 22.7596 | .001931 | .0439 |
| **519** | 269361 | 22.7816 | .001927 | .0439 |
| **520** | 270400 | 22.8035 | .001923 | .0439 |
| **521** | 271441 | 22.8254 | .001919 | .0438 |
| **522** | 272484 | 22.8473 | .001916 | .0438 |
| **523** | 273529 | 22.8692 | .001912 | .0437 |
| **524** | 274576 | 22.8910 | .001908 | .0437 |
| **525** | 275625 | 22.9129 | .001905 | .0436 |
| **526** | 276676 | 22.9347 | .001901 | .0436 |
| **527** | 277729 | 22.9565 | .001898 | .0436 |
| **528** | 278784 | 22.9783 | .001894 | .0435 |
| **529** | 279841 | 23.0000 | .001890 | .0435 |
| **530** | 280900 | 23.0217 | .001887 | .0434 |
| **531** | 281961 | 23.0434 | .001883 | .0434 |
| **532** | 283024 | 23.0651 | .001880 | .0434 |
| **533** | 284089 | 23.0868 | .001876 | .0433 |
| **534** | 285156 | 23.1084 | .001873 | .0433 |
| **535** | 286225 | 23.1301 | .001869 | .0432 |
| **536** | 287296 | 23.1517 | .001866 | .0432 |
| **537** | 288369 | 23.1733 | .001862 | .0432 |
| **538** | 289444 | 23.1948 | .001859 | .0431 |
| **539** | 290521 | 23.2164 | .001855 | .0431 |
| **540** | 291600 | 23.2379 | .001852 | .0430 |
| **541** | 292681 | 23.2594 | .001848 | .0430 |
| **542** | 293764 | 23.2809 | .001845 | .0430 |
| **543** | 294849 | 23.3024 | .001842 | .0429 |
| **544** | 295936 | 23.3238 | .001838 | .0429 |
| **545** | 297025 | 23.3452 | .001835 | .0428 |
| **546** | 298116 | 23.3666 | .001832 | .0428 |
| **547** | 299209 | 23.3880 | .001828 | .0428 |
| **548** | 300304 | 23.4094 | .001825 | .0427 |
| **549** | 301401 | 23.4307 | .001821 | .0427 |
| **550** | 302500 | 23.4521 | .001818 | .0426 |

**APPENDIX B** *(Continued)*

| $n$ | $n^2$ | $\sqrt{n}$ | $\frac{1}{n}$ | $\frac{1}{\sqrt{n}}$ |
|---|---|---|---|---|
| **551** | 303601 | 23.4734 | .001815 | .0426 |
| **552** | 304704 | 23.4947 | .001812 | .0426 |
| **553** | 305809 | 23.5160 | .001808 | .0425 |
| **554** | 306916 | 23.5372 | .001805 | .0425 |
| **555** | 308025 | 23.5584 | .001802 | .0424 |
| **556** | 309136 | 23.5797 | .001799 | .0424 |
| **557** | 310249 | 23.6008 | .001795 | .0424 |
| **558** | 311364 | 23.6220 | .001792 | .0423 |
| **559** | 312481 | 23.6432 | .001789 | .0423 |
| **560** | 313600 | 23.6643 | .001786 | .0423 |
| **561** | 314721 | 23.6854 | .001783 | .0422 |
| **562** | 315844 | 23.7065 | .001779 | .0422 |
| **563** | 316969 | 23.7276 | .001776 | .0421 |
| **564** | 318096 | 23.7487 | .001773 | .0421 |
| **565** | 319225 | 23.7697 | .001770 | .0421 |
| **566** | 320356 | 23.7908 | .001767 | .0420 |
| **567** | 321489 | 23.8118 | .001764 | .0420 |
| **568** | 322624 | 23.8328 | .001761 | .0420 |
| **569** | 323761 | 23.8537 | .001757 | .0419 |
| **570** | 324900 | 23.8747 | .001754 | .0419 |
| **571** | 326041 | 23.8956 | .001751 | .0418 |
| **572** | 327184 | 23.9165 | .001748 | .0418 |
| **573** | 328329 | 23.9374 | .001745 | .0418 |
| **574** | 329476 | 23.9583 | .001742 | .0417 |
| **575** | 330625 | 23.9792 | .001739 | .0417 |
| **576** | 331776 | 24.0000 | .001736 | .0417 |
| **577** | 332929 | 24.0208 | .001733 | .0416 |
| **578** | 334084 | 24.0416 | .001730 | .0416 |
| **579** | 335241 | 24.0624 | .001727 | .0416 |
| **580** | 336400 | 24.0832 | .001724 | .0415 |
| **581** | 337561 | 24.1039 | .001721 | .0415 |
| **582** | 338724 | 24.1247 | .001718 | .0415 |
| **583** | 339889 | 24.1454 | .001715 | .0414 |
| **584** | 341056 | 24.1661 | .001712 | .0414 |
| **585** | 342225 | 24.1868 | .001709 | .0413 |
| **586** | 343396 | 24.2074 | .001706 | .0413 |
| **587** | 344569 | 24.2281 | .001704 | .0413 |
| **588** | 345744 | 24.2487 | .001701 | .0412 |
| **589** | 346921 | 24.2693 | .001698 | .0412 |
| **590** | 348100 | 24.2899 | .001695 | .0412 |

**APPENDIX B** *(Continued)*

| $n$ | $n^2$ | $\sqrt{n}$ | $\frac{1}{n}$ | $\frac{1}{\sqrt{n}}$ |
|---|---|---|---|---|
| **591** | 349281 | 24.3105 | .001692 | .0411 |
| **592** | 350464 | 24.3311 | .001689 | .0411 |
| **593** | 351649 | 24.3516 | .001686 | .0411 |
| **594** | 352836 | 24.3721 | .001684 | .0410 |
| **595** | 354025 | 24.3926 | .001681 | .0410 |
| **596** | 355216 | 24.4131 | .001678 | .0410 |
| **597** | 356409 | 24.4336 | .001675 | .0409 |
| **598** | 357604 | 24.4540 | .001672 | .0409 |
| **599** | 358801 | 24.4745 | .001669 | .0409 |
| **600** | 360000 | 24.4949 | .001667 | .0408 |
| **601** | 361201 | 24.5153 | .001664 | .0408 |
| **602** | 362404 | 24.5357 | .001661 | .0408 |
| **603** | 363609 | 24.5561 | .001658 | .0407 |
| **604** | 364816 | 24.5764 | .001656 | .0407 |
| **605** | 366025 | 24.5967 | .001653 | .0407 |
| **606** | 367236 | 24.6171 | .001650 | .0406 |
| **607** | 368449 | 24.6374 | .001647 | .0406 |
| **608** | 369664 | 24.6577 | .001645 | .0406 |
| **609** | 370881 | 24.6779 | .001642 | .0405 |
| **610** | 372100 | 24.6982 | .001639 | .0405 |
| **611** | 373321 | 24.7184 | .001637 | .0405 |
| **612** | 374544 | 24.7386 | .001634 | .0404 |
| **613** | 375769 | 24.7588 | .001631 | .0404 |
| **614** | 376996 | 24.7790 | .001629 | .0404 |
| **615** | 378225 | 24.7992 | .001626 | .0403 |
| **616** | 379456 | 24.8193 | .001623 | .0403 |
| **617** | 380689 | 24.8395 | .001621 | .0403 |
| **618** | 381924 | 24.8596 | .001618 | .0402 |
| **619** | 383161 | 24.8797 | .001616 | .0402 |
| **620** | 384400 | 24.8998 | .001613 | .0402 |
| **621** | 385641 | 24.9199 | .001610 | .0401 |
| **622** | 386884 | 24.9399 | .001608 | .0401 |
| **623** | 388129 | 24.9600 | .001605 | .0401 |
| **624** | 389376 | 24.9800 | .001603 | .0400 |
| **625** | 390625 | 25.0000 | .001600 | .0400 |
| **626** | 391876 | 25.0200 | .001597 | .0400 |
| **627** | 393129 | 25.0400 | .001595 | .0399 |
| **628** | 394384 | 25.0599 | .001592 | .0399 |
| **629** | 395641 | 25.0799 | .001590 | .0399 |
| **630** | 396900 | 25.0998 | .001587 | .0398 |

**APPENDIX B** *(Continued)*

| $n$ | $n^2$ | $\sqrt{n}$ | $\frac{1}{n}$ | $\frac{1}{\sqrt{n}}$ |
|---|---|---|---|---|
| **631** | 398161 | 25.1197 | .001585 | .0398 |
| **632** | 399424 | 25.1396 | .001582 | .0398 |
| **633** | 400689 | 25.1595 | .001580 | .0397 |
| **634** | 401956 | 25.1794 | .001577 | .0397 |
| **635** | 403225 | 25.1992 | .001575 | .0397 |
| **636** | 404496 | 25.2190 | .001572 | .0397 |
| **637** | 405769 | 25.2389 | .001570 | .0396 |
| **638** | 407044 | 25.2587 | .001567 | .0396 |
| **639** | 408321 | 25.2784 | .001565 | .0396 |
| **640** | 409600 | 25.2982 | .001563 | .0395 |
| **641** | 410881 | 25.3180 | .001560 | .0395 |
| **642** | 412164 | 25.3377 | .001558 | .0395 |
| **643** | 413449 | 25.3574 | .001555 | .0394 |
| **644** | 414736 | 25.3772 | .001553 | .0394 |
| **645** | 416025 | 25.3969 | .001550 | .0394 |
| **646** | 417316 | 25.4165 | .001548 | .0393 |
| **647** | 418609 | 25.4362 | .001546 | .0393 |
| **648** | 419904 | 25.4558 | .001543 | .0393 |
| **649** | 421201 | 25.4755 | .001541 | .0393 |
| **650** | 422500 | 25.4951 | .001538 | .0392 |
| **651** | 423801 | 25.5147 | .001536 | .0392 |
| **652** | 425104 | 25.5343 | .001534 | .0392 |
| **653** | 426409 | 25.5539 | .001531 | .0391 |
| **654** | 427716 | 25.5734 | .001529 | .0391 |
| **655** | 429025 | 25.5930 | .001527 | .0391 |
| **656** | 430336 | 25.6125 | .001524 | .0390 |
| **657** | 431649 | 25.6320 | .001522 | .0390 |
| **658** | 432964 | 25.6515 | .001520 | .0390 |
| **659** | 434281 | 25.6710 | .001517 | .0390 |
| **660** | 435600 | 25.6905 | .001515 | .0389 |
| **661** | 436921 | 25.7099 | .001513 | .0389 |
| **662** | 438244 | 25.7294 | .001511 | .0389 |
| **663** | 439569 | 25.7488 | .001508 | .0388 |
| **664** | 440896 | 25.7682 | .001506 | .0388 |
| **665** | 442225 | 25.7876 | .001504 | .0388 |
| **666** | 443556 | 25.8070 | .001502 | .0387 |
| **667** | 444889 | 25.8263 | .001499 | .0387 |
| **668** | 446224 | 25.8457 | .001497 | .0387 |
| **669** | 447561 | 25.8650 | .001495 | .0387 |
| **670** | 448900 | 25.8844 | .001493 | .0386 |

**APPENDIX B** *(Continued)*

| $n$ | $n^2$ | $\sqrt{n}$ | $\frac{1}{n}$ | $\frac{1}{\sqrt{n}}$ |
|---|---|---|---|---|
| **671** | 450241 | 25.9037 | .001490 | .0386 |
| **672** | 451584 | 25.9230 | .001488 | .0386 |
| **673** | 452929 | 25.9422 | .001486 | .0385 |
| **674** | 454276 | 25.9615 | .001484 | .0385 |
| **675** | 455625 | 25.9808 | .001481 | .0385 |
| **676** | 456976 | 26.0000 | .001479 | .0385 |
| **677** | 458329 | 26.0192 | .001477 | .0384 |
| **678** | 459684 | 26.0384 | .001475 | .0384 |
| **679** | 461041 | 26.0576 | .001473 | .0384 |
| **680** | 462400 | 26.0768 | .001471 | .0383 |
| **681** | 463761 | 26.0960 | .001468 | .0383 |
| **682** | 465124 | 26.1151 | .001466 | .0383 |
| **683** | 466489 | 26.1343 | .001464 | .0383 |
| **684** | 467856 | 26.1534 | .001462 | .0382 |
| **685** | 469225 | 26.1725 | .001460 | .0382 |
| **686** | 470596 | 26.1916 | .001458 | .0382 |
| **687** | 471969 | 26.2107 | .001456 | .0382 |
| **688** | 473344 | 26.2298 | .001453 | .0381 |
| **689** | 474721 | 26.2488 | .001451 | .0381 |
| **690** | 476100 | 26.2679 | .001449 | .0381 |
| **691** | 477481 | 26.2869 | .001447 | .0380 |
| **692** | 478864 | 26.3059 | .001445 | .0380 |
| **693** | 480249 | 26.3249 | .001443 | .0380 |
| **694** | 481636 | 26.3439 | .001441 | .0380 |
| **695** | 483025 | 26.3629 | .001439 | .0379 |
| **696** | 484416 | 26.3818 | .001437 | .0379 |
| **697** | 485809 | 26.4008 | .001435 | .0379 |
| **698** | 487204 | 26.4197 | .001433 | .0379 |
| **699** | 488601 | 26.4386 | .001431 | .0378 |
| **700** | 490000 | 26.4575 | .001429 | .0378 |
| **701** | 491401 | 26.4764 | .001427 | .0378 |
| **702** | 492804 | 26.4953 | .001425 | .0377 |
| **703** | 494209 | 26.5141 | .001422 | .0377 |
| **704** | 495616 | 26.5330 | .001420 | .0377 |
| **705** | 497025 | 26.5518 | .001418 | .0377 |
| **706** | 498436 | 26.5707 | .001416 | .0376 |
| **707** | 499849 | 26.5895 | .001414 | .0376 |
| **708** | 501264 | 26.6083 | .001412 | .0376 |
| **709** | 502681 | 26.6271 | .001410 | .0376 |
| **710** | 504100 | 26.6458 | .001408 | .0375 |

**APPENDIX B** *(Continued)*

| $n$ | $n^2$ | $\sqrt{n}$ | $\frac{1}{n}$ | $\frac{1}{\sqrt{n}}$ |
|---|---|---|---|---|
| **711** | 505521 | 26.6646 | .001406 | .0375 |
| **712** | 506944 | 26.6833 | .001404 | .0375 |
| **713** | 508369 | 26.7021 | .001403 | .0375 |
| **714** | 509796 | 26.7208 | .001401 | .0374 |
| **715** | 511225 | 26.7395 | .001399 | .0374 |
| **716** | 512656 | 26.7582 | .001397 | .0374 |
| **717** | 514089 | 26.7769 | .001395 | .0373 |
| **718** | 515524 | 26.7955 | .001393 | .0373 |
| **719** | 516961 | 26.8142 | .001391 | .0373 |
| **720** | 518400 | 26.8328 | .001389 | .0373 |
| **721** | 519841 | 26.8514 | .001387 | .0372 |
| **722** | 521284 | 26.8701 | .001385 | .0372 |
| **723** | 522729 | 26.8887 | .001383 | .0372 |
| **724** | 524176 | 26.9072 | .001381 | .0372 |
| **725** | 525625 | 26.9258 | .001379 | .0371 |
| **726** | 527076 | 26.9444 | .001377 | .0371 |
| **727** | 528529 | 26.9629 | .001376 | .0371 |
| **728** | 529984 | 26.9815 | .001374 | .0371 |
| **729** | 531441 | 27.0000 | .001372 | .0370 |
| **730** | 532900 | 27.0185 | .001370 | .0370 |
| **731** | 534361 | 27.0370 | .001368 | .0370 |
| **732** | 535824 | 27.0555 | .001366 | .0370 |
| **733** | 537289 | 27.0740 | .001364 | .0369 |
| **734** | 538756 | 27.0924 | .001362 | .0369 |
| **735** | 540225 | 27.1109 | .001361 | .0369 |
| **736** | 541696 | 27.1293 | .001359 | .0369 |
| **737** | 543169 | 27.1477 | .001357 | .0368 |
| **738** | 544644 | 27.1662 | .001355 | .0368 |
| **739** | 546121 | 27.1846 | .001353 | .0368 |
| **740** | 547600 | 27.2029 | .001351 | .0368 |
| **741** | 549081 | 27.2213 | .001350 | .0367 |
| **742** | 550564 | 27.2397 | .001348 | .0367 |
| **743** | 552049 | 27.2580 | .001346 | .0367 |
| **744** | 553536 | 27.2764 | .001344 | .0367 |
| **745** | 555025 | 27.2947 | .001342 | .0366 |
| **746** | 556516 | 27.3130 | .001340 | .0366 |
| **747** | 558009 | 27.3313 | .001339 | .0366 |
| **748** | 559504 | 27.3496 | .001337 | .0366 |
| **749** | 561001 | 27.3679 | .001335 | .0365 |
| **750** | 562500 | 27.3861 | .001333 | .0365 |

**APPENDIX B** *(Continued)*

| $n$ | $n^2$ | $\sqrt{n}$ | $\frac{1}{n}$ | $\frac{1}{\sqrt{n}}$ |
|---|---|---|---|---|
| **751** | 564001 | 27.4044 | .001332 | .0365 |
| **752** | 565504 | 27.4226 | .001330 | .0365 |
| **753** | 567009 | 27.4408 | .001328 | .0364 |
| **754** | 568516 | 27.4591 | .001326 | .0364 |
| **755** | 570025 | 27.4773 | .001325 | .0364 |
| **756** | 571536 | 27.4955 | .001323 | .0364 |
| **757** | 573049 | 27.5136 | .001321 | .0363 |
| **758** | 574564 | 27.5318 | .001319 | .0363 |
| **759** | 576081 | 27.5500 | .001318 | .0363 |
| **760** | 577600 | 27.5681 | .001316 | .0363 |
| **761** | 579121 | 27.5862 | .001314 | .0363 |
| **762** | 580644 | 27.6043 | .001312 | .0362 |
| **763** | 582169 | 27.6225 | .001311 | .0362 |
| **764** | 583696 | 27.6405 | .001309 | .0362 |
| **765** | 585225 | 27.6586 | .001307 | .0362 |
| **766** | 586756 | 27.6767 | .001305 | .0361 |
| **767** | 588289 | 27.6948 | .001304 | .0361 |
| **768** | 589824 | 27.7128 | .001302 | .0361 |
| **769** | 591361 | 27.7308 | .001300 | .0361 |
| **770** | 592900 | 27.7489 | .001299 | .0360 |
| **771** | 594441 | 27.7669 | .001297 | .0360 |
| **772** | 595984 | 27.7849 | .001295 | .0360 |
| **773** | 597529 | 27.8029 | .001294 | .0360 |
| **774** | 599076 | 27.8209 | .001292 | .0359 |
| **775** | 600625 | 27.8388 | .001290 | .0359 |
| **776** | 602176 | 27.8568 | .001289 | .0359 |
| **777** | 603729 | 27.8747 | .001287 | .0359 |
| **778** | 605284 | 27.8927 | .001285 | .0359 |
| **779** | 606841 | 27.9106 | .001284 | .0358 |
| **780** | 608400 | 27.9285 | .001282 | .0358 |
| **781** | 609961 | 27.9464 | .001280 | .0358 |
| **782** | 611524 | 27.9643 | .001279 | .0358 |
| **783** | 613089 | 27.9821 | .001277 | .0357 |
| **784** | 614656 | 28.0000 | .001276 | .0357 |
| **785** | 616225 | 28.0179 | .001274 | .0357 |
| **786** | 617796 | 28.0357 | .001272 | .0357 |
| **787** | 619369 | 28.0535 | .001271 | .0356 |
| **788** | 620944 | 28.0713 | .001269 | .0356 |
| **789** | 622521 | 28.0891 | .001267 | .0356 |
| **790** | 624100 | 28.1069 | .001266 | .0356 |

**APPENDIX B** *(Continued)*

| $n$ | $n^2$ | $\sqrt{n}$ | $\frac{1}{n}$ | $\frac{1}{\sqrt{n}}$ |
|---|---|---|---|---|
| **791** | 625681 | 28.1247 | .001264 | .0356 |
| **792** | 627264 | 28.1425 | .001263 | .0355 |
| **793** | 628849 | 28.1603 | .001261 | .0355 |
| **794** | 630436 | 28.1780 | .001259 | .0355 |
| **795** | 632025 | 28.1957 | .001258 | .0355 |
| **796** | 633616 | 28.2135 | .001256 | .0354 |
| **797** | 635209 | 28.2312 | .001255 | .0354 |
| **798** | 636804 | 28.2489 | .001253 | .0354 |
| **799** | 638401 | 28.2666 | .001252 | .0354 |
| **800** | 640000 | 28.2843 | .001250 | .0354 |
| **801** | 641601 | 28.3019 | .001248 | .0353 |
| **802** | 643204 | 28.3196 | .001247 | .0353 |
| **803** | 644809 | 28.3373 | .001245 | .0353 |
| **804** | 646416 | 28.3549 | .001244 | .0353 |
| **805** | 648025 | 28.3725 | .001242 | .0352 |
| **806** | 649636 | 28.3901 | .001241 | .0352 |
| **807** | 651249 | 28.4077 | .001239 | .0352 |
| **808** | 652864 | 28.4253 | .001238 | .0352 |
| **809** | 654481 | 28.4429 | .001236 | .0352 |
| **810** | 656100 | 28.4605 | .001235 | .0351 |
| **811** | 657721 | 28.4781 | .001233 | .0351 |
| **812** | 659344 | 28.4956 | .001232 | .0351 |
| **813** | 660969 | 28.5132 | .001230 | .0351 |
| **814** | 662596 | 28.5307 | .001229 | .0351 |
| **815** | 664225 | 28.5482 | .001227 | .0350 |
| **816** | 665856 | 28.5657 | .001225 | .0350 |
| **817** | 667489 | 28.5832 | .001224 | .0350 |
| **818** | 669124 | 28.6007 | .001222 | .0350 |
| **819** | 670761 | 28.6182 | .001221 | .0349 |
| **820** | 672400 | 28.6356 | .001220 | .0349 |
| **821** | 674041 | 28.6531 | .001218 | .0349 |
| **822** | 675684 | 28.6705 | .001217 | .0349 |
| **823** | 677329 | 28.6880 | .001215 | .0349 |
| **824** | 678976 | 28.7054 | .001214 | .0348 |
| **825** | 680625 | 28.7228 | .001212 | .0348 |
| **826** | 682276 | 28.7402 | .001211 | .0348 |
| **827** | 683929 | 28.7576 | .001209 | .0348 |
| **828** | 685584 | 28.7750 | .001208 | .0348 |
| **829** | 687241 | 28.7924 | .001206 | .0347 |
| **830** | 688900 | 28.8097 | .001205 | .0347 |

**APPENDIX B** *(Continued)*

| $n$ | $n^2$ | $\sqrt{n}$ | $\frac{1}{n}$ | $\frac{1}{\sqrt{n}}$ |
|---|---|---|---|---|
| **831** | 690561 | 28.8271 | .001203 | .0347 |
| **832** | 692224 | 28.8444 | .001202 | .0347 |
| **833** | 693889 | 28.8617 | .001200 | .0346 |
| **834** | 695556 | 28.8791 | .001199 | .0346 |
| **835** | 697225 | 28.8964 | .001198 | .0346 |
| **836** | 698896 | 28.9137 | .001196 | .0346 |
| **837** | 700569 | 28.9310 | .001195 | .0346 |
| **838** | 702244 | 28.9482 | .001193 | .0345 |
| **839** | 703921 | 28.9655 | .001192 | .0345 |
| **840** | 705600 | 28.9828 | .001190 | .0345 |
| **841** | 707281 | 29.0000 | .001189 | .0345 |
| **842** | 708964 | 29.0172 | .001188 | .0345 |
| **843** | 710649 | 29.0345 | .001186 | .0344 |
| **844** | 712336 | 29.0517 | .001185 | .0344 |
| **845** | 714025 | 29.0689 | .001183 | .0344 |
| **846** | 715716 | 29.0861 | .001182 | .0344 |
| **847** | 717409 | 29.1033 | .001181 | .0344 |
| **848** | 719104 | 29.1204 | .001179 | .0343 |
| **849** | 720801 | 29.1376 | .001178 | .0343 |
| **850** | 722500 | 29.1548 | .001176 | .0343 |
| **851** | 724201 | 29.1719 | .001175 | .0343 |
| **852** | 725904 | 29.1890 | .001174 | .0343 |
| **853** | 727609 | 29.2062 | .001172 | .0342 |
| **854** | 729316 | 29.2233 | .001171 | .0342 |
| **855** | 731025 | 29.2404 | .001170 | .0342 |
| **856** | 732736 | 29.2575 | .001168 | .0342 |
| **857** | 734449 | 29.2746 | .001167 | .0342 |
| **858** | 736164 | 29.2916 | .001166 | .0341 |
| **859** | 737881 | 29.3087 | .001164 | .0341 |
| **860** | 739600 | 29.3258 | .001163 | .0341 |
| **861** | 741321 | 29.3428 | .001161 | .0341 |
| **862** | 743044 | 29.3598 | .001160 | .0341 |
| **863** | 744769 | 29.3769 | .001159 | .0340 |
| **864** | 746496 | 29.3939 | .001157 | .0340 |
| **865** | 748225 | 29.4109 | .001156 | .0340 |
| **866** | 749956 | 29.4279 | .001155 | .0340 |
| **867** | 751689 | 29.4449 | .001153 | .0340 |
| **868** | 753424 | 29.4618 | .001152 | .0339 |
| **869** | 755161 | 29.4788 | .001151 | .0339 |
| **870** | 756900 | 29.4958 | .001149 | .0339 |

**APPENDIX B** *(Continued)*

| $n$ | $n^2$ | $\sqrt{n}$ | $\frac{1}{n}$ | $\frac{1}{\sqrt{n}}$ |
|---|---|---|---|---|
| **871** | 758641 | 29.5127 | .001148 | .0339 |
| **872** | 760384 | 29.5296 | .001147 | .0339 |
| **873** | 762129 | 29.5466 | .001145 | .0338 |
| **874** | 763876 | 29.5635 | .001144 | .0338 |
| **875** | 765625 | 29.5804 | .001143 | .0338 |
| **876** | 767376 | 29.5973 | .001142 | .0338 |
| **877** | 769129 | 29.6142 | .001140 | .0338 |
| **878** | 770884 | 29.6311 | .001139 | .0337 |
| **879** | 772641 | 29.6479 | .001138 | .0337 |
| **880** | 774400 | 29.6648 | .001136 | .0337 |
| **881** | 776161 | 29.6816 | .001135 | .0337 |
| **882** | 777924 | 29.6985 | .001134 | .0337 |
| **883** | 779689 | 29.7153 | .001133 | .0337 |
| **884** | 781456 | 29.7321 | .001131 | .0336 |
| **885** | 783225 | 29.7489 | .001130 | .0336 |
| **886** | 784996 | 29.7658 | .001129 | .0336 |
| **887** | 786769 | 29.7825 | .001127 | .0336 |
| **888** | 788544 | 29.7993 | .001126 | .0336 |
| **889** | 790321 | 29.8161 | .001125 | .0335 |
| **890** | 792100 | 29.8329 | .001124 | .0335 |
| **891** | 793881 | 29.8496 | .001122 | .0335 |
| **892** | 795664 | 29.8664 | .001121 | .0335 |
| **893** | 797449 | 29.8831 | .001120 | .0335 |
| **894** | 799236 | 29.8998 | .001119 | .0334 |
| **895** | 801025 | 29.9166 | .001117 | .0334 |
| **896** | 802816 | 29.9333 | .001116 | .0334 |
| **897** | 804609 | 29.9500 | .001115 | .0334 |
| **898** | 806404 | 29.9666 | .001114 | .0334 |
| **899** | 808201 | 29.9833 | .001112 | .0334 |
| **900** | 810000 | 30.0000 | .001111 | .0333 |
| **901** | 811801 | 30.0167 | .001110 | .0333 |
| **902** | 813604 | 30.0333 | .001109 | .0333 |
| **903** | 815409 | 30.0500 | .001107 | .0333 |
| **904** | 817216 | 30.0666 | .001106 | .0333 |
| **905** | 819025 | 30.0832 | .001105 | .0332 |
| **906** | 820836 | 30.0998 | .001104 | .0332 |
| **907** | 822649 | 30.1164 | .001103 | .0332 |
| **908** | 824464 | 30.1330 | .001101 | .0332 |
| **909** | 826281 | 20.1496 | .001100 | 0332 |
| **910** | 828100 | 30.1662 | .001099 | .0331 |

**APPENDIX B** *(Continued)*

| $n$ | $n^2$ | $\sqrt{n}$ | $\frac{1}{n}$ | $\frac{1}{\sqrt{n}}$ |
|---|---|---|---|---|
| **911** | 829921 | 30.1828 | .001098 | .0331 |
| **912** | 831744 | 30.1993 | .001096 | .0331 |
| **913** | 833569 | 30.2159 | .001095 | .0331 |
| **914** | 835396 | 30.2324 | .001094 | .0331 |
| **915** | 837225 | 30.2490 | .001093 | .0331 |
| **916** | 839056 | 30.2655 | .001092 | .0330 |
| **917** | 840889 | 30.2820 | .001091 | .0330 |
| **918** | 842724 | 30.2985 | .001089 | .0330 |
| **919** | 844561 | 30.3150 | .001088 | .0330 |
| **920** | 846400 | 30.3315 | .001087 | .0330 |
| **921** | 848241 | 30.3480 | .001086 | .0330 |
| **922** | 850084 | 30.3645 | .001085 | .0329 |
| **923** | 851929 | 30.3809 | .001083 | .0329 |
| **924** | 853776 | 30.3974 | .001082 | .0329 |
| **925** | 855625 | 30.4138 | .001081 | .0329 |
| **926** | 857476 | 30.4302 | .001080 | .0329 |
| **927** | 859329 | 30.4467 | .001079 | .0328 |
| **928** | 861184 | 30.4631 | .001078 | .0328 |
| **929** | 863041 | 30.4795 | .001076 | .0328 |
| **930** | 864900 | 30.4959 | .001075 | .0328 |
| **931** | 866761 | 30.5123 | .001074 | .0328 |
| **932** | 868624 | 30.5287 | .001073 | .0328 |
| **933** | 870489 | 30.5450 | .001072 | .0327 |
| **934** | 872356 | 30.5614 | .001071 | .0327 |
| **935** | 874225 | 30.5778 | .001070 | .0327 |
| **936** | 876096 | 30.5941 | .001068 | .0327 |
| **937** | 877969 | 30.6105 | .001067 | .0327 |
| **938** | 879844 | 30.6268 | .001066 | .0327 |
| **939** | 881721 | 30.6431 | .001065 | .0326 |
| **940** | 883600 | 30.6594 | .001064 | .0326 |
| **941** | 885481 | 30.6757 | .001063 | .0326 |
| **942** | 887364 | 30.6920 | .001062 | .0326 |
| **943** | 889249 | 30.7083 | .001060 | .0326 |
| **944** | 891136 | 30.7246 | .001059 | .0325 |
| **945** | 893025 | 30.7409 | .001058 | .0325 |
| **946** | 894916 | 30.7571 | .001057 | .0325 |
| **947** | 896809 | 30.7734 | .001056 | .0325 |
| **948** | 898704 | 30.7896 | .001055 | .0325 |
| **949** | 900601 | 30.8058 | .001054 | .0325 |
| **950** | 902500 | 30.8221 | .001053 | .0324 |

**APPENDIX B** *(Continued)*

| $n$ | $n^2$ | $\sqrt{n}$ | $\frac{1}{n}$ | $\frac{1}{\sqrt{n}}$ |
|---|---|---|---|---|
| **951** | 904401 | 30.8383 | .001052 | .0324 |
| **952** | 906304 | 30.8545 | .001050 | .0324 |
| **953** | 908209 | 30.8707 | .001049 | .0324 |
| **954** | 910116 | 30.8869 | .001048 | .0324 |
| **955** | 912025 | 30.9031 | .001047 | .0324 |
| **956** | 913936 | 30.9192 | .001046 | .0323 |
| **957** | 915849 | 30.9354 | .001045 | .0323 |
| **958** | 917764 | 30.9516 | .001044 | .0323 |
| **959** | 919681 | 30.9677 | .001043 | .0323 |
| **960** | 921600 | 30.9839 | .001042 | .0323 |
| **961** | 923521 | 31.0000 | .001041 | .0323 |
| **962** | 925444 | 31.0161 | .001040 | .0322 |
| **963** | 927369 | 31.0322 | .001038 | .0322 |
| **964** | 929296 | 31.0483 | .001037 | .0322 |
| **965** | 931225 | 31.0644 | .001036 | .0322 |
| **966** | 933156 | 31.0805 | .001035 | .0322 |
| **967** | 935089 | 31.0966 | .001034 | .0322 |
| **968** | 937024 | 31.1127 | .001033 | .0321 |
| **969** | 938961 | 31.1288 | .001032 | .0321 |
| **970** | 940900 | 31.1448 | .001031 | .0321 |
| **971** | 942841 | 31.1609 | .001030 | .0321 |
| **972** | 944784 | 31.1769 | .001029 | .0321 |
| **973** | 946729 | 31.1929 | .001028 | .0321 |
| **974** | 948676 | 31.2090 | .001027 | .0320 |
| **975** | 950625 | 31.2250 | .001026 | .0320 |
| **976** | 952576 | 31.2410 | .001025 | .0320 |
| **977** | 954529 | 31.2570 | .001024 | .0320 |
| **978** | 956484 | 31.2730 | .001022 | .0320 |
| **979** | 958441 | 31.2890 | .001021 | .0320 |
| **980** | 960400 | 31.3050 | .001020 | .0319 |
| **981** | 962361 | 31.3209 | .001019 | .0319 |
| **982** | 964324 | 31.3369 | .001018 | .0319 |
| **983** | 966289 | 31.3528 | .001017 | .0319 |
| **984** | 968256 | 31.3688 | .001016 | .0319 |
| **985** | 970225 | 31.3847 | .001015 | .0319 |
| **986** | 972196 | 31.4006 | .001014 | .0318 |
| **987** | 974169 | 31.4166 | .001013 | .0318 |
| **988** | 976144 | 31.4325 | .001012 | .0318 |
| **989** | 978121 | 31.4484 | .001011 | .0318 |
| **990** | 980100 | 31.4643 | .001010 | .0318 |

**APPENDIX B** *(Continued)*

| $n$ | $n^2$ | $\sqrt{n}$ | $\frac{1}{n}$ | $\frac{1}{\sqrt{n}}$ |
|---|---|---|---|---|
| **991** | 982081 | 31.4802 | .001009 | .0318 |
| **992** | 984064 | 31.4960 | .001008 | .0318 |
| **993** | 986049 | 31.5119 | .001007 | .0317 |
| **994** | 988036 | 31.5278 | .001006 | .0317 |
| **995** | 990025 | 31.5436 | .001005 | .0317 |
| **996** | 992016 | 31.5595 | .001004 | .0317 |
| **997** | 994009 | 31.5753 | .001003 | .0317 |
| **998** | 996004 | 31.5911 | .001002 | .0317 |
| **999** | 998001 | 31.6070 | .001001 | .0316 |
| **1000** | 1000000 | 31.6228 | .001000 | .0316 |

# APPENDIX C

# AREAS OF THE NORMAL CURVE IN TERMS OF $z$

| (1) | (2) | (3) | (4) | (1) | (2) | (3) | (4) |
|---|---|---|---|---|---|---|---|
| $z$ | *A* AREA FROM MEAN TO $z$ | *B* AREA IN LARGER PORTION | *C* AREA IN SMALLER PORTION | $z$ | *A* AREA FROM MEAN TO $z$ | *B* AREA IN LARGER PORTION | *C* AREA IN SMALLER PORTION |
| 0.00 | .0000 | .5000 | .5000 | 0.30 | .1179 | .6179 | .3821 |
| 0.01 | .0040 | .5040 | .4960 | 0.31 | .1217 | .6217 | .3783 |
| 0.02 | .0080 | .5080 | .4920 | 0.32 | .1255 | .6255 | .3745 |
| 0.03 | .0120 | .5120 | .4880 | 0.33 | .1293 | .6293 | .3707 |
| 0.04 | .0160 | .5160 | .4840 | 0.34 | .1331 | .6331 | .3669 |
| 0.05 | .0199 | .5199 | .4801 | 0.35 | .1368 | .6368 | .3632 |
| 0.06 | .0239 | .5239 | .4761 | 0.36 | .1406 | .6406 | .3594 |
| 0.07 | .0279 | .5279 | .4721 | 0.37 | .1443 | .6443 | .3557 |
| 0.08 | .0319 | .5319 | .4681 | 0.38 | .1480 | .6480 | .3520 |
| 0.09 | .0359 | .5359 | .4641 | 0.39 | .1517 | .6517 | .3483 |
| 0.10 | .0398 | .5398 | .4602 | 0.40 | .1554 | .6554 | .3446 |
| 0.11 | .0438 | .5438 | .4562 | 0.41 | .1591 | .6591 | .3409 |
| 0.12 | .0478 | .5478 | .4522 | 0.42 | .1628 | .6628 | .3372 |
| 0.13 | .0517 | .5517 | .4483 | 0.43 | .1664 | .6664 | .3336 |
| 0.14 | .0557 | .5557 | .4443 | 0.44 | .1700 | .6700 | .3300 |
| 0.15 | .0596 | .5596 | .4404 | 0.45 | .1736 | .6736 | .3264 |
| 0.16 | .0636 | .5636 | .4364 | 0.46 | .1772 | .6772 | .3228 |
| 0.17 | .0675 | .5675 | .4325 | 0.47 | .1808 | .6808 | .3192 |
| 0.18 | .0714 | .5714 | .4286 | 0.48 | .1844 | .6844 | .3156 |
| 0.19 | .0753 | .5753 | .4247 | 0.49 | .1879 | .6879 | .3121 |
| 0.20 | .0793 | .5793 | .4207 | 0.50 | .1915 | .6915 | .3085 |
| 0.21 | .0832 | .5832 | .4168 | 0.51 | .1950 | .6950 | .3050 |
| 0.22 | .0871 | .5871 | .4129 | 0.52 | .1985 | .6985 | .3015 |
| 0.23 | .0910 | .5910 | .4090 | 0.53 | .2019 | .7019 | .2981 |
| 0.24 | .0948 | .5948 | .4052 | 0.54 | .2054 | .7054 | .2946 |
| 0.25 | .0987 | .5987 | .4013 | 0.55 | .2088 | .7088 | .2912 |
| 0.26 | .1026 | .6026 | .3974 | 0.56 | .2123 | .7123 | .2877 |
| 0.27 | .1064 | .6064 | .3936 | 0.57 | .2157 | .7157 | .2843 |
| 0.28 | .1103 | .6103 | .3897 | 0.58 | .2190 | .7190 | .2810 |
| 0.29 | .1141 | .6141 | .3859 | 0.59 | .2224 | .7224 | .2776 |

**APPENDIX C** *(Continued)*

| (1) | (2) | (3) | (4) | (1) | (2) | (3) | (4) |
|---|---|---|---|---|---|---|---|
| | *A* | *B* | *C* | | *A* | *B* | *C* |
| *z* | AREA FROM MEAN TO *z* | AREA IN LARGER PORTION | AREA IN SMALLER PORTION | *z* | AREA FROM MEAN TO *z* | AREA IN LARGER PORTION | AREA IN SMALLER PORTION |
| 0.60 | .2257 | .7257 | .2743 | 0.95 | .3289 | .8289 | .1711 |
| 0.61 | .2291 | .7291 | .2709 | 0.96 | .3315 | .8315 | .1685 |
| 0.62 | .2324 | .7324 | .2676 | 0.97 | .3340 | .8340 | .1660 |
| 0.63 | .2357 | .7357 | .2643 | 0.98 | .3365 | .8365 | .1635 |
| 0.64 | .2389 | .7389 | .2611 | 0.99 | .3389 | .8389 | .1611 |
| 0.65 | .2422 | .7422 | .2578 | 1.00 | .3413 | .8413 | .1587 |
| 0.66 | .2454 | .7454 | .2546 | 1.01 | .3438 | .8438 | .1562 |
| 0.67 | .2486 | .7486 | .2514 | 1.02 | .3461 | .8461 | .1539 |
| 0.68 | .2517 | .7517 | .2483 | 1.03 | .3485 | .8485 | .1515 |
| 0.69 | .2549 | .7549 | .2451 | 1.04 | .3508 | .8508 | .1492 |
| 0.70 | .2580 | .7580 | .2420 | 1.05 | .3531 | .8531 | .1469 |
| 0.71 | .2611 | .7611 | .2389 | 1.06 | .3554 | .8554 | .1446 |
| 0.72 | .2642 | .7642 | .2358 | 1.07 | .3577 | .8577 | .1423 |
| 0.73 | .2673 | .7673 | .2327 | 1.08 | .3599 | .8599 | .1401 |
| 0.74 | .2704 | .7704 | .2296 | 1.09 | .3621 | .8621 | .1379 |
| 0.75 | .2734 | .7734 | .2266 | 1.10 | .3643 | .8643 | .1357 |
| 0.76 | .2764 | .7764 | .2236 | 1.11 | .3665 | .8665 | .1335 |
| 0.77 | .2794 | .7794 | .2206 | 1.12 | .3686 | .8686 | .1314 |
| 0.78 | .2823 | .7823 | .2177 | 1.13 | .3708 | .8708 | .1292 |
| 0.79 | .2852 | .7852 | .2148 | 1.14 | .3729 | .8729 | .1271 |
| 0.80 | .2881 | .7881 | .2119 | 1.15 | .3749 | .8749 | .1251 |
| 0.81 | .2910 | .7910 | .2090 | 1.16 | .3770 | .8770 | .1230 |
| 0.82 | .2939 | .7939 | .2061 | 1.17 | .3790 | .8790 | .1210 |
| 0.83 | .2967 | .7967 | .2033 | 1.18 | .3810 | .8810 | .1190 |
| 0.84 | .2995 | .7995 | .2005 | 1.19 | .3830 | .8830 | .1170 |
| 0.85 | .3023 | .8023 | .1977 | 1.20 | .3849 | .8849 | .1151 |
| 0.86 | .3051 | .8051 | .1949 | 1.21 | .3869 | .8869 | .1131 |
| 0.87 | .3078 | .8078 | .1922 | 1.22 | .3888 | .8888 | .1112 |
| 0.88 | .3106 | .8106 | .1894 | 1.23 | .3907 | .8907 | .1093 |
| 0.89 | .3133 | .8133 | .1867 | 1.24 | .3925 | .8925 | .1075 |
| 0.90 | .3159 | .8159 | .1841 | 1.25 | .3944 | .8944 | .1056 |
| 0.91 | .3186 | .8186 | .1814 | 1.26 | .3962 | .8962 | .1038 |
| 0.92 | .3212 | .8212 | .1788 | 1.27 | .3980 | .8980 | .1020 |
| 0.93 | .3238 | .8238 | .1762 | 1.28 | .3997 | .8997 | .1003 |
| 0.94 | .3264 | .8264 | .1736 | 1.29 | .4015 | .9015 | .0985 |

**APPENDIX C** *(Continued)*

| (1) | (2) | (3) | (4) | (1) | (2) | (3) | (4) |
|---|---|---|---|---|---|---|---|
| | *A* AREA FROM MEAN TO $z$ | *B* AREA IN LARGER PORTION | *C* AREA IN SMALLER PORTION | | *A* AREA FROM MEAN TO $z$ | *B* AREA IN LARGER PORTION | *C* AREA IN SMALLER PORTION |
| $z$ | | | | $z$ | | | |
| 1.30 | .4032 | .9032 | .0968 | 1.65 | .4505 | .9505 | .0495 |
| 1.31 | .4049 | .9049 | .0951 | 1.66 | .4515 | .9515 | .0485 |
| 1.32 | .4066 | .9066 | .0934 | 1.67 | .4525 | .9525 | .0475 |
| 1.33 | .4082 | .9082 | .0918 | 1.68 | .4535 | .9535 | .0465 |
| 1.34 | .4099 | .9099 | .0901 | 1.69 | .4545 | .9545 | .0455 |
| 1.35 | .4115 | .9115 | .0885 | 1.70 | .4554 | .9554 | .0446 |
| 1.36 | .4131 | .9131 | .0869 | 1.71 | .4564 | .9564 | .0436 |
| 1.37 | .4147 | .9147 | .0853 | 1.72 | .4573 | .9573 | .0427 |
| 1.38 | .4162 | .9162 | .0838 | 1.73 | .4582 | .9582 | .0418 |
| 1.39 | .4177 | .9177 | .0823 | 1.74 | .4591 | .9591 | .0409 |
| 1.40 | .4192 | .9192 | .0808 | 1.75 | .4599 | .9599 | .0401 |
| 1.41 | .4207 | .9207 | .0793 | 1.76 | .4608 | .9608 | .0392 |
| 1.42 | .4222 | .9222 | .0778 | 1.77 | .4616 | .9616 | .0384 |
| 1.43 | .4236 | .9236 | .0764 | 1.78 | .4625 | .9625 | .0375 |
| 1.44 | .4251 | .9251 | .0749 | 1.79 | .4633 | .9633 | .0367 |
| 1.45 | .4265 | .9265 | .0735 | 1.80 | .4641 | .9641 | .0359 |
| 1.46 | .4279 | .9279 | .0721 | 1.81 | .4649 | .9649 | .0351 |
| 1.47 | .4292 | .9292 | .0708 | 1.82 | .4656 | .9656 | .0344 |
| 1.48 | .4306 | .9306 | .0694 | 1.83 | .4664 | .9664 | .0336 |
| 1.49 | .4319 | .9319 | .0681 | 1.84 | .4671 | .9671 | .0329 |
| 1.50 | .4332 | .9332 | .0668 | 1.85 | .4678 | .9678 | .0322 |
| 1.51 | .4345 | .9345 | .0655 | 1.86 | .4686 | .9686 | .0314 |
| 1.52 | .4357 | .9357 | .0643 | 1.87 | .4693 | .9693 | .0307 |
| 1.53 | .4370 | .9370 | .0630 | 1.88 | .4699 | .9699 | .0301 |
| 1.54 | .4382 | .9382 | .0618 | 1.89 | .4706 | .9706 | .0294 |
| 1.55 | .4394 | .9394 | .0606 | 1.90 | .4713 | .9713 | .0287 |
| 1.56 | .4406 | .9406 | .0594 | 1.91 | .4719 | .9719 | .0281 |
| 1.57 | .4418 | .9418 | .0582 | 1.92 | .4726 | .9726 | .0274 |
| 1.58 | .4429 | .9429 | .0571 | 1.93 | .4732 | .9732 | .0268 |
| 1.59 | .4441 | .9441 | .0559 | 1.94 | .4738 | .9738 | .0262 |
| 1.60 | .4452 | .9452 | .0548 | 1.95 | .4744 | .9744 | .0256 |
| 1.61 | .4463 | .9463 | .0537 | 1.96 | .4750 | .9750 | .0250 |
| 1.62 | .4474 | .9474 | .0526 | 1.97 | .4756 | .9756 | .0244 |
| 1.63 | .4484 | .9484 | .0516 | 1.98 | .4761 | .9761 | .0239 |
| 1.64 | .4495 | .9495 | .0505 | 1.99 | .4767 | .9767 | .0233 |

**APPENDIX C** (*Continued*)

| (1) | (2) | (3) | (4) | (1) | (2) | (3) | (4) |
|---|---|---|---|---|---|---|---|
| z | A<br>AREA FROM MEAN TO z | B<br>AREA IN LARGER PORTION | C<br>AREA IN SMALLER PORTION | z | A<br>AREA FROM MEAN TO z | B<br>AREA IN LARGER PORTION | C<br>AREA IN SMALLER PORTION |
| 2.00 | .4772 | .9772 | .0228 | 2.35 | .4906 | .9906 | .0094 |
| 2.01 | .4778 | .9778 | .0222 | 2.36 | .4909 | .9909 | .0091 |
| 2.02 | .4783 | .9783 | .0217 | 2.37 | .4911 | .9911 | .0089 |
| 2.03 | .4788 | .9788 | .0212 | 2.38 | .4913 | .9913 | .0087 |
| 2.04 | .4793 | .9793 | .0207 | 2.39 | .4916 | .9916 | .0084 |
| 2.05 | .4798 | .9798 | .0202 | 2.40 | .4918 | .9918 | .0082 |
| 2.06 | .4803 | .9803 | .0197 | 2.41 | .4920 | .9920 | .0080 |
| 2.07 | .4808 | .9808 | .0192 | 2.42 | .4922 | .9922 | .0078 |
| 2.08 | .4812 | .9812 | .0188 | 2.43 | .4925 | .9925 | .0075 |
| 2.09 | .4817 | .9817 | .0183 | 2.44 | .4927 | .9927 | .0073 |
| 2.10 | .4821 | .9821 | .0179 | 2.45 | .4929 | .9929 | .0071 |
| 2.11 | .4826 | .9826 | .0174 | 2.46 | .4931 | .9931 | .0069 |
| 2.12 | .4830 | .9830 | .0170 | 2.47 | .4932 | .9932 | .0068 |
| 2.13 | .4834 | .9834 | .0166 | 2.48 | .4934 | .9934 | .0066 |
| 2.14 | .4838 | .9838 | .0162 | 2.49 | .4936 | .9936 | .0064 |
| 2.15 | .4842 | .9842 | .0158 | 2.50 | .4938 | .9938 | .0062 |
| 2.16 | .4846 | .9846 | .0154 | 2.51 | .4940 | .9940 | .0060 |
| 2.17 | .4850 | .9850 | .0150 | 2.52 | .4941 | .9941 | .0059 |
| 2.18 | .4854 | .9854 | .0146 | 2.53 | .4943 | .9943 | .0057 |
| 2.19 | .4857 | .9857 | .0143 | 2.54 | .4945 | .9945 | .0055 |
| 2.20 | .4861 | .9861 | .0139 | 2.55 | .4946 | .9946 | .0054 |
| 2.21 | .4864 | .9864 | .0136 | 2.56 | .4948 | .9948 | .0052 |
| 2.22 | .4868 | .9868 | .0132 | 2.57 | .4949 | .9949 | .0051 |
| 2.23 | .4871 | .9871 | .0129 | 2.58 | .4951 | .9951 | .0049 |
| 2.24 | .4875 | .9875 | .0125 | 2.59 | .4952 | .9952 | .0048 |
| 2.25 | .4878 | .9878 | .0122 | 2.60 | .4953 | .9953 | .0047 |
| 2.26 | .4881 | .9881 | .0119 | 2.61 | .4955 | .9955 | .0045 |
| 2.27 | .4884 | .9884 | .0116 | 2.62 | .4956 | .9956 | .0044 |
| 2.28 | .4887 | .9887 | .0113 | 2.63 | .4957 | .9957 | .0043 |
| 2.29 | .4890 | .9890 | .0110 | 2.64 | .4959 | .9959 | .0041 |
| 2.30 | .4893 | .9893 | .0107 | 2.65 | .4960 | .9960 | .0040 |
| 2.31 | .4896 | .9896 | .0104 | 2.66 | .4961 | .9961 | .0039 |
| 2.32 | .4898 | .9898 | .0102 | 2.67 | .4962 | .9962 | .0038 |
| 2.33 | .4901 | .9901 | .0099 | 2.68 | .4963 | .9963 | .0037 |
| 2.34 | .4904 | .9904 | .0096 | 2.69 | .4964 | .9964 | .0036 |

**APPENDIX C** *(Continued)*

| (1) | (2) | (3) | (4) | (1) | (2) | (3) | (4) |
|---|---|---|---|---|---|---|---|
| | *A* AREA FROM MEAN TO *z* | *B* AREA IN LARGER PORTION | *C* AREA IN SMALLER PORTION | | *A* AREA FROM MEAN TO *z* | *B* AREA IN LARGER PORTION | *C* AREA IN SMALLER PORTION |
| *z* | | | | *z* | | | |
| 2.70 | .4965 | .9965 | .0035 | 3.00 | .4987 | .9987 | .0013 |
| 2.71 | .4966 | .9966 | .0034 | 3.01 | .4987 | .9987 | .0013 |
| 2.72 | .4967 | .9967 | .0033 | 3.02 | .4987 | .9987 | .0013 |
| 2.73 | .4968 | .9968 | .0032 | 3.03 | .4988 | .9988 | .0012 |
| 2.74 | .4969 | .9969 | .0031 | 3.04 | .4988 | .9988 | .0012 |
| 2.75 | .4970 | .9970 | .0030 | 3.05 | .4989 | .9989 | .0011 |
| 2.76 | .4971 | .9971 | .0029 | 3.06 | .4989 | .9989 | .0011 |
| 2.77 | .4972 | .9972 | .0028 | 3.07 | .4989 | .9989 | .0011 |
| 2.78 | .4973 | .9973 | .0027 | 3.08 | .4990 | .9990 | .0010 |
| 2.79 | .4974 | .9974 | .0026 | 3.09 | .4990 | .9990 | .0010 |
| 2.80 | .4974 | .9974 | .0026 | 3.10 | .4990 | .9990 | .0010 |
| 2.81 | .4975 | .9975 | .0025 | 3.11 | .4991 | .9991 | .0009 |
| 2.82 | .4976 | .9976 | .0024 | 3.12 | .4991 | .9991 | .0009 |
| 2.83 | .4977 | .9977 | .0023 | 3.13 | .4991 | .9991 | .0009 |
| 2.84 | .4977 | .9977 | .0023 | 3.14 | .4992 | .9992 | .0008 |
| 2.85 | .4978 | .9978 | .0022 | 3.15 | .4992 | .9992 | .0008 |
| 2.86 | .4979 | .9979 | .0021 | 3.16 | .4992 | .9992 | .0008 |
| 2.87 | .4979 | .9979 | .0021 | 3.17 | .4992 | .9992 | .0008 |
| 2.88 | .4980 | .9980 | .0020 | 3.18 | .4993 | .9993 | .0007 |
| 2.89 | .4981 | .9981 | .0019 | 3.19 | .4993 | .9993 | .0007 |
| 2.90 | .4981 | .9981 | .0019 | 3.20 | .4993 | .9993 | .0007 |
| 2.91 | .4982 | .9982 | .0018 | 3.21 | .4993 | .9993 | .0007 |
| 2.92 | .4982 | .9982 | .0018 | 3.22 | .4994 | .9994 | .0006 |
| 2.93 | .4983 | .9983 | .0017 | 3.23 | .4994 | .9994 | .0006 |
| 2.94 | .4984 | .9984 | .0016 | 3.24 | .4994 | .9994 | .0006 |
| 2.95 | .4984 | .9984 | .0016 | 3.30 | .4995 | .9995 | .0005 |
| 2.96 | .4985 | .9985 | .0015 | 3.40 | .4997 | .9997 | .0003 |
| 2.97 | .4985 | .9985 | .0015 | 3.50 | .4998 | .9998 | .0002 |
| 2.98 | .4986 | .9986 | .0014 | 3.60 | .4998 | .9998 | .0002 |
| 2.99 | .4986 | .9986 | .0014 | 3.70 | .4999 | .9999 | .0001 |

## APPENDIX D

# DISTRIBUTION OF $\chi^2$

| | Levels of Significance | | | | |
|---|---|---|---|---|---|
| *df* | .10 | .05 | .02 | .01 | .001 |
| 1 | 2.706 | 3.841 | 5.412 | 6.635 | 10.827 |
| 2 | 4.605 | 5.991 | 7.824 | 9.210 | 13.815 |
| 3 | 6.251 | 7.815 | 9.837 | 11.345 | 16.268 |
| 4 | 7.779 | 9.488 | 11.668 | 13.277 | 18.465 |
| 5 | 9.236 | 11.070 | 13.388 | 15.086 | 20.517 |
| 6 | 10.645 | 12.592 | 15.033 | 16.812 | 22.457 |
| 7 | 12.017 | 14.067 | 16.622 | 18.475 | 24.322 |
| 8 | 13.362 | 15.507 | 18.168 | 20.090 | 26.125 |
| 9 | 14.684 | 16.919 | 19.679 | 21.666 | 27.877 |
| 10 | 15.987 | 18.307 | 21.161 | 23.209 | 29.588 |
| 11 | 17.275 | 19.675 | 22.618 | 24.725 | 31.264 |
| 12 | 18.549 | 21.026 | 24.054 | 26.217 | 32.909 |
| 13 | 19.812 | 22.362 | 25.472 | 27.688 | 34.528 |
| 14 | 21.064 | 23.685 | 26.873 | 29.141 | 36.123 |
| 15 | 22.307 | 24.996 | 28.259 | 30.578 | 37.697 |
| 16 | 23.542 | 26.296 | 29.633 | 32.000 | 39.252 |
| 17 | 24.769 | 27.587 | 30.995 | 33.409 | 40.790 |
| 18 | 25.989 | 28.869 | 32.346 | 34.805 | 42.312 |
| 19 | 27.204 | 30.144 | 33.687 | 36.191 | 43.820 |
| 20 | 28.412 | 31.410 | 35.020 | 37.566 | 45.315 |

SOURCE: From Table IV of R. A. Fisher and F. Yates. *Statistical Tables for Biological, Agricultural, and Medical Research,* published by Longman Group Ltd., London (previously published by Oliver and Boyd Ltd., Edinburgh). Reprinted with permission of the authors and publishers.

**APPENDIX D** *(Continued)*

| | Levels of Significance | | | | |
|---|---|---|---|---|---|
| *df* | .10 | .05 | .02 | .01 | .001 |
| 21 | 29.615 | 32.671 | 36.343 | 38.932 | 46.797 |
| 22 | 30.813 | 33.924 | 37.659 | 40.289 | 48.268 |
| 23 | 32.007 | 35.172 | 38.968 | 41.638 | 49.728 |
| 24 | 33.196 | 36.415 | 40.270 | 42.980 | 51.179 |
| 25 | 34.382 | 37.652 | 41.566 | 44.314 | 52.620 |
| 26 | 35.563 | 38.885 | 42.856 | 45.642 | 54.052 |
| 27 | 36.741 | 40.113 | 44.140 | 46.963 | 55.476 |
| 28 | 37.916 | 41.337 | 45.419 | 48.278 | 56.893 |
| 29 | 39.087 | 42.557 | 46.693 | 49.588 | 58.302 |
| 30 | 40.256 | 43.773 | 47.962 | 50.892 | 59.703 |

# APPENDIX E

# 5 PERCENT* AND 1 PERCENT† POINTS FOR THE DISTRIBUTION OF *F*

**$f_1$ Degrees of Freedom for**

$f_2$ Degrees of Freedom for the Denominator

| $f_2$ | 1 | 2 | 3 | 4 | 5 | 6 | 7 | 8 | 9 | 10 | 11 | 12 |
|---|---|---|---|---|---|---|---|---|---|---|---|---|
| 1 | 161 | 200 | 216 | 225 | 230 | 234 | 237 | 239 | 241 | 242 | 243 | 244 |
| | **4,052** | **4,999** | **5,403** | **5,625** | **5,764** | **5,859** | **5,928** | **5,981** | **6,022** | **6,056** | **6,082** | **6,106** |
| 2 | 18.51 | 19.00 | 19.16 | 19.25 | 19.30 | 19.33 | 19.36 | 19.37 | 19.38 | 19.39 | 19.40 | 19.41 |
| | **98.49** | **99.00** | **99.17** | **99.25** | **99.30** | **99.33** | **99.34** | **99.36** | **99.38** | **99.40** | **99.41** | **99.42** |
| 3 | 10.13 | 9.55 | 9.28 | 9.12 | 9.01 | 8.94 | 8.88 | 8.84 | 8.81 | 8.78 | 8.76 | 8.74 |
| | **34.12** | **30.82** | **29.46** | **28.71** | **28.24** | **27.91** | **27.67** | **27.49** | **27.34** | **27.23** | **27.13** | **27.05** |
| 4 | 7.71 | 6.94 | 6.59 | 6.39 | 6.26 | 6.16 | 6.09 | 6.04 | 6.00 | 5.96 | 5.93 | 5.91 |
| | **21.20** | **18.00** | **16.69** | **15.98** | **15.52** | **15.21** | **14.98** | **14.80** | **14.66** | **14.54** | **14.45** | **14.37** |
| 5 | 6.61 | 5.79 | 5.41 | 5.19 | 5.05 | 4.95 | 4.88 | 4.82 | 4.78 | 4.74 | 4.70 | 4.68 |
| | **16.26** | **13.27** | **12.06** | **11.39** | **10.97** | **10.67** | **10.45** | **10.27** | **10.15** | **10.05** | **9.96** | **9.89** |
| 6 | 5.99 | 5.14 | 4.76 | 4.53 | 4.39 | 4.28 | 4.21 | 4.15 | 4.10 | 4.06 | 4.03 | 4.00 |
| | **13.74** | **10.92** | **9.78** | **9.15** | **8.75** | **8.47** | **8.26** | **8.10** | **7.98** | **7.87** | **7.79** | **7.72** |
| 7 | 5.59 | 4.74 | 4.35 | 4.12 | 3.97 | 3.87 | 3.79 | 3.73 | 3.68 | 3.63 | 3.60 | 3.57 |
| | **12.25** | **9.55** | **8.45** | **7.85** | **7.46** | **7.19** | **7.00** | **6.84** | **6.71** | **6.62** | **6.54** | **6.47** |
| 8 | 5.32 | 4.46 | 4.07 | 3.84 | 3.69 | 3.58 | 3.50 | 3.44 | 3.39 | 3.34 | 3.31 | 3.28 |
| | **11.26** | **8.65** | **7.59** | **7.01** | **6.63** | **6.37** | **6.19** | **6.03** | **5.91** | **5.82** | **5.74** | **5.67** |
| 9 | 5.12 | 4.26 | 3.86 | 3.63 | 3.48 | 3.37 | 3.29 | 3.23 | 3.18 | 3.13 | 3.10 | 3.07 |
| | **10.56** | **8.02** | **6.99** | **6.42** | **6.06** | **5.80** | **5.62** | **5.47** | **5.35** | **5.26** | **5.18** | **5.11** |
| 10 | 4.96 | 4.10 | 3.71 | 3.48 | 3.33 | 3.22 | 3.14 | 3.07 | 3.02 | 2.97 | 2.94 | 2.91 |
| | **10.04** | **7.56** | **6.55** | **5.99** | **5.64** | **5.39** | **5.21** | **5.06** | **4.95** | **4.85** | **4.78** | **4.71** |

* Lightface type = $\alpha$ = .05 † Boldface type = $\alpha$ = .01

SOURCE: Reprinted by permission from George W. Snedecor and William G. Cochran. *Statisti Methods*, 6th ed., © 1967 by Iowa State University Press, Ames, Iowa.

the Numerator

| 14 | 16 | 20 | 24 | 30 | 40 | 50 | 75 | 100 | 200 | 500 | ∞ | $f_2$ |
|---|---|---|---|---|---|---|---|---|---|---|---|---|
| 245 | 246 | 248 | 249 | 250 | 251 | 252 | 253 | 253 | 254 | 254 | 254 | 1 |
| **6,142** | **6,169** | **6,208** | **6,234** | **6,258** | **6,286** | **6,302** | **6,323** | **6,344** | **6,352** | **6,361** | **6,366** | |
| 19.42 | 19.43 | 19.44 | 1.945 | 19.46 | 19.47 | 19.47 | 19.48 | 19.49 | 19.49 | 19.50 | 19.50 | 2 |
| **99.43** | **99.44** | **99.45** | **99.46** | **99.47** | **99.48** | **99.48** | **99.49** | **99.49** | **99.49** | **99.50** | **99.50** | |
| 8.71 | 8.69 | 8.66 | 8.64 | 8.62 | 8.60 | 8.58 | 8.57 | 8.56 | 8.54 | 8.54 | 8.53 | 3 |
| **26.92** | **26.83** | **26.69** | **26.60** | **26.50** | **26.41** | **26.35** | **26.27** | **26.23** | **26.18** | **26.14** | **26.12** | |
| 5.87 | 5.84 | 5.80 | 5.77 | 5.74 | 5.71 | 5.70 | 5.68 | 5.66 | 5.65 | 5.64 | 5.63 | 4 |
| **14.24** | **14.15** | **14.02** | **13.93** | **13.83** | **13.74** | **13.69** | **13.61** | **13.57** | **13.52** | **13.48** | **13.46** | |
| 4.64 | 4.60 | 4.56 | 4.53 | 4.50 | 4.46 | 4.44 | 4.42 | 4.40 | 4.38 | 4.37 | 4.36 | 5 |
| **9.77** | **9.68** | **9.55** | **9.47** | **9.38** | **9.29** | **9.24** | **9.17** | **9.13** | **9.07** | **9.04** | **9.02** | |
| 3.96 | 3.92 | 3.87 | 3.84 | 3.81 | 3.77 | 3.75 | 3.72 | 3.71 | 3.69 | 3.68 | 3.67 | 6 |
| **7.60** | **7.52** | **7.39** | **7.31** | **7.23** | **7.14** | **7.09** | **7.02** | **6.99** | **6.94** | **6.90** | **6.88** | |
| 3.52 | 3.49 | 3.44 | 3.41 | 3.38 | 3.34 | 3.32 | 3.29 | 3.28 | 3.25 | 3.24 | 3.23 | 7 |
| **6.35** | **6.27** | **6.15** | **6.07** | **5.98** | **5.90** | **5.85** | **5.78** | **5.75** | **5.70** | **5.67** | **5.65** | |
| 3.23 | 3.20 | 3.15 | 3.12 | 3.08 | 3.05 | 3.03 | 3.00 | 2.98 | 2.96 | 2.94 | 2.93 | 8 |
| **5.56** | **5.48** | **5.36** | **5.28** | **5.20** | **5.11** | **5.06** | **5.00** | **4.96** | **4.91** | **4.88** | **4.86** | |
| 3.02 | 2.98 | 2.93 | 2.90 | 2.86 | 2.82 | 2.80 | 2.77 | 2.76 | 2.73 | 2.72 | 2.71 | 9 |
| **5.00** | **4.92** | **4.80** | **4.73** | **4.64** | **4.56** | **4.51** | **4.45** | **4.41** | **4.36** | **4.33** | **4.31** | |
| 2.86 | 2.82 | 2.77 | 2.74 | 2.70 | 2.67 | 2.64 | 2.61 | 2.59 | 2.56 | 2.55 | 2.54 | 10 |
| **4.60** | **4.52** | **4.41** | **4.33** | **4.25** | **4.17** | **4.12** | **4.05** | **4.01** | **3.96** | **3.93** | **3.91** | |

**APPENDIX E** *(Continued)*

$f_2$ Degrees of Freedom for the Denominator

| | $f_1$ Degrees of Freedom for | | | | | | | | | | | |
|---|---|---|---|---|---|---|---|---|---|---|---|---|
| $f_2$ | 1 | 2 | 3 | 4 | 5 | 6 | 7 | 8 | 9 | 10 | 11 | 12 |
| 11 | 4.84 | 3.98 | 3.59 | 3.36 | 3.20 | 3.09 | 3.01 | 2.95 | 2.90 | 2.86 | 2.82 | 2.79 |
| | **9.65** | **7.20** | **6.22** | **5.67** | **5.32** | **5.07** | **4.88** | **4.74** | **4.63** | **4.54** | **4.46** | **4.40** |
| 12 | 4.75 | 3.88 | 3.49 | 3.26 | 3.11 | 3.00 | 2.92 | 2.85 | 2.80 | 2.76 | 2.72 | 2.69 |
| | **9.33** | **6.93** | **5.95** | **5.41** | **5.06** | **4.82** | **4.65** | **4.50** | **4.39** | **4.30** | **4.22** | **4.16** |
| 13 | 4.67 | 3.80 | 3.41 | 3.18 | 3.02 | 2.92 | 2.84 | 2.77 | 2.72 | 2.67 | 2.63 | 2.60 |
| | **9.07** | **6.70** | **5.74** | **5.20** | **4.86** | **4.62** | **4.44** | **4.30** | **4.19** | **4.10** | **4.02** | **3.96** |
| 14 | 4.60 | 3.74 | 3.34 | 3.11 | 2.96 | 2.85 | 2.77 | 2.70 | 2.65 | 2.60 | 2.56 | 2.53 |
| | **8.86** | **6.51** | **5.56** | **5.03** | **4.69** | **4.46** | **4.28** | **4.14** | **4.03** | **3.94** | **3.86** | **3.80** |
| 15 | 4.54 | 3.68 | 3.29 | 3.06 | 2.90 | 2.79 | 2.70 | 2.64 | 2.59 | 2.55 | 2.51 | 2.48 |
| | **8.68** | **6.36** | **5.42** | **4.89** | **4.56** | **4.32** | **4.14** | **4.00** | **3.89** | **3.80** | **3.73** | **3.67** |
| 16 | 4.49 | 3.63 | 3.24 | 3.01 | 2.85 | 2.74 | 2.66 | 2.59 | 2.54 | 2.49 | 2.45 | 2.42 |
| | **8.53** | **6.23** | **5.29** | **4.77** | **4.44** | **4.20** | **4.03** | **3.89** | **3.78** | **3.69** | **3.61** | **3.55** |
| 17 | 4.45 | 3.59 | 3.20 | 2.96 | 2.81 | 2.70 | 2.62 | 2.55 | 2.50 | 2.45 | 2.41 | 2.38 |
| | **8.40** | **6.11** | **5.18** | **4.67** | **4.34** | **4.10** | **3.93** | **3.79** | **3.68** | **3.59** | **3.52** | **3.45** |
| 18 | 4.41 | 3.55 | 3.16 | 2.93 | 2.77 | 2.66 | 2.58 | 2.51 | 2.46 | 2.41 | 2.37 | 2.34 |
| | **8.28** | **6.01** | **5.09** | **4.58** | **4.25** | **4.01** | **3.85** | **3.71** | **3.60** | **3.51** | **3.44** | **3.37** |
| 19 | 4.38 | 3.52 | 3.13 | 2.90 | 2.74 | 2.63 | 2.55 | 2.48 | 2.43 | 2.38 | 2.34 | 2.31 |
| | **8.18** | **5.93** | **5.01** | **4.50** | **4.17** | **3.94** | **3.77** | **3.63** | **3.52** | **3.43** | **3.36** | **3.30** |
| 20 | 4.35 | 3.49 | 3.10 | 2.87 | 2.71 | 2.60 | 2.52 | 2.45 | 2.40 | 2.35 | 2.31 | 2.28 |
| | **8.10** | **5.85** | **4.94** | **4.43** | **4.10** | **3.87** | **3.71** | **3.56** | **3.45** | **3.37** | **3.30** | **3.23** |
| 21 | 4.32 | 3.47 | 3.07 | 2.84 | 2.68 | 2.57 | 2.49 | 2.42 | 2.37 | 2.32 | 2.28 | 2.25 |
| | **8.02** | **5.78** | **4.87** | **4.37** | **4.04** | **3.81** | **3.65** | **3.51** | **3.40** | **3.31** | **3.24** | **3.17** |
| 22 | 4.30 | 3.44 | 3.05 | 2.82 | 2.66 | 2.55 | 2.47 | 2.40 | 2.35 | 2.30 | 2.26 | 2.23 |
| | **7.94** | **5.72** | **4.82** | **4.31** | **3.99** | **3.76** | **3.59** | **3.45** | **3.35** | **3.26** | **3.18** | **3.12** |
| 23 | 4.28 | 3.42 | 3.03 | 2.80 | 2.64 | 2.53 | 2.45 | 2.38 | 2.32 | 2.28 | 2.24 | 2.20 |
| | **7.88** | **5.66** | **4.76** | **4.26** | **3.94** | **3.71** | **3.54** | **3.41** | **3.30** | **3.21** | **3.14** | **3.07** |
| 24 | 4.26 | 3.40 | 3.01 | 2.78 | 2.62 | 2.51 | 2.43 | 2.36 | 2.30 | 2.26 | 2.22 | 2.18 |
| | **7.82** | **5.61** | **4.72** | **4.22** | **3.90** | **3.67** | **3.50** | **3.36** | **3.25** | **3.17** | **3.09** | **3.03** |
| 25 | 4.24 | 3.38 | 2.99 | 2.76 | 2.60 | 2.49 | 2.41 | 2.34 | 2.28 | 2.24 | 2.20 | 2.16 |
| | **7.77** | **5.57** | **4.68** | **4.18** | **3.86** | **3.63** | **3.46** | **3.32** | **3.21** | **3.13** | **3.05** | **2.99** |

the Numerator

| 14 | 16 | 20 | 24 | 30 | 40 | 50 | 75 | 100 | 200 | 500 | ∞ | $f_2$ |
|---|---|---|---|---|---|---|---|---|---|---|---|---|
| 2.74 | 2.70 | 2.65 | 2.61 | 2.57 | 2.53 | 2.50 | 2.47 | 2.45 | 2.42 | 2.41 | 2.40 | 11 |
| **4.29** | **4.21** | **4.10** | **4.02** | **3.94** | **3.86** | **3.80** | **3.74** | **3.70** | **3.66** | **3.62** | **3.60** | |
| 2.64 | 2.60 | 2.54 | 2.50 | 2.46 | 2.42 | 2.40 | 2.36 | 2.35 | 2.32 | 2.31 | 2.30 | 12 |
| **4.05** | **3.98** | **3.86** | **3.78** | **3.70** | **3.61** | **3.56** | **3.49** | **3.46** | **3.41** | **3.38** | **3.36** | |
| 2.55 | 2.51 | 2.46 | 2.42 | 2.38 | 2.34 | 2.32 | 2.28 | 2.26 | 2.24 | 2.22 | 2.21 | 13 |
| **3.85** | **3.78** | **3.67** | **3.59** | **3.51** | **3.42** | **3.37** | **3.30** | **3.27** | **3.21** | **3.18** | **3.16** | |
| 2.48 | 2.44 | 2.39 | 2.35 | 2.31 | 2.27 | 2.24 | 2.21 | 2.19 | 2.16 | 2.14 | 2.13 | 14 |
| **3.70** | **3.62** | **3.51** | **3.43** | **3.34** | **3.26** | **3.21** | **3.14** | **3.11** | **3.06** | **3.02** | **3.00** | |
| 2.43 | 2.39 | 2.33 | 2.29 | 2.25 | 2.21 | 2.18 | 2.15 | 2.12 | 2.10 | 2.08 | 2.07 | 15 |
| **3.56** | **3.48** | **3.36** | **3.29** | **3.20** | **3.12** | **3.07** | **3.00** | **2.97** | **2.92** | **2.89** | **2.87** | |
| 2.37 | 2.33 | 2.28 | 2.24 | 2.20 | 2.16 | 2.13 | 2.09 | 2.07 | 2.04 | 2.02 | 2.01 | 16 |
| **3.45** | **3.37** | **3.25** | **3.18** | **3.10** | **3.01** | **2.96** | **2.89** | **2.86** | **2.80** | **2.77** | **2.75** | |
| 2.33 | 2.29 | 2.23 | 2.19 | 2.15 | 2.11 | 2.08 | 2.04 | 2.02 | 1.99 | 1.97 | 1.96 | 17 |
| **3.35** | **3.27** | **3.16** | **3.08** | **3.00** | **2.92** | **2.86** | **2.79** | **2.76** | **2.70** | **2.67** | **2.65** | |
| 2.29 | 2.25 | 2.19 | 2.15 | 2.11 | 2.07 | 2.04 | 2.00 | 1.98 | 1.95 | 1.93 | 1.92 | 18 |
| **3.27** | **3.19** | **3.07** | **3.00** | **2.91** | **2.83** | **2.78** | **2.71** | **2.68** | **2.62** | **2.59** | **2.57** | |
| 2.26 | 2.21 | 2.15 | 2.11 | 2.07 | 2.02 | 2.00 | 1.96 | 1.94 | 1.91 | 1.90 | 1.88 | 19 |
| **3.19** | **3.12** | **3.00** | **2.92** | **2.84** | **2.76** | **2.70** | **2.63** | **2.60** | **2.54** | **2.51** | **2.49** | |
| 2.23 | 2.18 | 2.12 | 2.08 | 2.04 | 1.99 | 1.96 | 1.92 | 1.90 | 1.87 | 1.85 | 1.84 | 20 |
| **3.13** | **3.05** | **2.94** | **2.86** | **2.77** | **2.69** | **2.63** | **2.56** | **2.53** | **2.47** | **2.44** | **2.42** | |
| 2.20 | 2.15 | 2.09 | 2.05 | 2.00 | 1.96 | 1.93 | 1.89 | 1.87 | 1.84 | 1.82 | 1.81 | 21 |
| **3.07** | **2.99** | **2.88** | **2.80** | **2.72** | **2.63** | **2.58** | **2.51** | **2.47** | **2.42** | **2.38** | **2.36** | |
| 2.18 | 2.13 | 2.07 | 2.03 | 1.98 | 1.93 | 1.91 | 1.87 | 1.84 | 1.81 | 1.80 | 1.78 | 22 |
| **3.02** | **2.94** | **2.83** | **2.75** | **2.67** | **2.58** | **2.53** | **2.46** | **2.42** | **2.37** | **2.33** | **2.31** | |
| 2.14 | 2.10 | 2.04 | 2.00 | 1.96 | 1.91 | 1.88 | 1.84 | 1.82 | 1.79 | 1.77 | 1.76 | 23 |
| **2.97** | **2.89** | **2.78** | **2.70** | **2.62** | **2.53** | **2.48** | **2.41** | **2.37** | **2.32** | **2.28** | **2.26** | |
| 2.13 | 2.09 | 2.02 | 1.98 | 1.94 | 1.89 | 1.86 | 1.82 | 1.80 | 1.76 | 1.74 | 1.73 | 24 |
| **2.93** | **2.85** | **2.74** | **2.66** | **2.58** | **2.49** | **2.44** | **2.36** | **2.33** | **2.27** | **2.23** | **2.21** | |
| 2.11 | 2.06 | 2.00 | 1.96 | 1.92 | 1.87 | 1.84 | 1.80 | 1.77 | 1.74 | 1.72 | 1.71 | 25 |
| **2.89** | **2.81** | **2.70** | **2.62** | **2.54** | **2.45** | **2.40** | **2.32** | **2.29** | **2.23** | **2.19** | **2.17** | |

**APPENDIX E** *(Continued)*

$f_1$ **Degrees of Freedom for**

| $f_2$ | 1 | 2 | 3 | 4 | 5 | 6 | 7 | 8 | 9 | 10 | 11 | 12 |
|---|---|---|---|---|---|---|---|---|---|---|---|---|
| 26 | 4.22 | 3.37 | 2.98 | 2.74 | 2.59 | 2.47 | 2.39 | 2.32 | 2.27 | 2.22 | 2.18 | 2.15 |
| | **7.72** | **5.53** | **4.64** | **4.14** | **3.82** | **3.59** | **3.42** | **3.29** | **3.17** | **3.09** | **3.02** | **2.96** |
| 27 | 4.21 | 3.35 | 2.96 | 2.73 | 2.57 | 2.46 | 2.37 | 2.30 | 2.25 | 2.20 | 2.16 | 2.13 |
| | **7.68** | **5.49** | **4.60** | **4.11** | **3.79** | **3.56** | **3.39** | **3.26** | **3.14** | **3.06** | **2.98** | **2.93** |
| 28 | 4.20 | 3.34 | 2.95 | 2.71 | 2.56 | 2.44 | 2.36 | 2.29 | 2.24 | 2.19 | 2.15 | 2.12 |
| | **7.64** | **5.45** | **4.57** | **4.07** | **3.76** | **3.53** | **3.36** | **3.23** | **3.11** | **3.03** | **2.95** | **2.90** |
| 29 | 4.18 | 3.33 | 2.93 | 2.70 | 2.54 | 2.43 | 2.35 | 2.28 | 2.22 | 2.18 | 2.14 | 2.10 |
| | **7.60** | **5.42** | **4.54** | **4.04** | **3.73** | **3.50** | **3.33** | **3.20** | **3.08** | **3.00** | **2.92** | **2.87** |
| 30 | 4.17 | 3.32 | 2.92 | 2.69 | 2.53 | 2.42 | 2.34 | 2.27 | 2.21 | 2.16 | 2.12 | 2.09 |
| | **7.56** | **5.39** | **4.51** | **4.02** | **3.70** | **3.47** | **3.30** | **3.17** | **3.06** | **2.98** | **2.90** | **2.84** |
| 32 | 4.15 | 3.30 | 2.90 | 2.67 | 2.51 | 2.40 | 2.32 | 2.25 | 2.19 | 2.14 | 2.10 | 2.07 |
| | **7.50** | **5.34** | **4.46** | **3.97** | **3.66** | **3.42** | **3.25** | **3.12** | **3.01** | **2.94** | **2.86** | **2.80** |
| 34 | 4.13 | 3.28 | 2.88 | 2.65 | 2.49 | 2.38 | 2.30 | 2.23 | 2.17 | 2.12 | 2.08 | 2.05 |
| | **7.44** | **5.29** | **4.42** | **3.93** | **3.61** | **3.38** | **3.21** | **3.08** | **2.97** | **2.89** | **2.82** | **2.76** |
| 36 | 4.11 | 3.26 | 2.86 | 2.63 | 2.48 | 2.36 | 2.28 | 2.21 | 2.15 | 2.10 | 2.06 | 2.03 |
| | **7.39** | **5.25** | **4.38** | **3.89** | **3.58** | **3.35** | **3.18** | **3.04** | **2.94** | **2.86** | **2.78** | **2.72** |
| 38 | 4.10 | 3.25 | 2.85 | 2.62 | 2.46 | 2.35 | 2.26 | 2.19 | 2.14 | 2.09 | 2.05 | 2.02 |
| | **7.35** | **5.21** | **4.34** | **3.86** | **3.54** | **3.32** | **3.15** | **3.02** | **2.91** | **2.82** | **2.75** | **2.69** |
| 40 | 4.08 | 3.23 | 2.84 | 2.61 | 2.45 | 2.34 | 2.25 | 2.18 | 2.12 | 2.07 | 2.04 | 2.00 |
| | **7.31** | **5.18** | **4.31** | **3.83** | **3.51** | **3.29** | **3.12** | **2.99** | **2.88** | **2.80** | **2.73** | **2.66** |
| 42 | 4.07 | 3.22 | 2.83 | 2.59 | 2.44 | 2.32 | 2.24 | 2.17 | 2.11 | 2.06 | 2.02 | 1.99 |
| | **7.27** | **5.15** | **4.29** | **3.80** | **3.49** | **3.26** | **3.10** | **2.96** | **2.86** | **2.77** | **2.70** | **2.64** |
| 44 | 4.06 | 3.21 | 2.82 | 2.58 | 2.43 | 2.31 | 2.23 | 2.16 | 2.10 | 2.05 | 2.01 | 1.98 |
| | **7.24** | **5.12** | **4.26** | **3.78** | **3.46** | **3.24** | **3.07** | **2.94** | **2.84** | **2.75** | **2.68** | **2.62** |
| 46 | 4.05 | 3.20 | 2.81 | 2.57 | 2.42 | 2.30 | 2.22 | 2.14 | 2.09 | 2.04 | 2.00 | 1.97 |
| | **7.21** | **5.10** | **4.24** | **3.76** | **3.44** | **3.22** | **3.05** | **2.92** | **2.82** | **2.73** | **2.66** | **2.60** |
| 48 | 4.04 | 3.19 | 2.80 | 2.56 | 2.41 | 2.30 | 2.21 | 2.14 | 2.08 | 2.03 | 1.99 | 1.96 |
| | **7.19** | **5.08** | **4.22** | **3.74** | **3.42** | **3.20** | **3.04** | **2.90** | **2.80** | **2.71** | **2.64** | **2.58** |
| 50 | 4.03 | 3.18 | 2.79 | 2.56 | 2.40 | 2.29 | 2.20 | 2.13 | 2.07 | 2.02 | 1.98 | 1.95 |
| | **7.17** | **5.06** | **4.20** | **3.72** | **3.41** | **3.18** | **3.02** | **2.88** | **2.78** | **2.70** | **2.62** | **2.56** |

$f_2$ **Degrees of Freedom for the Denominator**

**the Numerator**

| 14 | 16 | 20 | 24 | 30 | 40 | 50 | 75 | 100 | 200 | 500 | ∞ | $f_2$ |
|---|---|---|---|---|---|---|---|---|---|---|---|---|
| 2.10 | 2.05 | 1.99 | 1.95 | 1.90 | 1.85 | 1.82 | 1.78 | 1.76 | 1.72 | 1.70 | 1.69 | 26 |
| **2.86** | **2.77** | **2.66** | **2.58** | **2.50** | **2.41** | **2.36** | **2.28** | **2.25** | **2.19** | **2.15** | **2.13** | |
| 2.08 | 2.03 | 1.97 | 1.93 | 1.88 | 1.84 | 1.80 | 1.76 | 1.74 | 1.71 | 1.68 | 1.67 | 27 |
| **2.83** | **2.74** | **2.63** | **2.55** | **2.47** | **2.38** | **2.33** | **2.25** | **2.21** | **2.16** | **2.12** | **2.10** | |
| 2.06 | 2.02 | 1.96 | 1.91 | 1.87 | 1.81 | 1.78 | 1.75 | 1.72 | 1.69 | 1.67 | 1.65 | 28 |
| **2.80** | **2.71** | **2.60** | **2.52** | **2.44** | **2.35** | **2.30** | **2.22** | **2.18** | **2.13** | **2.09** | **2.06** | |
| 2.05 | 2.00 | 1.94 | 1.90 | 1.85 | 1.80 | 1.77 | 1.73 | 1.71 | 1.68 | 1.65 | 1.64 | 29 |
| **2.77** | **2.68** | **2.57** | **2.49** | **2.41** | **2.32** | **2.27** | **2.19** | **2.15** | **2.10** | **2.06** | **2.03** | |
| 2.04 | 1.99 | 1.93 | 1.89 | 1.84 | 1.79 | 1.76 | 1.72 | 1.69 | 1.66 | 1.64 | 1.62 | 30 |
| **2.74** | **2.66** | **2.55** | **2.47** | **2.38** | **2.29** | **2.24** | **2.16** | **2.13** | **2.07** | **2.03** | **2.01** | |
| 2.02 | 1.97 | 1.91 | 1.86 | 1.82 | 1.76 | 1.74 | 1.69 | 1.67 | 1.64 | 1.61 | 1.59 | 32 |
| **2.70** | **2.62** | **2.51** | **2.42** | **2.34** | **2.25** | **2.20** | **2.12** | **2.08** | **2.02** | **1.98** | **1.96** | |
| 2.00 | 1.95 | 1.89 | 1.84 | 1.80 | 1.74 | 1.71 | 1.67 | 1.64 | 1.61 | 1.59 | 1.57 | 34 |
| **2.66** | **2.58** | **2.47** | **2.38** | **2.30** | **2.21** | **2.15** | **2.08** | **2.04** | **1.98** | **1.94** | **1.91** | |
| 1.98 | 1.93 | 1.87 | 1.82 | 1.78 | 1.72 | 1.69 | 1.65 | 1.62 | 1.59 | 1.56 | 1.55 | 36 |
| **2.62** | **2.54** | **2.43** | **2.35** | **2.26** | **2.17** | **2.12** | **2.04** | **2.00** | **1.94** | **1.90** | **1.87** | |
| 1.96 | 1.92 | 1.85 | 1.80 | 1.76 | 1.71 | 1.67 | 1.63 | 1.60 | 1.57 | 1.54 | 1.53 | 38 |
| **2.59** | **2.51** | **2.40** | **2.32** | **2.22** | **2.14** | **2.08** | **2.00** | **1.97** | **1.90** | **1.86** | **1.84** | |
| 1.95 | 1.90 | 1.84 | 1.79 | 1.74 | 1.69 | 1.66 | 1.61 | 1.59 | 1.55 | 1.53 | 1.51 | 40 |
| **2.56** | **2.49** | **2.37** | **2.29** | **2.20** | **2.11** | **2.05** | **1.97** | **1.94** | **1.88** | **1.84** | **1.81** | |
| 1.94 | 1.89 | 1.82 | 1.78 | 1.73 | 1.68 | 1.64 | 1.60 | 1.57 | 1.54 | 1.51 | 1.49 | 42 |
| **2.54** | **2.46** | **2.35** | **2.26** | **2.17** | **2.08** | **2.02** | **1.94** | **1.91** | **1.85** | **1.80** | **1.78** | |
| 1.92 | 1.88 | 1.81 | 1.76 | 1.72 | 1.66 | 1.63 | 1.58 | 1.56 | 1.52 | 1.50 | 1.48 | 44 |
| **2.52** | **2.44** | **2.32** | **2.24** | **2.15** | **2.06** | **2.00** | **1.92** | **1.88** | **1.82** | **1.78** | **1.75** | |
| 1.91 | 1.87 | 1.80 | 1.75 | 1.71 | 1.65 | 1.62 | 1.57 | 1.54 | 1.51 | 1.48 | 1.46 | 46 |
| **2.50** | **2.42** | **2.30** | **2.22** | **2.13** | **2.04** | **1.98** | **1.90** | **1.86** | **1.80** | **1.76** | **1.72** | |
| 1.90 | 1.86 | 1.79 | 1.74 | 1.70 | 1.64 | 1.61 | 1.56 | 1.53 | 1.50 | 1.47 | 1.45 | 48 |
| **2.48** | **2.40** | **2.28** | **2.20** | **2.11** | **2.02** | **1.96** | **1.88** | **1.84** | **1.78** | **1.73** | **1.70** | |
| 1.90 | 1.85 | 1.78 | 1.74 | 1.69 | 1.63 | 1.60 | 1.55 | 1.52 | 1.48 | 1.46 | 1.44 | 50 |
| **2.46** | **2.39** | **2.26** | **2.18** | **2.10** | **2.00** | **1.94** | **1.86** | **1.82** | **1.76** | **1.71** | **1.68** | |

**APPENDIX E** *(Continued)*

$f_2$ **Degrees of Freedom for the Denominator**

| | **$f_1$ Degrees of Freedom for** | | | | | | | | | | | |
|---|---|---|---|---|---|---|---|---|---|---|---|---|
| $f_2$ | 1 | 2 | 3 | 4 | 5 | 6 | 7 | 8 | 9 | 10 | 11 | 12 |
| 55 | 4.02 | 3.17 | 2.78 | 2.54 | 2.38 | 2.27 | 2.18 | 2.11 | 2.05 | 2.00 | 1.97 | 1.93 |
| | **7.12** | **5.01** | **4.16** | **3.68** | **3.37** | **3.15** | **2.98** | **2.85** | **2.75** | **2.66** | **2.59** | **2.53** |
| 60 | 4.00 | 3.15 | 2.76 | 2.52 | 2.37 | 2.25 | 2.17 | 2.10 | 2.04 | 1.99 | 1.95 | 1.92 |
| | **7.08** | **4.98** | **4.13** | **3.65** | **3.34** | **3.12** | **2.95** | **2.82** | **2.72** | **2.63** | **2.56** | **2.50** |
| 65 | 3.99 | 3.14 | 2.75 | 2.51 | 2.36 | 2.24 | 2.15 | 2.08 | 2.02 | 1.98 | 1.94 | 1.90 |
| | **7.04** | **4.95** | **4.10** | **3.62** | **3.31** | **3.09** | **2.93** | **2.79** | **2.70** | **2.61** | **2.54** | **2.47** |
| 70 | 3.98 | 3.13 | 2.74 | 2.50 | 2.35 | 2.23 | 2.14 | 2.07 | 2.01 | 1.97 | 1.93 | 1.89 |
| | **7.01** | **4.92** | **4.08** | **3.60** | **3.29** | **3.07** | **2.91** | **2.77** | **2.67** | **2.59** | **2.51** | **2.45** |
| 80 | 3.96 | 3.11 | 2.72 | 2.48 | 2.33 | 2.21 | 2.12 | 2.05 | 1.99 | 1.95 | 1.91 | 1.88 |
| | **6.96** | **4.88** | **4.04** | **3.56** | **3.25** | **3.04** | **2.87** | **2.74** | **2.64** | **2.55** | **2.48** | **2.41** |
| 100 | 3.94 | 3.09 | 2.70 | 2.46 | 2.30 | 2.19 | 2.10 | 2.03 | 1.97 | 1.92 | 1.88 | 1.85 |
| | **6.90** | **4.82** | **3.98** | **3.51** | **3.20** | **2.99** | **2.82** | **2.69** | **2.59** | **2.51** | **2.43** | **2.36** |
| 125 | 3.92 | 3.07 | 2.68 | 2.44 | 2.29 | 2.17 | 2.08 | 2.01 | 1.95 | 1.90 | 1.86 | 1.83 |
| | **6.84** | **4.78** | **3.94** | **3.47** | **3.17** | **2.95** | **2.79** | **2.65** | **2.56** | **2.47** | **2.40** | **2.33** |
| 150 | 3.91 | 3.06 | 2.67 | 2.43 | 2.27 | 2.16 | 2.07 | 2.00 | 1.94 | 1.89 | 1.85 | 1.82 |
| | **6.81** | **4.75** | **3.91** | **3.44** | **3.14** | **2.92** | **2.76** | **2.62** | **2.53** | **2.44** | **2.37** | **2.30** |
| 200 | 3.89 | 3.04 | 2.65 | 2.41 | 2.26 | 2.14 | 2.05 | 1.98 | 1.92 | 1.87 | 1.83 | 1.80 |
| | **6.76** | **4.71** | **3.88** | **3.41** | **3.11** | **2.90** | **2.73** | **2.60** | **2.50** | **2.41** | **2.34** | **2.28** |
| 400 | 3.86 | 3.02 | 2.62 | 2.39 | 2.23 | 2.12 | 2.03 | 1.96 | 1.90 | 1.85 | 1.81 | 1.78 |
| | **6.70** | **4.66** | **3.83** | **3.36** | **3.06** | **2.85** | **2.69** | **2.55** | **2.46** | **2.37** | **2.29** | **2.23** |
| 1000 | 3.85 | 3.00 | 2.61 | 2.38 | 2.22 | 2.10 | 2.02 | 1.95 | 1.89 | 1.84 | 1.80 | 1.76 |
| | **6.66** | **4.62** | **3.80** | **3.34** | **3.04** | **2.82** | **2.66** | **2.53** | **2.43** | **2.34** | **2.26** | **2.20** |
| ∞ | 3.84 | 2.99 | 2.60 | 2.37 | 2.21 | 2.09 | 2.01 | 1.94 | 1.88 | 1.83 | 1.79 | 1.75 |
| | **6.64** | **4.60** | **3.78** | **3.32** | **3.02** | **2.80** | **2.64** | **2.51** | **2.41** | **2.32** | **2.24** | **2.18** |

the Numerator

| 14 | 16 | 20 | 24 | 30 | 40 | 50 | 75 | 100 | 200 | 500 | ∞ | $f_2$ |
|---|---|---|---|---|---|---|---|---|---|---|---|---|
| 1.88 | 1.83 | 1.76 | 1.72 | 1.67 | 1.61 | 1.58 | 1.52 | 1.50 | 1.46 | 1.43 | 1.41 | 55 |
| **2.43** | **2.35** | **2.23** | **2.15** | **2.06** | **1.96** | **1.90** | **1.82** | **1.78** | **1.71** | **1.66** | **1.64** | |
| 1.86 | 1.81 | 1.75 | 1.70 | 1.65 | 1.59 | 1.56 | 1.50 | 1.48 | 1.44 | 1.41 | 1.39 | 60 |
| **2.40** | **2.32** | **2.20** | **2.12** | **2.03** | **1.93** | **1.87** | **1.79** | **1.74** | **1.68** | **1.63** | **1.60** | |
| 1.85 | 1.80 | 1.73 | 1.68 | 1.63 | 1.57 | 1.54 | 1.49 | 1.46 | 1.42 | 1.39 | 1.37 | 65 |
| **2.37** | **2.30** | **2.18** | **2.09** | **2.00** | **1.90** | **1.84** | **1.76** | **1.71** | **1.64** | **1.60** | **1.56** | |
| 1.84 | 1.79 | 1.72 | 1.67 | 1.62 | 1.56 | 1.53 | 1.47 | 1.45 | 1.40 | 1.37 | 1.35 | 70 |
| **2.35** | **2.28** | **2.15** | **2.07** | **1.98** | **1.88** | **1.82** | **1.74** | **1.69** | **1.62** | **1.56** | **1.53** | |
| 1.82 | 1.77 | 1.70 | 1.65 | 1.60 | 1.54 | 1.51 | 1.45 | 1.42 | 1.38 | 1.35 | 1.32 | 80 |
| **2.32** | **2.24** | **2.11** | **2.03** | **1.94** | **1.84** | **1.78** | **1.70** | **1.65** | **1.57** | **1.52** | **1.49** | |
| 1.79 | 1.75 | 1.68 | 1.63 | 1.57 | 1.51 | 1.48 | 1.42 | 1.39 | 1.34 | 1.30 | 1.28 | 100 |
| **2.26** | **2.19** | **2.06** | **1.98** | **1.89** | **1.79** | **1.73** | **1.64** | **1.59** | **1.51** | **1.46** | **1.43** | |
| 1.77 | 1.72 | 1.65 | 1.60 | 1.55 | 1.49 | 1.45 | 1.39 | 1.36 | 1.31 | 1.27 | 1.25 | 125 |
| **2.23** | **2.15** | **2.03** | **1.94** | **1.85** | **1.75** | **1.68** | **1.59** | **1.54** | **1.46** | **1.40** | **1.37** | |
| 1.76 | 1.71 | 1.64 | 1.59 | 1.54 | 1.47 | 1.44 | 1.37 | 1.34 | 1.29 | 1.25 | 1.22 | 150 |
| **2.20** | **2.12** | **2.00** | **1.91** | **1.83** | **1.72** | **1.66** | **1.56** | **1.51** | **1.43** | **1.37** | **1.33** | |
| 1.74 | 1.69 | 1.62 | 1.57 | 1.52 | 1.45 | 1.42 | 1.35 | 1.32 | 1.26 | 1.22 | 1.19 | 200 |
| **2.17** | **2.09** | **1.97** | **1.88** | **1.79** | **1.69** | **1.62** | **1.53** | **1.48** | **1.39** | **1.33** | **1.28** | |
| 1.72 | 1.67 | 1.60 | 1.54 | 1.49 | 1.42 | 1.38 | 1.32 | 1.28 | 1.22 | 1.16 | 1.13 | 400 |
| **2.12** | **2.04** | **1.92** | **1.84** | **1.74** | **1.64** | **1.57** | **1.47** | **1.42** | **1.32** | **1.24** | **1.19** | |
| 1.70 | 1.65 | 1.58 | 1.53 | 1.47 | 1.41 | 1.36 | 1.30 | 1.26 | 1.19 | 1.13 | 1.08 | 1000 |
| **2.09** | **2.01** | **1.89** | **1.81** | **1.71** | **1.61** | **1.54** | **1.44** | **1.38** | **1.28** | **1.19** | **1.11** | |
| 1.69 | 1.64 | 1.57 | 1.52 | 1.46 | 1.40 | 1.35 | 1.28 | 1.24 | 1.17 | 1.11 | 1.00 | ∞ |
| **2.07** | **1.99** | **1.87** | **1.79** | **1.69** | **1.59** | **1.52** | **1.41** | **1.36** | **1.25** | **1.15** | **1.00** | |

## APPENDIX E (Continued)

# .1 PERCENT POINTS FOR THE DISTRIBUTION OF $F$ ($\alpha = .001$)

| *df* for the Denominator | 1 | 2 | 3 | 4 | 5 | *df* for the 6 |
|---|---|---|---|---|---|---|
| 4 | 74.14 | 61.25 | 56.18 | 53.44 | 51.71 | 50.53 |
| 5 | 47.18 | 37.12 | 33.20 | 31.09 | 29.75 | 28.84 |
| 6 | 35.51 | 27.00 | 23.70 | 21.92 | 20.81 | 20.03 |
| 7 | 29.25 | 21.69 | 18.77 | 17.19 | 16.21 | 15.52 |
| 8 | 25.42 | 18.49 | 15.83 | 14.39 | 13.49 | 12.86 |
| 9 | 22.86 | 16.39 | 13.90 | 12.56 | 11.71 | 11.13 |
| 10 | 21.04 | 14.91 | 12.55 | 11.28 | 10.48 | 9.92 |
| 11 | 19.69 | 13.81 | 11.56 | 10.35 | 9.58 | 9.05 |
| 12 | 18.64 | 12.97 | 10.80 | 9.63 | 8.89 | 8.38 |
| 13 | 17.81 | 12.31 | 10.21 | 9.07 | 8.35 | 7.86 |
| 14 | 17.14 | 11.78 | 9.73 | 8.62 | 7.92 | 7.43 |
| 15 | 16.59 | 11.34 | 9.34 | 8.25 | 7.57 | 7.09 |
| 16 | 16.11 | 10.97 | 9.00 | 7.95 | 7.27 | 6.81 |
| 17 | 15.72 | 10.66 | 8.73 | 7.68 | 7.02 | 6.56 |
| 18 | 15.38 | 10.39 | 8.49 | 7.46 | 6.81 | 6.35 |
| 19 | 15.08 | 10.16 | 8.28 | 7.26 | 6.62 | 6.18 |
| 20 | 14.82 | 9.95 | 8.10 | 7.10 | 6.46 | 6.02 |
| 21 | 14.59 | 9.77 | 7.94 | 6.95 | 6.32 | 5.88 |
| 22 | 14.38 | 9.61 | 7.80 | 6.81 | 6.19 | 5.76 |
| 23 | 14.19 | 9.47 | 7.67 | 6.69 | 6.08 | 5.65 |
| 24 | 14.03 | 9.34 | 7.55 | 6.59 | 5.98 | 5.55 |
| 25 | 13.88 | 9.22 | 7.45 | 6.49 | 5.88 | 5.46 |
| 26 | 13.74 | 9.12 | 7.36 | 6.41 | 5.80 | 5.38 |
| 27 | 13.61 | 9.02 | 7.27 | 6.33 | 5.73 | 5.31 |
| 28 | 13.50 | 8.93 | 7.19 | 6.25 | 5.66 | 5.24 |
| 29 | 13.39 | 8.85 | 7.12 | 6.19 | 5.59 | 5.18 |
| 30 | 13.29 | 8.77 | 7.05 | 6.12 | 5.53 | 5.12 |
| 40 | 12.61 | 8.25 | 6.60 | 5.70 | 5.13 | 4.73 |
| 60 | 11.97 | 7.76 | 6.17 | 5.31 | 4.76 | 4.37 |
| 120 | 11.38 | 7.32 | 5.79 | 4.95 | 4.42 | 4.04 |
| ∞ | 10.83 | 6.91 | 5.42 | 4.62 | 4.10 | 3.74 |

SOURCE: Adapted from "Percentage Points, *F*-Distribution," *Handbook of Tables for Mathematicians,* © CRC Press, Cleveland, Ohio, 1970, p. 626. Used by permission of CRC Press, Inc.

| Numerator 7 | 8 | 9 | 10 | 20 | 40 | ∞ | |
|---|---|---|---|---|---|---|---|
| 49.66 | 49.00 | 48.47 | 48.05 | 46.10 | 45.00 | 44.05 | 4 |
| 28.16 | 27.64 | 27.24 | 26.92 | 25.39 | 24.60 | 23.79 | 5 |
| 19.46 | 19.03 | 18.69 | 18.41 | 17.12 | 16.44 | 15.75 | 6 |
| 15.02 | 14.63 | 14.33 | 14.08 | 12.93 | 12.33 | 11.70 | 7 |
| 12.40 | 12.04 | 11.77 | 11.54 | 10.48 | 9.92 | 9.33 | 8 |
| 10.70 | 10.37 | 10.11 | 9.89 | 8.90 | 8.37 | 7.81 | 9 |
| 9.52 | 9.20 | 8.96 | 8.75 | 7.80 | 7.30 | 6.76 | 10 |
| 8.66 | 8.35 | 8.12 | 7.92 | 7.01 | 6.52 | 6.00 | 11 |
| 8.00 | 7.71 | 7.48 | 7.29 | 6.40 | 5.93 | 5.42 | 12 |
| 7.49 | 7.21 | 6.98 | 6.80 | 5.93 | 5.47 | 4.97 | 13 |
| 7.08 | 6.80 | 6.58 | 6.40 | 5.56 | 5.10 | 4.60 | 14 |
| 6.75 | 6.47 | 6.26 | 6.08 | 5.25 | 4.80 | 4.31 | 15 |
| 6.46 | 6.19 | 5.98 | 5.81 | 4.99 | 4.54 | 4.06 | 16 |
| 6.22 | 5.96 | 5.75 | 5.58 | 4.78 | 4.33 | 3.85 | 17 |
| 6.02 | 5.76 | 5.56 | 5.39 | 4.59 | 4.15 | 3.67 | 18 |
| 5.85 | 5.59 | 5.39 | 5.22 | 4.43 | 3.99 | 3.51 | 19 |
| 5.69 | 5.44 | 5.24 | 5.08 | 4.29 | 3.86 | 3.38 | 20 |
| 5.56 | 5.31 | 5.11 | 4.95 | 4.17 | 3.74 | 3.26 | 21 |
| 5.44 | 5.19 | 4.99 | 4.83 | 4.06 | 3.63 | 3.15 | 22 |
| 5.33 | 5.09 | 4.89 | 4.73 | 3.96 | 3.53 | 3.05 | 23 |
| 5.23 | 4.99 | 4.80 | 4.64 | 3.87 | 3.45 | 2.97 | 24 |
| 5.15 | 4.91 | 4.71 | 4.56 | 3.79 | 3.37 | 2.89 | 25 |
| 5.07 | 4.83 | 4.64 | 4.48 | 3.72 | 3.30 | 2.82 | 26 |
| 5.00 | 4.76 | 4.57 | 4.41 | 3.66 | 3.23 | 2.75 | 27 |
| 4.93 | 4.69 | 4.50 | 4.35 | 3.60 | 3.18 | 2.69 | 28 |
| 4.87 | 4.64 | 4.45 | 4.29 | 3.54 | 3.12 | 2.64 | 29 |
| 4.82 | 4.58 | 4.39 | 4.24 | 3.49 | 3.07 | 2.59 | 30 |
| 4.44 | 4.21 | 4.02 | 3.87 | 3.15 | 2.73 | 2.23 | 40 |
| 4.09 | 3.87 | 3.69 | 3.54 | 2.83 | 2.41 | 1.89 | 60 |
| 3.77 | 3.55 | 3.38 | 3.24 | 2.53 | 2.11 | 1.54 | 120 |
| 3.47 | 3.27 | 3.10 | 2.96 | 2.27 | 1.84 | 1.00 | ∞ |

# APPENDIX F

# DISTRIBUTION OF $t$

| | Levels of Significance ($\alpha$) | | | |
|---|---|---|---|---|
| | DIRECTIONAL TEST | | | |
| | .05 | .025 | .005 | .0005 |
| | NONDIRECTIONAL TEST | | | |
| *df* | .1 | .05 | .01 | .001 |
| 1 | 6.314 | 12.706 | 63.657 | 636.619 |
| 2 | 2.920 | 4.303 | 9.925 | 31.598 |
| 3 | 2.353 | 3.182 | 5.841 | 12.941 |
| 4 | 2.132 | 2.776 | 4.604 | 8.610 |
| 5 | 2.015 | 2.571 | 4.032 | 6.859 |
| 6 | 1.943 | 2.447 | 3.707 | 5.959 |
| 7 | 1.895 | 2.365 | 3.499 | 5.405 |
| 8 | 1.860 | 2.306 | 3.355 | 5.041 |
| 9 | 1.833 | 2.262 | 3.250 | 4.781 |
| 10 | 1.812 | 2.228 | 3.169 | 4.587 |
| 11 | 1.796 | 2.201 | 3.106 | 4.437 |
| 12 | 1.782 | 2.179 | 3.055 | 4.318 |
| 13 | 1.771 | 2.160 | 3.012 | 4.221 |
| 14 | 1.761 | 2.145 | 2.977 | 4.140 |
| 15 | 1.753 | 2.131 | 2.947 | 4.073 |
| 16 | 1.746 | 2.120 | 2.921 | 4.015 |
| 17 | 1.740 | 2.110 | 2.898 | 3.965 |
| 18 | 1.734 | 2.101 | 2.878 | 3.922 |
| 19 | 1.729 | 2.093 | 2.861 | 3.883 |
| 20 | 1.725 | 2.086 | 2.845 | 3.850 |
| 21 | 1.721 | 2.080 | 2.831 | 3.819 |
| 22 | 1.717 | 2.074 | 2.819 | 3.792 |
| 23 | 1.714 | 2.069 | 2.807 | 3.767 |
| 24 | 1.711 | 2.064 | 2.797 | 3.745 |
| 25 | 1.708 | 2.060 | 2.787 | 3.725 |

SOURCE: Abridged from Table III of R. A. Fisher and F. Yates. *Statistical Tables for Biological, Agricultural, and Medical Research,* published by Longman Group Ltd., London (previously published by Oliver and Boyd, Ltd., Edinburgh). Reprinted with permission of the authors and publishers.

**APPENDIX F** *(Continued)*

| | Levels of Significance ($\alpha$) | | | |
|---|---|---|---|---|
| | DIRECTIONAL TEST | | | |
| | .05 | .025 | .005 | .0005 |
| | NONDIRECTIONAL TEST | | | |
| *df* | .1 | .05 | .01 | .001 |
| 26 | 1.706 | 2.056 | 2.779 | 3.707 |
| 27 | 1.703 | 2.052 | 2.771 | 3.690 |
| 28 | 1.701 | 2.048 | 2.763 | 3.674 |
| 29 | 1.699 | 2.045 | 2.756 | 3.659 |
| 30 | 1.697 | 2.042 | 2.750 | 3.646 |
| 40 | 1.684 | 2.021 | 2.704 | 3.551 |
| 60 | 1.671 | 2.000 | 2.660 | 3.460 |
| 120 | 1.658 | 1.980 | 2.617 | 3.373 |
| $\infty$ | 1.645 | 1.960 | 2.576 | 3.291 |

## APPENDIX G

# VALUES OF $r$ FOR DIFFERENT LEVELS OF SIGNIFICANCE (TWO-TAILED TESTS)

| *df* | .1 | .05 | .02 | .01 | .001 |
|---|---|---|---|---|---|
| 1 | .98769 | .99692 | .999507 | .999877 | .9999988 |
| 2 | .90000 | .95000 | .98000 | .990000 | .99900 |
| 3 | .8054 | .8783 | .93433 | .95873 | .99116 |
| 4 | .7293 | .8114 | .8822 | .91720 | .97406 |
| 5 | .6694 | .7545 | .8329 | .8745 | .95074 |
| 6 | .6215 | .7067 | .7887 | .8343 | .92493 |
| 7 | .5822 | .6664 | .7498 | .7977 | .8982 |
| 8 | .5494 | .6319 | .7155 | .7646 | .8721 |
| 9 | .5214 | .6021 | .6851 | .7348 | .8471 |
| 10 | .4973 | .5760 | .6581 | .7079 | .8233 |
| 11 | .4762 | .5529 | .6339 | .6835 | .8010 |
| 12 | .4575 | .5324 | .6120 | .6614 | .7800 |
| 13 | .4409 | .5139 | .5923 | .6411 | .7603 |
| 14 | .4259 | .4973 | .5742 | .6226 | .7420 |
| 15 | .4124 | .4821 | .5577 | .6055 | .7246 |
| 16 | .4000 | .4683 | .5425 | .5897 | .7084 |
| 17 | .3887 | .4555 | .5285 | .5751 | .6932 |
| 18 | .3783 | .4438 | .5155 | .5614 | .6787 |
| 19 | .3687 | .4329 | .5034 | .5487 | .6652 |
| 20 | .3598 | .4227 | .4921 | .5368 | .6524 |
| 25 | .3233 | .3809 | .4451 | .4869 | .5974 |
| 30 | .2960 | .3494 | .4093 | .4487 | .5541 |
| 35 | .2746 | .3246 | .3810 | .4182 | .5189 |
| 40 | .2573 | .3044 | .3578 | .3932 | .4896 |
| 45 | .2428 | .2875 | .3384 | .3721 | .4648 |

SOURCE: Reprinted from Table VI of R. A. Fisher and F. Yates. *Statistical Tables for Biological, Agricultural, and Medical Research,* published by Longman Group Ltd., London (previously published by Oliver and Boyd Ltd., Edinburgh). Reprinted with permission of the authors and publishers.

**APPENDIX G** *(Continued)*

| *df* | .1 | .05 | .02 | .01 | .001 |
|---|---|---|---|---|---|
| 50 | .2306 | .2732 | .3218 | .3541 | .4433 |
| 60 | .2108 | .2500 | .2948 | .3248 | .4078 |
| 70 | .1954 | .2319 | .2737 | .3017 | .3799 |
| 80 | .1829 | .2172 | .2565 | .2830 | .3568 |
| 90 | .1726 | .2050 | .2422 | .2673 | .3375 |
| 100 | .1638 | .1946 | .2301 | .2540 | .3211 |

# APPENDIX H

# TABLE OF Z VALUES FOR $r$*

| $r$ | $Z$ | $r$ | $Z$ | $r$ | $Z$ | $r$ | $Z$ |
|---|---|---|---|---|---|---|---|
| .20† | .203 | .40 | .424 | .60 | .693 | .80 | 1.099 |
| .21 | .213 | .41 | .436 | .61 | .709 | .81 | 1.127 |
| .22 | .224 | .42 | .448 | .62 | .725 | .82 | 1.157 |
| .23 | .234 | .43 | .460 | .63 | .741 | .83 | 1.188 |
| .24 | .245 | .44 | .472 | .64 | .758 | .84 | 1.221 |
| .25 | .255 | .45 | .485 | .65 | .775 | .85 | 1.256 |
| .26 | .266 | .46 | .497 | .66 | .793 | .86 | 1.293 |
| .27 | .277 | .47 | .510 | .67 | .811 | .87 | 1.333 |
| .28 | .288 | .48 | .523 | .68 | .829 | .88 | 1.376 |
| .29 | .299 | .49 | .536 | .69 | .848 | .89 | 1.422 |
| .30 | .310 | .50 | .549 | .70 | .867 | .90 | 1.472 |
| .31 | .321 | .51 | .563 | .71 | .887 | .91 | 1.528 |
| .32 | .332 | .52 | .577 | .72 | .908 | .92 | 1.589 |
| .33 | .343 | .53 | .590 | .73 | .929 | .93 | 1.658 |
| .34 | .354 | .54 | .604 | .74 | .950 | .94 | 1.738 |
| .35 | .365 | .55 | .618 | .75 | .973 | .95 | 1.832 |
| .36 | .377 | .56 | .633 | .76 | .996 | .96 | 1.946 |
| .37 | .388 | .57 | .648 | .77 | 1.020 | .97 | 2.092 |
| .38 | .400 | .58 | .662 | .78 | 1.045 | .98 | 2.298 |
| .39 | .412 | .59 | .678 | .79 | 1.071 | .99 | 2.647 |

* Based on the solution of the formula $z = \frac{1}{2}[\log_e (1 + r) - \log_e (1 - r)]$.

† For values below .20, $r$ and $Z$ are practically identical.

## APPENDIX I

# CRITICAL VALUES OF $T$ IN THE WILCOXON MATCHED-PAIRS SIGNED RANKS TEST

| | LEVEL OF SIGNIFICANCE FOR ONE-TAILED TEST | | |
|---|---|---|---|
| | .025 | .01 | .005 |
| | LEVEL OF SIGNIFICANCE FOR TWO-TAILED TEST | | |
| $N$ | .05 | .02 | .01 |
| 6 | 0 | — | — |
| 7 | 2 | 0 | — |
| 8 | 4 | 2 | 0 |
| 9 | 6 | 3 | 2 |
| 10 | 8 | 5 | 3 |
| 11 | 11 | 7 | 5 |
| 12 | 14 | 10 | 7 |
| 13 | 17 | 13 | 10 |
| 14 | 21 | 16 | 13 |
| 15 | 25 | 20 | 16 |
| 16 | 30 | 24 | 20 |
| 17 | 35 | 28 | 23 |
| 18 | 40 | 33 | 28 |
| 19 | 46 | 38 | 32 |
| 20 | 52 | 43 | 38 |
| 21 | 59 | 49 | 43 |
| 22 | 66 | 56 | 49 |
| 23 | 73 | 62 | 55 |
| 24 | 81 | 69 | 61 |
| 25 | 89 | 77 | 68 |

SOURCE: Adapted from Table I of F. Wilcoxon. *Some Rapid Approximate Statistical Procedures*. New York: American Cyanamid Company, 1949, p. 13. Reproduced by permission of the Lederle Laboratories, a division of the American Cyanamid Company.

# APPENDIX J

# CRITICAL VALUES OF *U* IN THE MANN-WHITNEY TEST

**(*a*) Critical Values of *U* for a One-Tailed Test at .001 or for a Two-Tailed Test at .002**

| $n_1$ \ $n_2$ | 9 | 10 | 11 | 12 | 13 | 14 | 15 | 16 | 17 | 18 | 19 | 20 |
|---|---|---|---|---|---|---|---|---|---|---|---|---|
| 1 | | | | | | | | | | | | |
| 2 | | | | | | | | | | | | |
| 3 | | | | | | | | | 0 | 0 | 0 | 0 |
| 4 | | 0 | 0 | 0 | 1 | 1 | 1 | 2 | 2 | 3 | 3 | 3 |
| 5 | 1 | 1 | 2 | 2 | 3 | 3 | 4 | 5 | 5 | 6 | 7 | 7 |
| 6 | 2 | 3 | 4 | 4 | 5 | 6 | 7 | 8 | 9 | 10 | 11 | 12 |
| 7 | 3 | 5 | 6 | 7 | 8 | 9 | 10 | 11 | 13 | 14 | 15 | 16 |
| 8 | 5 | 6 | 8 | 9 | 11 | 12 | 14 | 15 | 17 | 18 | 20 | 21 |
| 9 | 7 | 8 | 10 | 12 | 14 | 15 | 17 | 19 | 21 | 23 | 25 | 26 |
| 10 | 8 | 10 | 12 | 14 | 17 | 19 | 21 | 23 | 25 | 27 | 29 | 32 |
| 11 | 10 | 12 | 15 | 17 | 20 | 22 | 24 | 27 | 29 | 32 | 34 | 37 |
| 12 | 13 | 14 | 17 | 20 | 23 | 25 | 28 | 31 | 34 | 37 | 40 | 42 |
| 13 | 14 | 17 | 20 | 23 | 26 | 29 | 32 | 35 | 38 | 42 | 45 | 48 |
| 14 | 15 | 19 | 22 | 25 | 29 | 32 | 36 | 39 | 43 | 46 | 50 | 54 |
| 15 | 17 | 21 | 24 | 28 | 32 | 36 | 40 | 43 | 47 | 51 | 55 | 59 |
| 16 | 19 | 23 | 27 | 31 | 35 | 39 | 43 | 48 | 52 | 56 | 60 | 65 |
| 17 | 21 | 25 | 29 | 34 | 38 | 43 | 47 | 52 | 57 | 61 | 66 | 70 |
| 18 | 23 | 27 | 32 | 37 | 42 | 46 | 51 | 56 | 61 | 66 | 71 | 76 |
| 19 | 25 | 29 | 34 | 40 | 45 | 50 | 55 | 60 | 66 | 71 | 77 | 82 |
| 20 | 26 | 32 | 37 | 42 | 48 | 54 | 59 | 65 | 70 | 76 | 82 | 88 |

SOURCE: Adapted and abridged from Tables 1, 3, 5, and 7 of D. Auble, "Extended Tables for the Mann-Whitney Statistic." *Bulletin of the Institute of Educational Research at Indiana University,* 1953, *1,* No. 2. Reprinted with permission of the publisher.

**(*b*) Critical Values of *U* for a One-Tailed Test at .01 or for a Two-Tailed Test at .02**

| $n_1$ \ $n_2$ | 9 | 10 | 11 | 12 | 13 | 14 | 15 | 16 | 17 | 18 | 19 | 20 |
|---|---|---|---|---|---|---|---|---|---|---|---|---|
| 1 | | | | | | | | | | | | |
| 2 | | | | | 0 | 0 | 0 | 0 | 0 | 0 | 1 | 1 |
| 3 | 1 | 1 | 1 | 2 | 2 | 2 | 3 | 3 | 4 | 4 | 4 | 5 |
| 4 | 3 | 3 | 4 | 5 | 5 | 6 | 7 | 7 | 8 | 9 | 9 | 10 |
| 5 | 5 | 6 | 7 | 8 | 9 | 10 | 11 | 12 | 13 | 14 | 15 | 16 |
| 6 | 7 | 8 | 9 | 11 | 12 | 13 | 15 | 16 | 18 | 19 | 20 | 22 |
| 7 | 9 | 11 | 12 | 14 | 16 | 17 | 19 | 21 | 23 | 24 | 26 | 28 |
| 8 | 11 | 13 | 15 | 17 | 20 | 22 | 24 | 26 | 28 | 30 | 32 | 34 |
| 9 | 14 | 16 | 18 | 21 | 23 | 26 | 28 | 31 | 33 | 36 | 38 | 40 |
| 10 | 16 | 19 | 22 | 24 | 27 | 30 | 33 | 36 | 38 | 41 | 44 | 47 |
| 11 | 18 | 22 | 25 | 28 | 31 | 34 | 37 | 41 | 44 | 47 | 50 | 53 |
| 12 | 21 | 24 | 28 | 31 | 35 | 38 | 42 | 46 | 49 | 53 | 56 | 60 |
| 13 | 23 | 27 | 31 | 35 | 39 | 43 | 47 | 51 | 55 | 59 | 63 | 67 |
| 14 | 26 | 30 | 34 | 38 | 43 | 47 | 51 | 56 | 60 | 65 | 69 | 73 |
| 15 | 28 | 33 | 37 | 42 | 47 | 51 | 56 | 61 | 66 | 70 | 75 | 80 |
| 16 | 31 | 36 | 41 | 46 | 51 | 56 | 61 | 66 | 71 | 76 | 82 | 87 |
| 17 | 33 | 38 | 44 | 49 | 55 | 60 | 66 | 71 | 77 | 82 | 88 | 93 |
| 18 | 36 | 41 | 47 | 53 | 59 | 65 | 70 | 76 | 82 | 88 | 94 | 100 |
| 19 | 38 | 44 | 50 | 56 | 63 | 69 | 75 | 82 | 88 | 94 | 101 | 107 |
| 20 | 40 | 47 | 53 | 60 | 67 | 73 | 80 | 87 | 93 | 100 | 107 | 114 |

**(c) Critical Values of *U* for a One-Tailed Test at .025 or for a Two-Tailed Test at .05**

| $n_1$ \ $n_2$ | 9 | 10 | 11 | 12 | 13 | 14 | 15 | 16 | 17 | 18 | 19 | 20 |
|---|---|---|---|---|---|---|---|---|---|---|---|---|
| 1 | | | | | | | | | | | | |
| 2 | 0 | 0 | 0 | 1 | 1 | 1 | 1 | 1 | 2 | 2 | 2 | 2 |
| 3 | 2 | 3 | 3 | 4 | 4 | 5 | 5 | 6 | 6 | 7 | 7 | 8 |
| 4 | 4 | 5 | 6 | 7 | 8 | 9 | 10 | 11 | 11 | 12 | 13 | 13 |
| 5 | 7 | 8 | 9 | 11 | 12 | 13 | 14 | 15 | 17 | 18 | 19 | 20 |
| 6 | 10 | 11 | 13 | 14 | 16 | 17 | 19 | 21 | 22 | 24 | 25 | 27 |
| 7 | 12 | 14 | 16 | 18 | 20 | 22 | 24 | 26 | 28 | 30 | 32 | 34 |
| 8 | 15 | 17 | 19 | 22 | 24 | 26 | 29 | 31 | 34 | 36 | 38 | 41 |
| 9 | 17 | 20 | 23 | 26 | 28 | 31 | 34 | 37 | 39 | 42 | 45 | 48 |
| 10 | 20 | 23 | 26 | 29 | 33 | 36 | 39 | 42 | 45 | 48 | 52 | 55 |
| 11 | 23 | 26 | 30 | 33 | 37 | 40 | 44 | 47 | 51 | 55 | 58 | 62 |
| 12 | 26 | 29 | 33 | 37 | 41 | 45 | 49 | 53 | 57 | 61 | 65 | 69 |
| 13 | 28 | 33 | 37 | 41 | 45 | 50 | 54 | 59 | 63 | 67 | 72 | 76 |
| 14 | 31 | 36 | 40 | 45 | 50 | 55 | 59 | 64 | 67 | 74 | 78 | 83 |
| 15 | 34 | 39 | 44 | 49 | 54 | 59 | 64 | 70 | 75 | 80 | 85 | 90 |
| 16 | 37 | 42 | 47 | 53 | 59 | 64 | 70 | 75 | 81 | 86 | 92 | 98 |
| 17 | 39 | 45 | 51 | 57 | 63 | 67 | 75 | 81 | 87 | 93 | 99 | 105 |
| 18 | 42 | 48 | 55 | 61 | 67 | 74 | 80 | 86 | 93 | 99 | 106 | 112 |
| 19 | 45 | 52 | 58 | 65 | 72 | 78 | 85 | 92 | 99 | 106 | 113 | 119 |
| 20 | 48 | 55 | 62 | 69 | 76 | 83 | 90 | 98 | 105 | 112 | 119 | 127 |

**APPENDIX J** *(Continued)*

**(*d*) Critical Values of *U* for a One-Tailed Test at .05 or for a Two-Tailed Test at .10**

| $n_1$ \ $n_2$ | 9 | 10 | 11 | 12 | 13 | 14 | 15 | 16 | 17 | 18 | 19 | 20 |
|---|---|---|---|---|---|---|---|---|---|---|---|---|
| 1 | | | | | | | | | | | 0 | 0 |
| 2 | 1 | 1 | 1 | 2 | 2 | 2 | 3 | 3 | 3 | 4 | 4 | 4 |
| 3 | 3 | 4 | 5 | 5 | 6 | 7 | 7 | 8 | 9 | 9 | 10 | 11 |
| 4 | 6 | 7 | 8 | 9 | 10 | 11 | 12 | 14 | 15 | 16 | 17 | 18 |
| 5 | 9 | 11 | 12 | 13 | 15 | 16 | 18 | 19 | 20 | 22 | 23 | 25 |
| 6 | 12 | 14 | 16 | 17 | 19 | 21 | 23 | 25 | 26 | 28 | 30 | 32 |
| 7 | 15 | 17 | 19 | 21 | 24 | 26 | 28 | 30 | 33 | 35 | 37 | 39 |
| 8 | 18 | 20 | 23 | 26 | 28 | 31 | 33 | 36 | 39 | 41 | 44 | 47 |
| 9 | 21 | 24 | 27 | 30 | 33 | 36 | 39 | 42 | 45 | 48 | 51 | 54 |
| 10 | 24 | 27 | 31 | 34 | 37 | 41 | 44 | 48 | 51 | 55 | 58 | 62 |
| 11 | 27 | 31 | 34 | 38 | 42 | 46 | 50 | 54 | 57 | 61 | 65 | 69 |
| 12 | 30 | 34 | 38 | 42 | 47 | 51 | 55 | 60 | 64 | 68 | 72 | 77 |
| 13 | 33 | 37 | 42 | 47 | 51 | 56 | 61 | 65 | 70 | 75 | 80 | 84 |
| 14 | 36 | 41 | 46 | 51 | 56 | 61 | 66 | 71 | 77 | 82 | 87 | 92 |
| 15 | 39 | 44 | 50 | 55 | 61 | 66 | 72 | 77 | 83 | 88 | 94 | 100 |
| 16 | 42 | 48 | 54 | 60 | 65 | 71 | 77 | 83 | 89 | 95 | 101 | 107 |
| 17 | 45 | 51 | 57 | 64 | 70 | 77 | 83 | 89 | 96 | 102 | 109 | 115 |
| 18 | 48 | 55 | 61 | 68 | 75 | 82 | 88 | 95 | 102 | 109 | 116 | 123 |
| 19 | 51 | 58 | 65 | 72 | 80 | 87 | 94 | 101 | 109 | 116 | 123 | 130 |
| 20 | 54 | 62 | 69 | 77 | 84 | 92 | 100 | 107 | 115 | 123 | 130 | 138 |

## APPENDIX K

# VALUES OF THE COEFFICIENT OF CONCORDANCE, *W*, SIGNIFICANT AT THE 20, 10, 5, AND 1 PERCENT LEVELS

| *m* | $\alpha$ | *n* 3 | 4 | 5 | 6 | 7 | 8 | 9 | 10 |
|---|---|---|---|---|---|---|---|---|---|
| | .20 | .78 | .60 | .53 | .49 | .47 | .46 | .45 | .44 |
| | .10 | | .73 | .62 | .58 | .55 | .53 | .52 | .51 |
| 3 | .05 | 1.00 | .82 | .71 | .65 | .62 | .60 | .58 | .56 |
| | .01 | | .96 | .84 | .77 | .73 | .70 | .67 | .65 |
| | .20 | .56 | .40 | .38 | .37 | .36 | .35 | .34 | .33 |
| | .10 | .75 | .52 | .47 | .44 | .42 | .41 | .40 | .39 |
| 4 | .05 | .81 | .65 | .54 | .51 | .48 | .46 | .45 | .44 |
| | .01 | 1.00 | .80 | .67 | .62 | .59 | .56 | .54 | .52 |
| | .20 | .36 | .34 | .30 | .29 | .28 | .28 | .27 | .27 |
| | .10 | .52 | .42 | .38 | .36 | .34 | .33 | .32 | .31 |
| 5 | .05 | .64 | .52 | .44 | .41 | .39 | .38 | .36 | .35 |
| | .01 | .84 | .66 | .56 | .52 | .49 | .46 | .44 | .43 |
| | .20 | .33 | .27 | .25 | .24 | .24 | .23 | .23 | .23 |
| | .10 | .44 | .36 | .32 | .30 | .29 | .28 | .27 | .26 |
| 6 | .05 | .58 | .42 | .37 | .35 | .33 | .32 | .31 | .30 |
| | .01 | .75 | .56 | .49 | .45 | .42 | .40 | .38 | .37 |
| | .20 | .27 | .23 | .22 | .21 | .20 | .20 | .20 | .19 |
| | .10 | .39 | .30 | .27 | .26 | .25 | .24 | .23 | .23 |
| 7 | .05 | .51 | .36 | .32 | .30 | .29 | .27 | .26 | .26 |
| | .01 | .63 | .48 | .43 | .39 | .36 | .34 | .33 | .32 |
| | .20 | .25 | .20 | .19 | .18 | .18 | .17 | .17 | .17 |
| | .10 | .33 | .26 | .24 | .23 | .22 | .21 | .20 | .20 |
| 8 | .05 | .39 | .32 | .29 | .27 | .25 | .24 | .23 | .23 |
| | .01 | .56 | .43 | .38 | .35 | .32 | .31 | .29 | .28 |

SOURCE: M. W. Tate and R. C. Clelland. *Nonparametric and Shortcut Statistics.* Danville, Ill.: The Interstate Printers and Publishers, 1957. Used by permission.

**APPENDIX K** *(Continued)*

| | | *n* | | | | | | | |
|---|---|---|---|---|---|---|---|---|---|
| ***m*** | $\alpha$ | 3 | 4 | 5 | 6 | 7 | 8 | 9 | 10 |
| | .20 | .20 | .18 | .17 | .16 | .16 | .16 | .15 | .15 |
| | .10 | .31 | .23 | .21 | .20 | .19 | .19 | .18 | .18 |
| 9 | .05 | .35 | .28 | .26 | .24 | .23 | .22 | .21 | .20 |
| | .01 | .48 | .38 | .34 | .31 | .29 | .27 | .26 | .25 |
| | .20 | .19 | .16 | .15 | .15 | .14 | .14 | .14 | .13 |
| | .10 | .25 | .21 | .19 | .18 | .17 | .17 | .16 | .16 |
| 10 | .05 | .31 | .25 | .23 | .21 | .20 | .20 | .19 | .18 |
| | .01 | .48 | .35 | .31 | .28 | .26 | .25 | .24 | .23 |
| | .20 | .14 | .13 | .13 | .12 | .12 | .12 | .11 | .11 |
| | .10 | .19 | .17 | .16 | .15 | .15 | .14 | .14 | .13 |
| 12 | .05 | .25 | .21 | .19 | .18 | .17 | .16 | .16 | .15 |
| | .01 | .36 | .30 | .26 | .24 | .22 | .21 | .20 | .19 |
| | .20 | .12 | .11 | .11 | .10 | .10 | .10 | .10 | .10 |
| | .10 | .17 | .15 | .14 | .13 | .13 | .12 | .12 | .12 |
| 14 | .05 | .21 | .18 | .17 | .16 | .15 | .14 | .14 | .13 |
| | .01 | .31 | .26 | .23 | .21 | .19 | .18 | .17 | .17 |
| | .20 | .10 | .10 | .09 | .09 | .09 | .09 | .09 | .08 |
| | .10 | .15 | .13 | .12 | .12 | .11 | .11 | .10 | .10 |
| 16 | .05 | .19 | .16 | .15 | .14 | .13 | .11 | .12 | .12 |
| | .01 | .28 | .23 | .20 | .18 | .17 | .16 | .15 | .15 |
| | .20 | .09 | .09 | .08 | .08 | .08 | .08 | .08 | .07 |
| | .10 | .13 | .12 | .11 | .10 | .10 | .09 | .09 | .09 |
| 18 | .05 | .17 | .14 | .13 | .12 | .11 | .11 | .11 | .10 |
| | .01 | .25 | .20 | .18 | .16 | .15 | .14 | .14 | .13 |
| | .20 | .08 | .08 | .07 | .07 | .07 | .07 | .07 | .07 |
| | .10 | .11 | .10 | .10 | .09 | .09 | .08 | .08 | .08 |
| 20 | .05 | .15 | .13 | .12 | .11 | .10 | .10 | .10 | .09 |
| | .01 | .22 | .18 | .16 | .15 | .14 | .13 | .12 | .11 |
| | .20 | .07 | .06 | .06 | .06 | .06 | .06 | .05 | .05 |
| | .10 | .09 | .08 | .08 | .07 | .07 | .07 | .07 | .06 |
| 25 | .05 | .12 | .10 | .09 | .09 | .08 | .08 | .08 | .07 |
| | .01 | .18 | .15 | .13 | .12 | .11 | .10 | .10 | .09 |
| | .20 | .05 | .05 | .05 | .05 | .05 | .05 | .05 | .04 |
| | .10 | .08 | .07 | .06 | .06 | .06 | .06 | .06 | .05 |
| 30 | .05 | .10 | .09 | .08 | .07 | .07 | .07 | .07 | .06 |
| | .01 | .15 | .12 | .11 | .10 | .09 | .09 | .08 | .08 |

## APPENDIX L

# VALUES OF $H$ FOR THREE SAMPLES SIGNIFICANT AT THE 10, 5, AND 1 PERCENT LEVELS

| Sample Sizes | | | Level | | |
|---|---|---|---|---|---|
| $N_1$ | $N_2$ | $N_3$ | .10 | .05 | .01 |
| 2 | 2 | 2 | 4.57 | | |
| 3 | 2 | 1 | 4.29 | | |
| 3 | 2 | 2 | 4.50 | 4.71 | |
| 3 | 3 | 1 | 4.57 | 5.14 | |
| 3 | 3 | 2 | 4.56 | 5.36 | 6.25 |
| 3 | 3 | 3 | 4.62 | 5.60 | 6.49 |
| 4 | 2 | 1 | 4.50 | | |
| 4 | 2 | 2 | 4.46 | 5.33 | |
| 4 | 3 | 1 | 4.06 | 5.21 | |
| 4 | 3 | 2 | 4.51 | 5.44 | 6.30 |
| 4 | 3 | 3 | 4.70 | 5.73 | 6.75 |
| 4 | 4 | 1 | 4.17 | 4.79 | 6.67 |
| 4 | 4 | 2 | 4.55 | 5.45 | 6.87 |
| 4 | 4 | 3 | 4.55 | 5.60 | 7.14 |
| 4 | 4 | 4 | 4.65 | 5.69 | 7.54 |
| 5 | 2 | 1 | 4.20 | 5.00 | |
| 5 | 2 | 2 | 4.37 | 5.16 | 6.53 |
| 5 | 3 | 1 | 4.02 | 4.96 | |
| 5 | 3 | 2 | 4.49 | 5.25 | 6.82 |
| 5 | 3 | 3 | 4.53 | 5.44 | 6.98 |
| 5 | 4 | 1 | 3.99 | 4.99 | 6.84 |
| 5 | 4 | 2 | 4.52 | 5.27 | 7.12 |
| 5 | 4 | 3 | 4.55 | 5.63 | 7.40 |
| 5 | 4 | 4 | 4.62 | 5.62 | 7.74 |
| 5 | 5 | 1 | 4.11 | 5.13 | 6.84 |
| 5 | 5 | 2 | 4.51 | 5.25 | 7.27 |
| 5 | 5 | 3 | 4.55 | 5.63 | 7.54 |
| 5 | 5 | 4 | 4.52 | 5.64 | 7.79 |
| 5 | 5 | 5 | 4.56 | 5.66 | 7.98 |

SOURCE: Abridged from Table 6.1 of W. H. Kruskal and W. A. Wallis, "Use of Ranks on One-Criterion Variance Analysis." *Journal of the American Statistical Association,* 1952, *47,* 584–621. Reproduced by permission of the Journal of the American Statistical Association.

# APPENDIX M

# CRITICAL VALUES OF *A*

For any given value of $N - 1$, the table shows the values of $A$ corresponding to various levels of probability. $A$ is significant at a given level if it is equal to or *less than* the value shown in the table.

| | LEVELS OF SIGNIFICANCE FOR A DIRECTIONAL TEST | | | | | |
|---|---|---|---|---|---|---|
| | .05 | .025 | .01 | .005 | .0005 | |
| | LEVELS OF SIGNIFICANCE FOR A NON-DIRECTIONAL TEST | | | | | |
| $N - 1$* | .10 | .05 | .02 | .01 | .001 | $N - 1$ |
| 1 | 0.5125 | 0.5031 | 0.50049 | 0.50012 | 0.5000012 | 1 |
| 2 | 0.412 | 0.369 | 0.347 | 0.340 | 0.334 | 2 |
| 3 | 0.385 | 0.324 | 0.286 | 0.272 | 0.254 | 3 |
| 4 | 0.376 | 0.304 | 0.257 | 0.238 | 0.211 | 4 |
| 5 | 0.372 | 0.293 | 0.240 | 0.218 | 0.184 | 5 |
| 6 | 0.370 | 0.286 | 0.230 | 0.205 | 0.167 | 6 |
| 7 | 0.369 | 0.281 | 0.222 | 0.196 | 0.155 | 7 |
| 8 | 0.368 | 0.278 | 0.217 | 0.190 | 0.146 | 8 |
| 9 | 0.368 | 0.276 | 0.213 | 0.185 | 0.139 | 9 |
| 10 | 0.368 | 0.274 | 0.210 | 0.181 | 0.134 | 10 |
| 11 | 0.368 | 0.273 | 0.207 | 0.178 | 0.130 | 11 |
| 12 | 0.368 | 0.271 | 0.205 | 0.176 | 0.126 | 12 |
| 13 | 0.368 | 0.270 | 0.204 | 0.174 | 0.124 | 13 |
| 14 | 0.368 | 0.270 | 0.202 | 0.172 | 0.121 | 14 |
| 15 | 0.368 | 0.269 | 0.201 | 0.170 | 0.119 | 15 |
| 16 | 0.368 | 0.268 | 0.200 | 0.169 | 0.117 | 16 |
| 17 | 0.368 | 0.268 | 0.199 | 0.168 | 0.116 | 17 |
| 18 | 0.368 | 0.267 | 0.198 | 0.167 | 0.114 | 18 |
| 19 | 0.368 | 0.267 | 0.197 | 0.166 | 0.113 | 19 |
| 20 | 0.368 | 0.266 | 0.197 | 0.165 | 0.112 | 20 |

* $N$ = the number of pairs.

SOURCE: J. Sandler. "A Test of Significance of the Difference Between the Means of Correlated Measures, Based upon a Simplification of Student's *t*. *British Journal of Psychology*, 1955, *46*, 225–226. Used by permission of Cambridge University Press.

**APPENDIX M** *(Continued)*

| | LEVELS OF SIGNIFICANCE FOR A DIRECTIONAL TEST | | | | | |
|---|---|---|---|---|---|---|
| | .05 | .025 | .01 | .005 | .0005 | |
| | LEVELS OF SIGNIFICANCE FOR A NON-DIRECTIONAL TEST | | | | | |
| $N-1$* | .10 | .05 | .02 | .01 | .001 | $N-1$ |
| 21 | 0.368 | 0.266 | 0.196 | 0.165 | 0.111 | 21 |
| 22 | 0.368 | 0.266 | 0.196 | 0.164 | 0.110 | 22 |
| 23 | 0.368 | 0.266 | 0.195 | 0.163 | 0.109 | 23 |
| 24 | 0.368 | 0.265 | 0.195 | 0.163 | 0.108 | 24 |
| 25 | 0.368 | 0.265 | 0.194 | 0.162 | 0.108 | 25 |
| 26 | 0.368 | 0.265 | 0.194 | 0.162 | 0.107 | 26 |
| 27 | 0.368 | 0.265 | 0.193 | 0.161 | 0.107 | 27 |
| 28 | 0.368 | 0.265 | 0.193 | 0.161 | 0.106 | 28 |
| 29 | 0.368 | 0.264 | 0.193 | 0.161 | 0.106 | 29 |
| 30 | 0.368 | 0.264 | 0.193 | 0.160 | 0.105 | 30 |
| 40 | 0.368 | 0.263 | 0.191 | 0.158 | 0.102 | 40 |
| 60 | 0.369 | 0.262 | 0.189 | 0.155 | 0.099 | 60 |
| 120 | 0.369 | 0.261 | 0.187 | 0.153 | 0.095 | 120 |
| ∞ | 0.370 | 0.260 | 0.185 | 0.151 | 0.092 | ∞ |

## APPENDIX N

# TABLE OF RANDOM NUMBERS

To use this table, select a digit or set of digits at random and then read in any direction or combination of directions. For example, assume that a sample of 25 is to be drawn from a population of 500 members and that the first set of digits, 848, are the three that occur in the last column (marked by an *) of the second page of this table. Since there is no 848 in this sample, the researcher continues up the page to the next number, 355. This person becomes the first member of the sample. Continuing, the next number is 428 and the individual with this number becomes the second member of the sample. And so it proceeds. If a number appears twice it is disregarded and all numbers above 500 are skipped. When the top of the column is reached the researcher may decide to go in any direction that he wishes from there. He continues until he has drawn his 25 cases.

| | | | | | | | |
|---|---|---|---|---|---|---|---|
| 80583 | 93944 | 52456 | 73766 | 06830 | 53656 | 95043 | 52628 |
| 18453 | 24065 | 08458 | 95366 | 53473 | 07541 | 45547 | 70808 |
| 60814 | 37777 | 10057 | 42332 | 63335 | 20483 | 31732 | 57254 |
| 07236 | 12152 | 05088 | 65825 | 64169 | 49022 | 86995 | 90328 |
| 71396 | 89215 | 30722 | 22102 | 39542 | 07772 | 35841 | 85721 |
| 26849 | 84547 | 14663 | 56346 | 70774 | 35439 | 46850 | 52341 |
| 60352 | 33049 | 53633 | 70863 | 95596 | 20094 | 69248 | 93446 |
| 92087 | 96294 | 43514 | 37481 | 38649 | 06343 | 14013 | 31711 |
| 15701 | 08337 | 98588 | 09495 | 92176 | 72535 | 56303 | 87352 |
| 85275 | 36898 | 71569 | 75673 | 81007 | 47749 | 81304 | 48557 |
| 26857 | 73156 | 46758 | 70472 | 66067 | 42792 | 70284 | 24320 |
| 14633 | 84924 | 73750 | 85788 | 54244 | 91030 | 90415 | 93615 |
| 15694 | 48297 | 57256 | 61342 | 30945 | 75789 | 39904 | 02192 |
| 80613 | 19019 | 93119 | 56077 | 69170 | 37403 | 88152 | 00077 |
| 45688 | 32486 | 40744 | 56974 | 08345 | 88975 | 45134 | 63538 |
| 75545 | 35247 | 18619 | 13674 | 86864 | 29901 | 14908 | 08830 |
| 81122 | 11724 | 74627 | 73707 | 69979 | 20288 | 87342 | 78818 |
| 38904 | 13141 | 32392 | 19763 | 93278 | 81757 | 52548 | 54091 |
| 13902 | 63742 | 78464 | 22501 | 68107 | 23621 | 71586 | 73417 |
| 08972 | 11598 | 62095 | 36787 | 63535 | 24170 | 64756 | 03324 |

**APPENDIX N** *(Continued)*

| | | | | | | | |
|---|---|---|---|---|---|---|---|
| 24152 | 00023 | 12302 | 80783 | 93584 | 72869 | 60096 | 21551 |
| 39024 | 00867 | 76378 | 41605 | 11303 | 22970 | 07855 | 39269 |
| 49458 | 74284 | 05041 | 49807 | 44035 | 52166 | 21282 | 21296 |
| 20158 | 34243 | 46978 | 35482 | 17395 | 96131 | 35947 | 67807 |
| 37051 | 93029 | 47665 | 64382 | 87648 | 85261 | 04986 | 83666 |
| 86015 | 46874 | 32444 | 48277 | 58303 | 29822 | 93174 | 93994 |
| 23108 | 88222 | 88570 | 74015 | 73515 | 90400 | 71148 | 43674 |
| 54880 | 87873 | 95160 | 59221 | 61222 | 60561 | 62326 | 18462 |
| 29748 | 12102 | 80580 | 41867 | 85496 | 57560 | 81604 | 18811 |
| 07944 | 05600 | 60478 | 03343 | 45875 | 21069 | 85644 | 47217 |
| 30389 | 87374 | 64278 | 58044 | 14924 | 39650 | 95294 | 00583 |
| 26870 | 76150 | 68476 | 64659 | 70312 | 05682 | 66986 | 34091 |
| 35124 | 67018 | 41361 | 82760 | 90850 | 64618 | 80620 | 51727 |
| 21375 | 05871 | 93823 | 43178 | 24781 | 89683 | 55411 | 85695 |
| 17714 | 53295 | 07706 | 17813 | 40902 | 05069 | 99083 | 06720 |
| 84618 | 97553 | 31223 | 08420 | 72682 | 07385 | 90726 | 57104 |
| 09604 | 60475 | 94119 | 01840 | 21443 | 41808 | 68984 | 83632 |
| 20466 | 68795 | 77762 | 20791 | 01176 | 28838 | 36421 | 16428 |
| 52781 | 76514 | 83483 | 47055 | 80582 | 71944 | 92638 | 40355 |
| 78494 | 72306 | 94541 | 37408 | 13177 | 55292 | 21036 | 82848* |
| 93692 | 25527 | 21785 | 41101 | 91178 | 10174 | 43708 | 66354 |
| 66146 | 63210 | 47458 | 64809 | 98189 | 81851 | 46062 | 27647 |
| 28992 | 63165 | 40405 | 68032 | 96717 | 54244 | 71171 | 15102 |
| 46182 | 49126 | 71209 | 92061 | 39448 | 93136 | 42175 | 88350 |
| 47269 | 15747 | 85561 | 29671 | 58137 | 17820 | 54358 | 64578 |
| 14738 | 86667 | 28825 | 35793 | 28976 | 66252 | 19715 | 94082 |
| 19056 | 13939 | 12843 | 82590 | 09815 | 93146 | 38793 | 85774 |
| 75814 | 85986 | 83824 | 52692 | 54130 | 55160 | 54196 | 34108 |
| 62448 | 46385 | 63011 | 98901 | 14974 | 40344 | 60014 | 07201 |
| 80395 | 81114 | 88325 | 80851 | 43667 | 70883 | 16315 | 53969 |
| 35075 | 33949 | 27767 | 43584 | 85301 | 88977 | 60365 | 94653 |
| 56623 | 34442 | 13025 | 14338 | 54066 | 15243 | 83799 | 42402 |
| 36409 | 83232 | 80217 | 26392 | 98525 | 24335 | 32960 | 07405 |
| 57620 | 52606 | 10875 | 62004 | 90391 | 61105 | 19322 | 53845 |
| 07399 | 37408 | 54127 | 57326 | 26629 | 19087 | 11220 | 94747 |
| 68980 | 05339 | 60311 | 42824 | 37301 | 42678 | 31751 | 57260 |
| 14454 | 04504 | 49739 | 71484 | 92003 | 98086 | 88492 | 99382 |
| 07481 | 83828 | 78626 | 51594 | 16453 | 94614 | 30934 | 47744 |
| 27499 | 98748 | 66692 | 13986 | 99837 | 00582 | 22888 | 48893 |
| 35902 | 91386 | 44071 | 28091 | 97362 | 97703 | 78212 | 16993 |

**APPENDIX N** *(Continued)*

| | | | | | | | |
|---|---|---|---|---|---|---|---|
| 96207 | 44156 | 32825 | 29527 | 04220 | 86304 | 03061 | 18072 |
| 59175 | 20695 | 51981 | 50654 | 94938 | 81997 | 07954 | 19814 |
| 05128 | 09719 | 47677 | 26269 | 62290 | 64464 | 06958 | 92983 |
| 13499 | 06319 | 20971 | 87749 | 90429 | 12272 | 99599 | 10507 |
| 64421 | 80814 | 66281 | 31003 | 00682 | 27398 | 43622 | 63147 |
| 74910 | 64345 | 78542 | 42785 | 13661 | 58873 | 34677 | 58300 |
| 61318 | 31855 | 81333 | 10591 | 40510 | 07893 | 45305 | 07521 |
| 76503 | 34513 | 81594 | 13628 | 51215 | 90290 | 59747 | 68277 |
| 11654 | 99892 | 61613 | 62269 | 50263 | 90212 | 16520 | 69676 |
| 92852 | 55866 | 00397 | 58391 | 12609 | 17646 | 68652 | 27376 |
| 01158 | 63267 | 41290 | 67312 | 71857 | 15957 | 79375 | 95220 |
| 55823 | 47641 | 05870 | 01119 | 92784 | 26340 | 33521 | 26665 |
| 66821 | 41576 | 82444 | 99005 | 04921 | 73701 | 59589 | 49067 |
| 96277 | 48257 | 20247 | 81759 | 45197 | 25332 | 20554 | 91409 |
| 43947 | 51680 | 48460 | 85558 | 15191 | 18782 | 59404 | 72059 |
| 01918 | 28316 | 60833 | 25983 | 01291 | 41349 | 42614 | 29297 |
| 70071 | 14736 | 43529 | 06318 | 38384 | 74761 | 34994 | 41374 |
| 11133 | 07586 | 88722 | 56736 | 66164 | 49431 | 99385 | 41600 |
| 78138 | 66559 | 94813 | 31900 | 54155 | 83436 | 66497 | 68646 |
| 27482 | 45476 | 74457 | 90561 | 72848 | 11834 | 48509 | 23929 |
| 88651 | 22596 | 25163 | 01889 | 70014 | 15021 | 15470 | 48355 |
| 74843 | 93413 | 65251 | 07629 | 37239 | 33295 | 20094 | 98977 |
| 28597 | 20405 | 36815 | 43625 | 18637 | 37509 | 73788 | 06533 |
| 74022 | 84617 | 64397 | 11692 | 05327 | 82162 | 60530 | 45128 |
| 65741 | 14014 | 04515 | 25624 | 95096 | 67946 | 44372 | 15486 |
| 58838 | 73859 | 83761 | 60873 | 43253 | 84145 | 36066 | 94850 |
| 52623 | 07992 | 14387 | 06345 | 80854 | 09279 | 74384 | 89342 |
| 07759 | 51777 | 51321 | 92246 | 60000 | 77074 | 38992 | 22815 |
| 27493 | 70939 | 72472 | 00008 | 40890 | 18002 | 36151 | 99073 |
| 11161 | 78576 | 05466 | 55306 | 93128 | 18464 | 57178 | 65762 |

# APPENDIX O

# LIST OF FORMULAS

| FORMULA NUMBER | | PAGE |
|---|---|---|
| (2.1) | Any centile $= ll + \dfrac{Np - cf}{f_1}(i)$ | 18 |
| (3.1) | $\bar{X} = \dfrac{\Sigma X}{N}$ | 24 |
| (3.2) | $\bar{X} = \dfrac{\Sigma fX}{N}$ | 25 |
| (3.3) | $\bar{X}_T = \dfrac{\Sigma N\bar{X}}{N_T}$ | 27 |
| (4.1a) | $S = \sqrt{\dfrac{\Sigma x^2}{N}}$ | 34 |
| (4.1b) | $\sigma = \sqrt{\dfrac{\Sigma x^2}{N}}$ | 34 |
| (4.2) | $\Sigma x^2 = SS = \Sigma X^2 - \dfrac{(\Sigma X)^2}{N}$ | 35 |
| (4.3) | $S = \dfrac{1}{N}\sqrt{N\Sigma X^2 - (\Sigma X)^2}$ | 36 |
| (4.4) | $s = \sqrt{\dfrac{\Sigma x^2}{N-1}} = \sqrt{\dfrac{N\Sigma X^2 - (\Sigma X)^2}{N(N-1)}}$ | 36 |
| (4.5) | $S^2 = \dfrac{\Sigma x^2}{N}$ or $s^2 = \dfrac{\Sigma x^2}{N-1}$ | 36 |
| (4.6) | Averaging standard deviations: $s_T = \sqrt{\dfrac{N_A(\bar{X}_A{}^2 + S_A{}^2) + N_B(\bar{X}_B{}^2 + s_B{}^2)}{N_A + N_B} - \bar{X}_T{}^2}$ | 38 |
| (4.7) | $Q = \dfrac{Q_3 - Q_1}{2}$ | 41 |
| (5.1) | $z = \dfrac{X - \bar{X}}{s} = \dfrac{x}{s}$ | 45 |

| FORMULA NUMBER | | PAGE |
|---|---|---|
| (5.2) | Transformed standard score: $z(\text{new } S) + \text{new } \bar{X}$ | 46 |
| (7.1) | Chi-square, general formula: $\chi^2 = \sum \frac{(O - E)^2}{E}$ | 74 |
| (7.2) | Degrees of freedom in chi-square: $df = (r - 1)(c - 1)$ | 78 |
| (7.3) | $\chi^2 = \sum \frac{O^2}{E} - N$ | 79 |
| (7.4) | Chi-square formula for a 2 × 2 table using the obtained frequencies: $\chi^2 = \frac{N[(ad) - (bc)]^2}{i\,j\,k\,l}$ | 80 |
| (7.5) | Pirie-Hamden correction for small frequencies in a 2 × 2 table: $\chi^2 = \frac{N[\lvert ad - bc \rvert - .5]^2}{i\,j\,k\,l}$ | 91 |
| (8.1) | Standard deviation of a sampling distribution: $\sigma_{\bar{X}} = \frac{\sigma}{\sqrt{N_{\bar{X}}}}$ | 100 |
| (8.2) | Standard error of the mean: $s_{\bar{X}} = \frac{s}{\sqrt{N_{\bar{X}}}}$ | 102 |
| (8.3) | $z$ equivalent of observed sample means in the sampling distribution: $z = \frac{\bar{X} - \mu_{\bar{X}}}{\sigma_{\bar{X}}}$ | 105 |
| (8.4) | $g$ confidence interval when $N > 60$: $\bar{X} \pm (z)(s_{\bar{X}})$ | 107 |
| (8.5) | Formula (8.3) for small samples: $t = \frac{\bar{X} - \mu_{\bar{X}}}{s_{\bar{X}}}$ | 108 |

| FORMULA NUMBER | | PAGE |
|---|---|---|
| (8.6) | $g$ confidence interval when $N \leqslant 60$: $\bar{X} \pm (t)(s_{\bar{X}})$ | 109 |
| (9.1) | Standard error of the difference between two means, population variances known: $\sigma_{\bar{X}_1-\bar{X}_2} = \sqrt{\sigma_{\bar{X}_1}{}^2 + \sigma_{\bar{X}_2}{}^2}$ | 128 |
| (9.2) | Standard error of the difference between two means, population variances unknown: $s_{\bar{X}_1-\bar{X}_2} = \sqrt{\dfrac{\Sigma x_1{}^2 + \Sigma x_2{}^2}{N_1 + N_2 - 2}\left(\dfrac{1}{N_1} + \dfrac{1}{N_2}\right)}$ | 128 |
| (9.3) | Formula (9.2) when $N_1 = N_2$: $s_{\bar{X}_1-\bar{X}_2} = \sqrt{\dfrac{\Sigma x_1{}^2 + \Sigma x_2{}^2}{N(N-1)}}$ | 128 |
| (9.4) | Two-sample $t$ tests, population variances unknown: $t = \dfrac{\bar{X}_1 - \bar{X}_2}{s_{\bar{X}_1-\bar{X}_2}}$ | 129 |
| (9.5) | Generalized computational formula: $t = \dfrac{\bar{X}_1 - \bar{X}_2}{\sqrt{\dfrac{\Sigma x_1{}^2 + \Sigma x_2{}^2}{N_1 + N_2 - 2}\left(\dfrac{1}{N_1} + \dfrac{1}{N_2}\right)}}$ | 129 |
| (9.6) | Two-sample test when the parameter means are *not* assumed equal: $t = \dfrac{(\bar{X}_1 - \bar{X}_2) - (\mu_1 - \mu_2)}{s_{\bar{X}_1-\bar{X}_2}}$ | 133 |
| (9.7) | $\bar{D} = \dfrac{\Sigma D}{N} = \bar{X}_1 - \bar{X}_2$ | 134 |
| (9.8) | $t$ test for two dependent population means: $t = \dfrac{\bar{X}_1 - \bar{X}_2}{\sqrt{\dfrac{\Sigma D^2 - (\Sigma D)^2/N}{N(N-1)}}}$ | 134 |
| (9.9) | $A = \dfrac{\Sigma D^2}{(\Sigma D)^2}$ | 138 |

| FORMULA NUMBER | | PAGE |
|---|---|---|
| (9.10) | Testing the significance of a proportion: $z = \dfrac{P - p}{\sqrt{\dfrac{pq}{N}}}$ | 140 |
| (9.11) | Test for the significance of the difference between two proportions: $z = \dfrac{P_1 - P_2}{\sqrt{P'Q'\left(\dfrac{1}{N_1} + \dfrac{1}{N_2}\right)}}$ | 141 |
| (9.12) | $P' = \dfrac{N_1P_1 + N_2P_2}{N_1 + N_2}$ | 141 |
| (9.13) | $Q' = 1 - P'$ | 141 |
| (9.14) | $F = \dfrac{s_i^2}{s_j^2}$ | 142 |
| (10.1) | $F = \dfrac{s_b^2}{s_w^2}$ | 154 |
| (10.2) | $df_b = k - 1$ | 155 |
| (10.3) | $df_w = k(N - 1)$ | 155 |
| (10.4) | Total or grand mean: $\bar{X}_T = \dfrac{\Sigma X_T}{N_T}$ | 156 |
| (10.5) | $s_b^2 = \dfrac{\Sigma x_b^2}{df_b} = \dfrac{SS_b}{df_b}$ | 156 |
| (10.6) | $s_w^2 = \dfrac{\Sigma x_w^2}{df_w} = \dfrac{SS_w}{df_w}$ | 156 |
| (10.7) | $F = \dfrac{SS_b/df_b}{SS_w/df_w}$ | 157 |
| (10.8) | $SS_b = N\Sigma(\bar{X}_k - \bar{X}_T)^2$ | 157 |
| (10.9) | $SS_b = \dfrac{\Sigma(\Sigma X_k)^2}{N} - \dfrac{(\Sigma X_T)^2}{N_T}$ | 157 |
| (10.10) | $SS_w = \Sigma(X_k - \bar{X}_k)^2$ | 157 |
| (10.11) | $SS_w = \Sigma X_T^2 - \dfrac{\Sigma(\Sigma X_k)^2}{N}$ | 157 |

| FORMULA NUMBER | | PAGE |
|---|---|---|
| (10.12) | $MS = s^2 = \Sigma x^2/df = SS/df$ | 159 |
| (10.13) | $F = \dfrac{MS_b}{MS_w} = \dfrac{s_b{}^2}{s_w{}^2} = \dfrac{SS_b/df_b}{SS_w/df_w}$ | 159 |
| (10.14) | $SS_T = \Sigma(X - \bar{X}_T)^2$ | 162 |
| (10.15) | Computational formula for (10.14): | |
| | $SS_T = \Sigma X_T{}^2 - \dfrac{(\Sigma X_T)^2}{N_T}$ | 162 |
| (10.16) | $SS_T = SS_b + SS_w$ | 162 |
| (10.17) | $df_T = N_T - 1$ | 162 |
| (10.18) | $df_T = df_b + df_w$ | 162 |
| (10.19) | Smallest significant difference, Scheffé method: | |
| | $ssd = \sqrt{\dfrac{(2N)(k-1)(F_\alpha)(MS_w)}{N^2}}$ | 166 |
| | or | |
| | $\dfrac{1}{N}\sqrt{(2N)(k-1)(F_\alpha)(MS_w)}$ | 166 |
| (11.1) | $SS_b = SS_{\text{rows}} + SS_{\text{cols}} + SS_{r \times c}$ | 173 |
| (11.2) | Computational formula for rows: | |
| | $SS_{\text{rows}} = \dfrac{\Sigma(\Sigma X_r)^2}{N_r} - \dfrac{(\Sigma X_T)^2}{N_T}$ | 174 |
| (11.3) | Computational formula for columns: | |
| | $SS_{\text{cols}} = \dfrac{\Sigma(\Sigma X_c)^2}{N_c} - \dfrac{(\Sigma X_T)^2}{N_T}$ | 174 |
| (11.4) | $SS_{r \times c} = SS_T - SS_w - SS_{\text{rows}} - SS_{\text{cols}}$ | 174 |
| (11.5) | $df_{\text{rows}} = r = 1$ | 174 |
| (11.6) | $df_{\text{cols}} = c - 1$ | 174 |
| (11.7) | $df_{r \times c} = (r - 1)(c - 1)$ | 174 |
| (12.1) | Pearson $r$, standard score formula: | |
| | $r = \dfrac{\Sigma z_x z_y}{N}$ | 194 |

| FORMULA NUMBER | | PAGE |
|---|---|---|
| (12.2) | $r = \dfrac{N\Sigma XY - (\Sigma X)(\Sigma Y)}{\sqrt{[N\Sigma X^2 - (\Sigma X)^2][N\Sigma Y^2 - (\Sigma Y)^2]}}$ | 194 |
| (12.3) | $t = \dfrac{r\sqrt{N-2}}{\sqrt{1-r^2}}$ or $t = \dfrac{r}{\sqrt{1-r^2}}\sqrt{N-2}$ | 198 |
| (12.4) | Testing the difference between two $Z$'s: $z = \dfrac{Z_1 - Z_2}{\sqrt{\dfrac{1}{N_1 - 3} + \dfrac{1}{N_2 - 3}}}$ | 200 |
| (12.5) | Standard error of Fisher's $Z$ statistic: $S_Z = \dfrac{1}{\sqrt{N-3}}$ | 200 |
| (13.1) | Equation of the straight line: $Y = a + bX$ | 205 |
| (13.2) | Equation of the straight line used in regression analysis: $\tilde{Y} = a + bX$ | 206 |
| (13.3) | $b$ coefficient for predicting $Y$ from $X$: $b_{yx} = \dfrac{\Sigma XY - [(\Sigma X)(\Sigma Y)/N]}{\Sigma X^2 - [(\Sigma X)^2/N]}$ or $b_{yx} = \dfrac{N\Sigma XY - (\Sigma X)(\Sigma Y)}{N\Sigma X^2 - (\Sigma X)^2}$ | 207 |
| (13.4) | $a$ coefficient used in predicting $Y$ from $X$: $a_{yx} = \overline{Y} - \overline{X}(b_{yx})$ | 207 |
| (13.5) | $b$ coefficient for predicting $X$ from $Y$ in both raw score and deviation form: $b_{xy} = \dfrac{\Sigma XY - [(\Sigma X)(\Sigma Y)/N]}{\Sigma Y^2 - [(\Sigma Y)^2/N]}$ or $b_{xy} = \dfrac{N\Sigma XY - (\Sigma X)(\Sigma Y)}{N\Sigma Y^2 - (\Sigma Y)^2}$ | 208 |
| (13.6) | $a$ coefficient used in predicting $X$ from $Y$: $a_{xy} = \overline{X} - (b_{xy})\overline{Y}$ | 208 |
| (13.7) | $b$ coefficient in terms of $r$ and $s$'s: $b_{yx} = r_{xy}\dfrac{s_y}{s_x}$ | 209 |

FORMULA NUMBER | PAGE

(13.8) Regression equation in terms of $r$, means, and $s$'s: 209

$$\tilde{Y} = \bar{Y} + r_{xy}\left(\frac{s_y}{s_x}\right)(X - \bar{X})$$

(13.9) An alternate formula for $r$: 210

$$r = \frac{S_X}{S_Y}(b_{YX})$$

(13.10) Standard error of estimate, raw score formula: 210

$$s_{Y\cdot X} = \sqrt{\frac{\Sigma(Y - \tilde{Y})^2}{N - 2}}$$

(13.11 is the same formula using data obtained from setting up the regression equation:) 211

(13.11)

$$s_{Y\cdot X} = \sqrt{[N\Sigma Y^2 - (\Sigma Y)^2] - \frac{[N\Sigma XY - (\Sigma X)(\Sigma Y)]^2}{N\Sigma X^2 - (\Sigma X)^2}\left[\frac{1}{N(N - 2)}\right]}$$

(13.12) Standard error of estimate, in terms of test statistics: 211

$$s_{Y\cdot X} = s_Y\sqrt{1 - r_{xy}{}^2}$$

(13.13) Correction formula for formula (13.12): 211

$$\sqrt{\frac{N - 1}{N - 2}}$$

(13.14) Relationship between $r$ and $k$: 213

$$r^2 + k^2 = 1$$

(14.1) Spearman rank-order correlation coefficient: 217

$$r_S = 1 - \frac{6\Sigma D^2}{N(N^2 - 1)}$$

(14.2) Testing significance of $r_S$: 218

$$t = \frac{r_S\sqrt{N - 2}}{\sqrt{1 - r_S{}^2}}$$

(14.3) Coefficient of concordance, $W$: 219

$$W = \frac{12\Sigma D^2}{m^2(n)(n^2 - 1)}$$

| FORMULA NUMBER | | PAGE |
|---|---|---|
| (14.4) | $T = \frac{P - Q}{N(N-1)/2}$ | 220 |
| (14.5) | Testing the significance of $T$: $z = \frac{T}{S_T} = \frac{T}{\sqrt{2(2N+5)/9N(N-1)}}$ | 223 |
| (14.6) | Point-biserial correlation coefficient: $r_{pb} = \frac{N_t(\Sigma f_r X) - (N_r)(\Sigma f X)}{\sqrt{N_r N_w [N_t(\Sigma f X^2) - (\Sigma f X)^2]}}$ | 224 |
| (14.7) | The fourfold or phi coefficient: $\phi = \frac{ad - bc}{\sqrt{ijkl}}$ or $\frac{ad - bc}{\sqrt{(a+b)(a+c)(b+d)(c+d)}}$ | 226 |
| (14.8) | Relationships of chi-square and phi: $\chi^2 = N\phi^2$ | 228 |
| (14.9) | Coefficient of contingency: $C = \sqrt{\frac{\chi^2}{N + \chi^2}}$ | 229 |
| (14.10) | Cramér's statistic: $\phi' = \sqrt{\frac{\chi^2}{N(L-1)}}$ | 230 |
| (14.11) | Multiple correlation coefficient, $R$, three variables: $R_{1 \cdot 23} = \sqrt{\frac{r_{12}^2 + r_{13}^2 - (2r_{12}r_{13}r_{23})}{1 - r_{23}^2}}$ | 232 |
| (14.12) | Standard error of multiple $R$: $S_{R_{1 \cdot 23}} = S_1\sqrt{1 - R^2}$ | 232 |
| (14.13) | Partial $r$: $r_{12 \cdot 3} = \frac{r_{12} - (r_{13}r_{23})}{\sqrt{(1 - r_{13}^2)(1 - r_{23}^2)}}$ | 233 |
| (15.1) | Check on sum of the ranks in the Mann-Whitney $U$ test: $\frac{N(N+1)}{2} = \Sigma R_1 + \Sigma R_2$ | 244 |

| FORMULA NUMBER | | PAGE |
|---|---|---|
| (15.2a) | Computation of $U_1$ in the Mann-Whitney $U$ test: $$U_1 = N_1N_2 + \frac{N_1(N_1 + 1)}{2} - \Sigma R_1$$ | 244 |
| (15.2b) | Computation of $U_2$: $$U_2 = N_1N_2 + \frac{N_2(N_2 + 1)}{2} - \Sigma R_2$$ | 244 |
| (15.3) | $z$ ratio for the Mann-Whitney $U$ test: $$z = \frac{U_1 - (N_1N_2/2)}{\sqrt{[N_1N_2(N_1 + N_2 + 1)]/12}} = \frac{U_1 - \mu_u}{\sigma_u}$$ | 245 |
| (15.4) | Kruskal-Wallis $H$ test: $$H = \frac{12}{N(N+1)}\left[\sum \frac{R_i^2}{N_k}\right] - 3(N + 1)$$ | 247 |
| (15.5) | Standard deviation of $T$ in the Wilcoxon matched-pairs signed rank test: $$\sigma_T = \frac{N(N + 1)(2N + 1)}{24}$$ | 250 |
| (15.6) | Test of significance for Wilcoxon test when $N > 25$: $$z = \frac{T - [N(N + 1)/4]}{\sqrt{\frac{N(N + 1)(2N + 1)}{24}}} = \frac{T - \mu_T}{\sigma_T}$$ | 250 |
| (16.1) | Components of a single score: $$X = X_t + X_e$$ | 254 |
| (16.2) | Relationship between obtained mean and true mean: $$\bar{X} = \bar{X}_t$$ | 255 |
| (16.3) | Breakdown of total test variance into component parts: $$S_X{}^2 = S_t{}^2 + S_e{}^2$$ | 255 |
| (16.4) | Two definitions of reliability: $$\frac{S_t^2}{S_x{}^2} = 1 - \frac{S_e{}^2}{S_x{}^2}$$ | 255 |
| (16.5) | Reliability equation: $$r_{xx} = \frac{S_t^2}{S_x{}^2}$$ | 255 |

| FORMULA NUMBER | | PAGE |
|---|---|---|
| (16.6) | Alternate equation for reliability: $$r_{xx} = 1 - \frac{S_e^2}{S_x^2}$$ | 255 |
| (16.7) | Standard error of measurement: $$S_e = S_x\sqrt{1 - r_{xx}}$$ | 256 |
| (16.8) | Spearman-Brown formula to use with the split-half reliability technique: $$r_{xx} = \frac{2r_{oe}}{1 - r_{oe}}$$ | 257 |
| (16.9) | Spearman-Brown formula, general case: $$r_{xx} = \frac{Nr}{1 + (N - 1)r}$$ | 258 |
| (16.10) | Kuder-Richardson reliability formula, No. 20: $$r_{xx} = \frac{k}{k-1}\left[1 - \frac{\Sigma PQ}{S_x^2}\right]$$ | 259 |
| (16.11) | Variance of a test: $$S^2 = \Sigma P_iQ_i + 2\Sigma r_{jj}\sqrt{P_iQ_iP_jQ_j}$$ | 259 |
| (16.12) | Correction for attenuation in both test and criterion scores: $$r_{x_ty_t} = \frac{r_{xy}}{\sqrt{r_{xx}r_{yy}}}$$ | 262 |
| (16.13) | Correction for attenuation in the criterion only: $$r_{x_ty_t} = \frac{r_{xy}}{\sqrt{r_{yy}}}$$ | 263 |

# ANSWERS TO EXERCISES

## CHAPTER 2

1. a. (1) 1 (2) 2 (3) 10
(4) .04 (5) 3.4 (6) 20

b.

| | LOWER LIMIT | MIDPOINT | UPPER LIMIT |
|---|---|---|---|
| (1) | .5 | 6.5 | 12.5 |
| (2) | 100.5 | 110.5 | 120.5 |
| (3) | 44.5 | 67 | 89.5 |
| (4) | 21.5 | 33 | 44.5 |
| (5) | 2.45 | 2.75 | 3.05 |
| (6) | 199.5 | 301 | 402.5 |

6. a. (1) 101.2 (2) 80.1 (3) 62.5 (4) 50.5
(5) 88.4 (6) 75.2 (7) 85.6 (8) 65.2

## CHAPTER 3

1.

| | MEAN | MEDIAN | MODE |
|---|---|---|---|
| a. | 17.6 | 17.5 | — |
| b. | 17.5 | 18.0 | 18 |
| c. | 25.6 | 26.0 | 26 |
| d. | 29.0 | 29.0 | 29 |
| e. | 26.7 | 26.5 | 24 |

2. $\bar{X} = 638.3$; $Mdn = 665$; $Mo = 570{,}740$
3. $\bar{X} = 11.1$; $Mdn = 11.2$; $Mo = 11$
4. 40.4

## CHAPTER 4

1.

| | $\bar{X}$ | $S$ | $s$ | $Q$ |
|---|---|---|---|---|
| a. | 15 | 3.6 | 3.8 | 2.75 |
| b. | 23.8 | 8.5 | 9.0 | 6 |
| c. | 27.2 | 10.1 | 10.8 | 9 |
| d. | 10 | 0 | 0 | 0 |
| e. | 8.9 | 2.6 | 2.7 | 1.3 |

2. $S = 7.2$; $\bar{X} = 29.9$
3. $\bar{X} = 72.9$; $S = 11.9$; $S^2 = 141.6$
4. 13.3
5. $Q = 2.1$
6. $\bar{X} = 640.3$; $S = 9.4$

## CHAPTER 5

1. a. (1) .00 (4) −2.14
   (2) −1.00 (5) 2.57
   (3) .86 (6) −1.64
   b. (1) 500 (4) 286
   (2) 400 (5) 757
   (3) 586 (6) 336
   c. (1) 15 (4) 8.5
   (2) 12 (5) 22.7
   (3) 17.6 (6) 10.1
2. a. 61
   b. 58
3. a. 49
4. a. 176; 650; 3
   b. 173; 460; 95
   c. .25; .93; .004
   d. $p$ = .2476 or 25 in 100
5. 75.4; 44.6; 70.1; 49.9
6. 20
7. a; e; f; g; h

## CHAPTER 7

1. $1.43 \not> 3.84$
2. $24.00 > 10.83$. (Note that the expected frequency of one-spots would be 1/6 (120) = 20 and non–one-spots 100.
3. $19.62 > 9.21$
4. $30.73 > 11.07$
5. $1.25 \not> 2.71$
6. $28.86 > 13.82$
7. $10.40 > 7.82$
8. $8.56 \not> 13.28$
9. $0.94 \not> 2.71$
10. $8.67 > 5.99$

## CHAPTER 8

6. a. $z = \pm 2.0$; probability = .0228 + .0228 = .05
   b. $z = \pm .26$; probability = .3974 + .3974 = .79
7. $z_u = +1.08$, $z_l = -1.33$; probability = .3599 + .4082 = .77
8. $z_u = +3.13$, $z_l = +.9$; percentage = 49.91 − 31.59 = 18.32%
9. a. Estimates: $\mu = 281.3$, $\sigma = 44.8$, $\mu_{\overline{X}} = 281.3$, $\sigma_{\overline{X}} = 4.86$
   b. $z = 1.96$; confidence interval = 271.77–290.83
10. b. $z = 3.30$; confidence interval = 7.07–7.33
11. $t = 2.98$; confidence interval = 4.48–6.32
12. $t = 2.015$; confidence interval = 130.4–135.0

## CHAPTER 9

1. $\sigma_{\overline{X}} = 2.13$
   a. $+2.35 > +1.96$
   b. $+2.35 > +2.33$
2. $\sigma_{\overline{X}} = 5.25$
   a. $-1.77 < -1.64$
   b. $-1.77 \not< -3.10$

3. $s_{\bar{X}} = .20; -1.96 < +1.50 < +1.96$
4. $s_{\bar{X}} = 7.56; -2.12 \nless -2.77$
5. $s_{\bar{X}} = .92$
   a. $-3.26 < -2.10$
   b. $-3.26 < -1.73$
6. $+2.74 > +1.74$
7. $-1.43 \nless -1.96$
8. $t_{obs} = 1.47$
9. $-3.51 < +2.29 < +3.51$. (This critical value was obtained through interpolation in the $t$ tables.)
10. a. $+23.08 > +2.03$
    b. $.03 < .264$ (Significant)
11. a. $+1.38 \ngtr +2.26$
    b. $.57 \nless .276$ (Not significant)
12. $\chi_{obs}{}^2 = 24.00; 24.00 > 6.63$
    $z_{obs} = -4.90; -4.90 < -2.58$
13. $-1.39 \nless -1.64; P' = .5$
14. $-1.37 \nless -2.33; P' = .22$
15. $2.76 > 2.12$
16. a. $\beta = .09$; power = .91
    b. Power = .46

## CHAPTER 10

1. $t$: $-3.89 < -3.71$
   $F$: $15.15 > 13.74$
   $t^2 = (-3.8929)^2 = 15.155$
   $F = 15.155$
2. All $SS$, $df$, $MS$, and $F$ values in Table 10.2 will be duplicated. The $ssd$ test will replicate that described in the text for Table 10.2 data.

3. b.

| SOURCE | *SS* | *df* | *MS* | *F* |
|---|---|---|---|---|
| **Between** | **339.50** | **2** | **169.75** | **28.97*** |
| **Within** | **52.75** | **9** | **5.86** | |
| **Total** | **392.25** | **11** | | |

* Significant at .05 level (28.97 > 4.26)

c. $ssd = 5.00$
   $\bar{X}_1 - \bar{X}_2 > 5.00$
   $\bar{X}_3 - \bar{X}_2 > 5.00$
   $\bar{X}_3 - \bar{X}_1 > 5.00$

4. b.

| SOURCE | *SS* | *df* | *MS* | *F* |
|---|---|---|---|---|
| **Between** | **1.76** | **4** | **.44** | **.06** |
| **Within** | **150.40** | **20** | **7.52** | |
| **Total** | **152.16** | **24** | | |

$.06 \ngtr 4.43$

c. *Post hoc* tests are not justified.

5. b.

| SOURCE | SS | df | MS | F |
|---|---|---|---|---|
| **Between** | **96.44** | **2** | **48.22** | **4.18*** |
| **Within** | **173.17** | **15** | **11.54** | |
| **Total** | **269.61** | **17** | | |

* Significant at .05 level (4.18 > 3.68)

c. $ssd = 5.32$
$\bar{X}_3 - \bar{X}_1 > 5.33$

## CHAPTER 11

1.

| SOURCE | SS | df | MS | F | CRITICAL VALUE |
|---|---|---|---|---|---|
| **Rows** | **10.125** | **1** | **10.125** | **9.00** ≯ | **21.2** |
| **Columns** | **28.125** | **1** | **28.125** | **25.00** > | **21.2** |
| **Interaction** | **.125** | **1** | **.125** | **.11** ≯ | **21.2** |
| **Within** | **4.50** | **4** | **1.125** | | |
| **Total** | **42.875** | **7** | | | |

2. d.

| SOURCE | SS | df | MS | F |
|---|---|---|---|---|
| **Between** | | | | |
| **Rows** | **5.63** | **1** | **5.63** | **7.82** |
| **Columns** | **82.40** | **2** | **41.20** | **57.22*** |
| **Interaction** | **1.07** | **2** | **.54** | **.75** |
| **Within** | **17.20** | **24** | **.72** | |
| **Total** | **106.30** | **29** | | |

* Significant at .001 level (57.22 > 9.34)
7.82 ≯ 14.03; 0.75 ≯ 9.34

4. b.

| SOURCE | SS | df | MS | F |
|---|---|---|---|---|
| **Between** | | | | |
| **Rows** | **183.00** | **2** | **91.50** | **61.00*** |
| **Columns** | **64.67** | **3** | **21.56** | **14.37*** |
| **Interaction** | **82.33** | **6** | **13.72** | **9.15*** |
| **Within** | **18.00** | **12** | **1.50** | |
| **Total** | **348.00** | **23** | | |

* Significant at .05 level (61.00 > 3.88; 14.37 > 3.49; 9.15 > 3.00)

5. c.

| SOURCE | *SS* | *df* | *MS* | *F* |
|---|---|---|---|---|
| **Between** | | | | |
| **Rows** | **232.07** | **2** | **116.04** | **4.87** |
| **Columns** | **1008.07** | **2** | **504.04** | **21.13*** |
| **Interaction** | **2.60** | **4** | **.65** | **.03** |
| **Within** | **429.33** | **18** | **23.85** | |
| **Total** | **1672.07** | **26** | | |

* Significant at .01 level (21.13 > 6.01)
$4.87 \not> 6.01$; $.03 \not> 4.58$

## CHAPTER 12

1. .85
2. .70
3. a. .80
   b. $\bar{X} = 611$; $S = .6$
5. a. $r = -.20$
6. a. $p < .02$
   b. $p > .05$
   c. $p < .001$
   d. $p > .05$
   e. $p < .05$
7. $z = .48$; $p > .05$
8. $t = 4.98$; $p < .0001$
9. 99%: .59–.875
   95%: .64–.855
10. .90

## CHAPTER 13

1. $\tilde{Y} = 20 + .33X$
   $\tilde{X} = 30 + .75Y$
2. a. $\tilde{Y} = -2.37 + 1.81X$
   b. 17.5; 21.2; 3.1
   c. 2.89 = 2.9
3. a. 5.0
   b. 4.0
   c. .46
4. 9.1
6. a. .30
   c. .70

## CHAPTER 14

1. $-.90$; $p < .01$
2. .95; $p < .01$
3. $-.79$
4. .89; $z = 4.68$; $p < .01$
5. .94; $p < .01$
6. .60; $p < .001$
7. .22; $p < .001$

8. a. $p > .05$
   b. $p < .05$
   c. $p < .01$
   d. $p < .05$
   e. $p < .01$
   f. $p > .05$
   g. $p < .05$
   h. $p < .001$
9. a. .47
   b. .38
10. .53
11. a. −.013
    b. .23

## CHAPTER 15

1. $\chi^2 = 4.95; p < .01$
2. $\chi^2 = 6.0; p > .05$
3. $U_2 = 13$; reject $H_0$ at 1% level
4. $z = 1.12; p > .05$
5. $H = 11.8; p < .01$
6. $H = 7.89; p < .01$
7. $H = 33.1; p < .01$
8. $T = -2; p < .01$
9. $T = 3; p < .01$

## CHAPTER 16

1. a. .93
   b. 1.3
   c. .89
2. a. .94
   b. .85
   c. 8
3. a. .98
   b. .91

## APPENDIX A

1. a. 33
   b. −20
   c. −16
   d. 9
   e. 31
2. a. 71
   b. 283
   c. −283
   d. −71
3. a. 801
   b. −300
   c. 783
   d. −132
4. a. 13
   b. −20.1
   c. 15
   d. −29.8
5. a. .43
   b. .01
   c. 8.85
   d. .03
   e. 256
   f. 124679.6
   g. .0024
6. a. 49.35
   b. .00003
   c. 118057
   d. 51.64
7. a. .05
   b. .34
   c. .96
8. a. .03
   b. .56
   c. −.25
9. a. .50
   b. .06
   c. .30
10. a. 1.20
    b. 1.31
    c. 9.63
11. a. 76.6
    b. 3.62
    c. .05

12. a. 1058
    b. 6084
    c. −162
    d. 4301
    e. 7374.1

13. a. .03
    b. 0
    c. .0044
    d. .141
    e. .14
    f. .046

14. a. 7.5
    b. 67.4
    c. 67.6
    d. 1.0
    e. .1
    f. 1.2

15. a. $Y = -.44$
    b. $Z = 6.67$
    c. $X = -6.1$

# REFERENCES


American Psychological Association. *Standards for Educational and Psychological Tests*. Washington, D.C.: American Psychological Association, 1974.

Edwards, A. L. *Statistical Methods for the Behavioral Sciences*. New York: Holt, Rinehart and Winston, 2nd ed., 1967.

Ferguson, G. A. *Statistical Analysis in Psychology and Education*. New York: McGraw-Hill, 3rd ed., 1971.

Guilford, J. P. and B. Fruchter. *Fundamental Statistics in Psychology and Education*. New York: McGraw-Hill, 5th ed., 1973.

Hays, W. L. *Statistics for the Social Sciences*. New York: Holt, Rinehart and Winston, 2nd ed., 1973.

Karabinus, R. A. "The *r*-Point Biserial Limitation." *Educational and Psychological Measurement,* 1975, *35,* 277–282.

Lewis, D. and C. J. Burke. "Use and Misuse of the Chi-Square Test." *Psychological Bulletin,* 1949, *46*, 433–489.

Marascuilo, L. A. *Statistical Methods for Behavioral Science Research,* New York: McGraw-Hill, 1971.

McNemar, Q. *Psychological Statistics*. New York, Wiley, 4th ed., 1969.

Pirie, W. R. and M. A. Hamden. "Some Revised Continuity Corrections for Discrete Data." *Biometrics*, 1972, *28,* 693–701.

Scheffé, H. *The Analysis of Variance*. New York: Wiley, 1957.

Siegel, S. *Nonparametric Statistics for the Behavioral Sciences*. New York: McGraw-Hill, 1956.

Winer, B. J. *Statistical Methods in Experimental Design*. New York: McGraw-Hall, 2nd ed., 1971.

# INDEX

77 78 79 80 9 8 7 6 5 4 3 2 1